Sustainable Technologies for Water and Wastewater Treatment

T0174208

Sustainable Technologies for Water and Wastewater Treatment

Edited by

*Noel Jacob Kaleekkal, Prasanna Kumar S. Mural,
Saravanamuthu Vigneswaran, and Upal Ghosh*

CRC Press
Taylor & Francis Group
Boca Raton London New York

CRC Press is an imprint of the
Taylor & Francis Group, an **informa** business

First edition published 2021
by CRC Press
6000 Broken Sound Parkway NW, Suite 300, Boca Raton, FL 33487-2742
and by CRC Press
2 Park Square, Milton Park, Abingdon, Oxon, OX14 4RN

© 2021 Taylor & Francis Group, LLC
CRC Press is an imprint of Taylor & Francis Group, LLC

Library of Congress Cataloging-in-Publication Data

Names: Kaleekkal, Noel Jacob, editor.
Title: Sustainable technologies for water and wastewater treatment/edited by Noel Jacob
     Kaleekkal, Prasanna Kumar S Mural, Saravanamuthu Vigneswaran, and Upal Ghosh.
Description: First edition. | Boca Raton, FL: CRC Press, 2021. | Includes bibliographical
     references and index.
Identifiers: LCCN 2020056208 (print) | LCCN 2020056209 (ebook) |
     ISBN 9780367510374 (hardcover) | ISBN 9781003052234 (ebook)
Subjects: LCSH: Sewage–Purification–Biological treatment. | Water–Purification–Membrane
     filtration. | Green chemistry.
Classification: LCC TD755.S857 2021 (print) | LCC TD755 (ebook) | DDC 628.1/62–dc23
LC record available at https://lccn.loc.gov/2020056208
LC ebook record available at https://lccn.loc.gov/2020056209

ISBN: 9780367510374 (hbk)
ISBN: 9780367552534 (pbk)
ISBN: 9781003052234 (ebk)

Typeset in Times LT Std
by KnowledgeWorks Global Ltd.

Contents

PART III   Energy-Water Nexus

PART IV   Nascent Technologies

PART V   Bio and Bio-Inspired Materials
for Water Reclamation

PART VI   Integrated Technologies for Water Treatment

Preface

The rapid population growth, incessant deforestation, rampant urbanization coupled with the depleting surface/groundwater resources, and global warming have led to acute water scarcity and flash flooding, which could render potable water a diminishing natural resource. The aquatic pollutants from domestic and industrial waste streams are detrimental to the ecosystem and can be dangerous for living beings. Apart from various seawater desalination technologies, other ways to recover usable water from wastewater are being explored worldwide by researchers.

The exponential demand for safe water, concomitant with the improvement of technology and government-imposed standards, have made conventional treatment processes obsolete. The increasing use of nanoscale material in various water treatment technologies is unprecedented. These materials have high surface areas, can be reused, and can possess a wide range of characteristics like photocatalytic ability, selective adsorption, and reactivity. These materials can be fabricated on the design-for-purpose concept and can be utilized in many types of technologies.

This book collates the advances and explores the environmental sustainability of various technologies for water and wastewater treatment. Further, it can serve as a one-stop reference book to a much wider audience and play a pivotal role in bridging various scattered sustainable technologies. The authors are experts who have incorporated different dimensions from the basics to state-of-the-art advancements for new technologies. This book would be a valuable resource for students, engineers, scientists, and industrial practitioners working in water treatment technologies.

The first section outlines the passive sampling techniques for detecting contaminants, the occurrence/removal of endocrine-disrupting pollutants, and the advancement of nanotechnology in water remediation. The second section discusses pressure-driven membrane processes (thin-film nanocomposite and mixed matrix membranes) and explores osmotic and vapour-pressure driven membrane applications. The role of electrospinning as a versatile tool for the development of nanofibrous membranes is discussed. The third section focuses on the energy-water nexus addresses the efficiency and challenges of microbial fuel cells. An outline of the various advanced oxidation processes is included.

The fourth section highlights some nascent sustainable technologies such as hydrodynamic cavitation and ultrasound to suppress contaminants in wastewater. The fifth section includes biological and bio-inspired materials for the recovery of water. Bioremediation, bioactive nanomaterials, biopolymers, and bioreactors are discussed along with the challenges in implementation. The last section deals with integrated sustainable technologies such as decentralized water treatment, membrane bioreactors (aerobic and anaerobic), and hybrid systems for wastewater treatment.

This book aims to provide sufficient knowledge and organize a comprehensive review of recent scientific literature that would give the reader a deeper understanding of the various technologies associated with wastewater treatment and desalination. This book also throws light on multiple novel technologies currently being developed in laboratories and helps understand the challenges associated with their commercialization. This book will pique the curiosity of researchers globally to explore attractive alternative solutions to address the global water scarcity of this century.

Acknowledgments

We thank the Ministry of Human Resource Development (Ministry of Education), Government of India, for the project under the Scheme for Promotion of Academic and Research Collaboration (SPARC), which has enabled increased cooperation and scientific information sharing.

We are greatly indebted to the authors who have dedicated their time and energy to bring out the latest developments in their field.

Dr. Noel Jacob Kaleekkal and Dr. Prasanna Kumar S. Mural are also grateful to the Department of Chemical Engineering, National Institute of Technology Calicut, Kerala, India, for providing them with the necessary facilities and support.

The doctoral students, Mr. Akash M. Chandran, Ms. Jenny N., and Mr. Varun S., are also appreciated for their timely assistance.

We also place on record the enthusiasm and support provided by the publisher to this book.

The editors are also thankful to their family members who have supported them in their journey.

And above all, we thank God Almighty for the success of this project!

About the Editors

Dr. Noel Jacob Kaleekkal is currently an Assistant Professor at the National Institute of Technology Calicut, India, and also leads the Membrane Separation Group. He holds a Chemical Engineering undergraduate degree from the National Institute of Technology Karnataka, India. He obtained his postgraduate degree (Chemical Engineering) and PhD (Technology) from Anna University, India. His research interests lie in the development of novel membranes for different applications His initial focus was on the development and evaluation of membranes for hemodialysis application. He has also worked on the pressure-driven membrane process for water treatment including removal of heavy metals, humic substances, separation of oil-water emulsion, etc. He has also coauthored peer-reviewed articles on proton exchange membranes for direct-methanol fuel cells. His most recent researches focus on low-pressure, novel membrane technologies that aim to address the concerns of diminishing water and energy resources. He has received funded projects for developing novel forward osmosis membranes as well as superhydrophobic membranes for membrane distillation for the recovery and reuse of wastewater and seawater.

Dr. Prasanna Kumar S. Mural has a bachelor's degree in Chemical Engineering from the National Institute of Technology Karnataka, India. He has completed his master's degree in Plastic Technology from CIPET, Bhubaneshwar, and MBA in Total Quality Management from Sikkim Manipal and a PhD from the Indian Institute of Science, Bangalore, India. Currently he is serving as an assistant professor and leads the Materials Chemistry and Polymer Technology Research Group in the Department of Chemical Engineering at the National Institute of Technology Calicut, Kerala, India. His main focus of research was on the development and evaluation of antibacterial membranes for water purification applications. Dr. Prasanna Kumar has developed a unique one-step method of membrane preparation via polymer blends. He also authored a peer-reviewed article on antibacterial water purification membranes. Presently, his research involves the application of polymer nanocomposite for water and energy application to address the water and energy crisis.

Prof. Saravanamuthu Vigneswaran is currently a Distinguished Professor of Environmental Engineering in the Faculty of Engineering and IT at the University of Technology, Sydney (UTS), Australia. He obtained his Doctor of Engineering and Dr.Sc from University of Montpellier and University of Toulouse, respectively. During the last twenty-five years, he has made significant contributions to the understanding of membrane systems in water reuse and desalination. He has several national and international projects in this area. He is a Distinguished Fellow of the International Water Association (IWA) and was the Vice-Chair of the IWA's specialist group on membrane technology and currently secretary of the rainwater harvesting group. He obtained IWA Project Innovation Awards in the Applied Research category

(Honour Award) three consecutive times, IWA Development Honour Award in 2019 and Kamal Fernando mentor award in 2018. He is the winner of Google Impact Challenge Technology against Poverty Prize, 2017. He has published more than 350 journal papers (with a Google Scholar index of 68). He has authored and edited three books on filtration, water treatment and membrane processes. The first two books on "Water, Wastewater and Sludge Filtration" and "Water Treatment:Simple Options" were published by CRC press, USA. The latest book he authored and edited was *Membrane Technology and Environmental Applications* with Zhang, T.C., Surampalli, R.Y., Vigneswaran, S., Tyagi, R.D., Ong, S.L., and Kao, C.M. (2012). This book was sponsored by the Membrane Technology Task Committee of the Environmental Council Environmental and Water Resources Institute (EWRI) of the American Society of Civil Engineers, ASCE Press, New York. In addition to that he was invited by UNESCO to write two volumes on "Water Recycle, Reuse and Reclamation" and "Waste-water Treatment Technologies."

Dr. Upal Ghosh is a professor in the Department of Chemical, Biochemical, and Environmental Engineering at the University of Maryland, Baltimore County. He has an undergraduate degree in Chemical Engineering and MSc and PhD degrees in Environmental Engineering. His group performs research in environmental engineering and science with a focus on the fate, effects, and remediation of toxic pollutants in the environment. They use multidisciplinary tools to investigate exposure and bioavailability of organic and metal pollutants to organisms and apply the new understanding to develop novel approaches for risk assessment and remediation. His research has contributed to the development and transition of novel sediment remediation technologies based on altering sediment geochemistry and enhancing biological degradation. Dr. Ghosh has also led the development of monitoring tools for pollutant bioavailability, especially work on passive sampling techniques for measuring freely dissolved concentrations in sediment porewater. His work has been published in the leading journals in the field and the technology development has led to several US patents. His research contributions have been recognized through multiple awards including the University System of Maryland Regents Award for Excellence in Scholarship, Research, and Creative Activity in 2016. He is an Associate Editor of the journal *Environmental Toxicology and Chemistry*. Dr. Ghosh also is the co-founder of two start-up companies that are transitioning emerging sediment remediation technologies to the field.

Contributors

J. Anandkumar
Department of Chemical Engineering
National Institute of Technology
Raipur, India

Vineet Aniya
Process Engineering and Technology
  Transfer Division
CSIR-Indian Institute of Chemical
  Technology
Hyderabad, India

Mandar Bokare
University of Maryland Baltimore County
Baltimore, MD, United States

Suryasarathi Bose
Department of Materials Engineering
Indian Institute of Science
Bangalore, India

Yuhe Cao
University of Arkansas
Fayetteville, AR, United States

Akash M. Chandran
Materials Chemistry and Polymer
  Technology Group
Department of Chemical Engineering
National Institute of Technology
Calicut, India

Narayan Chandra Das
Rubber Technology Centre
Indian Institute of Technology-Kharagpur,
  721302
West Bengal, India

Upal Ghosh
University of Maryland Baltimore County
Baltimore, MD, United States

Joel Parayil Jacob
Membrane Separation Group
Department of Chemical Engineering
National Institute of Technology
Calicut, India

J. Jayapriya
Department of Applied Science and
  Technology
Anna University
Chennai, India

K. L. Jesintha
Centre for Research
Department of Biotechnology
Kamaraj College of Engineering
  & Technology
India

J. Juliana
Department of Civil Engineering
National Institute of Technology
Trichy, India

Gautham B. Jegadeesan
School of Chemical & Biotechnology
SASTRA Deemed University
Thanjavur, India

M. A. H. Johir
University of Technology Sydney (UTS)
Broadway, Ultimo, NSW, Australia

Noel Jacob Kaleekkal
Membrane Separation Group
Department of Chemical Engineering
National Institute of Technology
Calicut, India

Aiswarya M.
Materials Chemistry and Polymer
    Technology Group
Department of Chemical Engineering
National Institute of Technology
Calicut, India

Joyabrata Mal
Department of Biotechnology
Motilal Nehru National Institute
    of Technology
Allahabad, Uttar Pradesh, India

S. Mariaamalraj
Centre for Research
Department of Biotechnology
Kamaraj College of Engineering
    & Technology
India

Aquib Mohammed
Materials Chemistry and Polymer
    Technology Group
Department of Chemical Engineering
National Institute of Technology
Calicut, India

Siddhartha Moulik
Cavitation and Dynamics Lab
Department of Process
Engineering & Technology Transfer
CSIR-Indian Institute of Chemical
    Technology
Hyderabad, India

Aditi Mullick
Cavitation and Dynamics Lab
Department of Process
Engineering & Technology Transfer
CSIR-Indian Institute of Chemical
    Technology
Hyderabad, India

Anupam Mukherjee
Water-Energy Nexus Lab
Department of Chemical Engineering
BITS Pilani Goa Campus
Goa, India

Prasanna Kumar S. Mural
Materials Chemistry and Polymer
    Technology Group
Department of Chemical Engineering
National Institute of Technology
Calicut, India

Harsha Nagar
Department of Chemical Engineering
Chaitanya Bharathi Institute
    of Technology
Hyderabad, India

A. R. Neelakandan
School of Biotechnology
National Institute of Technology
Calicut, India

V. C. Padmanaban
Centre for Research
Department of Biotechnology
Kamaraj College of Engineering
    & Technology
India

Nirenkumar Pathak
University of Technology Sydney (UTS)
Broadway, Ultimo, NSW, Australia

V. Ponnusam
Biomass Conversion and Bioproducts
    Laboratory
Center for Bioenergy
School of Chemical & Biotechnology
SASTRA Deemed University
Thanjavur, India

M. S. Priyanka
National Institute of Technology
Calicut, India

G. K. Rajanikant
School of Biotechnology
National Institute of Technology
Calicut, India

S. Rangabhashiyam
Department of Biotechnology
School of Chemical and Biotechnology
SASTRA Deemed University
Thanjavur, India

Sanjay Remanan
Rubber Technology Centre
Indian Institute of Technology-Kharagpur,
    721302
West Bengal, India

Anirban Roy
Water-Energy Nexus Lab
Department of Chemical Engineering
BITS Pilani Goa Campus
Goa, India

Biju Prava Sahariah
University Teaching Department
Chhattisgarh Swami Vivekanand
    Technical University
India

Paresh Kumar Samantaray
International Institute for Nanocomposites
    Manufacturing (IINM)
WMG, University of Warwick
U.K.

M. Saravanan
Biomass Conversion and Bioproducts
    Laboratory
Center for Bioenergy
School of Chemical & Biotechnology
SASTRA Deemed University
Thanjavur, India

Monalisa Satapathy
Department of Chemical Engineering
National Institute of Technology
Raipur, India

Hokyong Shon
University of Technology Sydney (UTS)
Broadway, Ultimo, NSW, Australia

Buddhima Siriweera
Environmental Engineering and
    Management Program
School of Environment
Resources and Development
Asian Institute of Technology
Thailand

N. Santosh Srinivas
School of Chemical & Biotechnology
SASTRA Deemed University
Thanjavur, India

Anoopa Ann Thomas
Membrane Separation Group
Department of Chemical Engineering
National Institute of Technology
Calicut, India

George K. Varghese
National Institute of Technology
Calicut, India

S. Varun
Materials Chemistry and Polymer
    Technology Group
Department of Chemical Engineering
National Institute of Technology
Calicut, India

Saravanamuthu Vigneswaran
Faculty of Engineering and IT
University of Technology
Sydney (UTS)
Broadway, Ultimo, NSW, Australia

Chettiyappan Visvanathan
Environmental Engineering and
    Management Program
School of Environment
Resources and Development
Asian Institute of Technology
Thailand

Ranil Wickramasinghe
University of Arkansas
Fayetteville, AR, United States

Aparna Yadu
Department of Chemical Engineering
National Institute of Technology
Raipur, India

Part I

Introduction

1 In Situ Sampling Techniques for Identification and Quantification of Trace Pollutants

Upal Ghosh* and Mandar Bokare
University of Maryland Baltimore County,
Baltimore, MD, United States

CONTENTS

1.1 INTRODUCTION

1.1.1 TRACE POLLUTANTS IN THE WATER ENVIRONMENT

Impairments to water quality can be caused by a range of toxic chemicals released from current or past anthropogenic activities. Broadly, the toxic chemicals fall under two classes based on chemical molecular properties and history of release: 1) water-soluble compounds that typically originate from ongoing releases, and 2) sparingly soluble hydrophobic organic compounds (HOCs) released in the past that strongly bind to solids and often bioaccumulate in organisms. The water-soluble compounds include nutrients like nitrogen, phosphorus, and oxygen depleting substances from wastewater. They also include soluble pharmaceutical compounds, small organic molecules such as benzene and phenols, and soluble metals. These chemicals weakly associate with solids and are thus flushed out with water flow into the ocean with impacts during their passage through the water body.

The sparingly soluble and hydrophobic compounds associate strongly with solids and predominantly settle out with the suspended solids into the bottom sediment. These contaminated sediments serve as long-term legacy repositories of HOCs released in the past. Chemicals such as dichlorodiphenyltrichloroethane (DDT) and polychlorinated biphenyls (PCBs) that have been banned for decades and metals like mercury, which are highly regulated, are still found in many contaminated sediment sites such as Grasse River (US EPA, 2019) and the Hudson River (US EPA, 2020a). Pollutants in legacy sediments slowly leach out into the water column, impacting existing ecosystems, although these compounds are not currently used. Some of these chemicals are semi-volatile and can exchange with the atmosphere and enter into long-distance atmospheric transport. The hydrophobic nature of these chemicals also makes them more conducive for uptake into the aquatic food web. Several rivers in the United States have fish-consumption advisories due to potential exposure to carcinogenic chemicals (PCBs) and neurotoxins (methylmercury).

1.1.2 THE IMPORTANCE OF FREELY DISSOLVED CONCENTRATIONS

Sediments often serve as a long-term source by discharging the legacy pollutants into the aqueous phase (e.g., porewater and overlying water). The freely dissolved concentration in the aqueous phase (C_{free}) is a useful measure of the amount of contaminant bioavailable to aquatic organisms. Bioaccumulation and toxicity of HOCs to benthic and pelagic organisms is well-characterized by C_{free} of the pollutants. For example, Werner et al. (2010) showed that PCB concentrations in bio-lipids of a marine species exposed to sediment for 28 days were well-predicted by the free aqueous phase in sediment porewater. However, PCB bioaccumulation was over predicted when using the solid phase concentrations. In another study, Kreitinger et al. (2007) measured toxicity of PAHs to *Hyalella azteca* after 28 days in 34 sediment samples collected from four manufactured-gas plant sites. The toxicity of PAHs was accurately predicted by the free aqueous phase concentrations in the sediment porewater, while toxicity threshold was difficult to determine when using the solid phase concentrations. Thus, accurate measurement of ultra-low C_{free} of HOCs in aqueous environmental media is critical for assessing bioaccumulation and toxicity.

Bioaccumulation of HOCs, such as PCBs in the aquatic food web, can harm the ecosystem and human health (Apitz et al., 2005). The high bioaccumulation potential of PCBs results in unacceptably high concentrations in fish even at low freely dissolved water phase concentrations. For example, the bioaccumulation factor-based target water concentration to limit the population level incremental cancer risk to 1 in a million is 0.064 ng/L for total PCBs and much lower concentrations for individual compounds US EPA (2020b). Like PCBs, polycyclic aromatic hydrocarbons

(PAHs) are strongly hydrophobic compounds, but due to the extensive metabolism of the compound in higher animals, they do not biomagnify in the food web. The toxic effects are primarily observed in the sediment-dwelling invertebrates. The bioaccumulation factors of PAHs are high in exposed organisms. The freely dissolved aqueous concentrations (sediment porewater for benthic animals) of PAHs need to be limited to very low concentrations.

Besides, the simultaneous measurement of C_{free} in sediment porewater and overlying water allows for assessing fluxes between sediment bed and water column (Eek and Breedveld, 2010). For example, when an engineered cap is used as part of site cleanup, the contaminant flux from the underlying contaminated sediment into the cap layers and into the overlying water needs to be determined before and after the remediation in order to evaluate the performance of the cap layer.

1.1.3 CONVENTIONAL APPROACHES FOR MEASURING OR ESTIMATING C_{free}

One technique for determining the C_{free} is the direct measurement. Direct measurements were previously conducted by collecting porewater/surface water using centrifugation and/or filtration of sediment samples (Cho et al., 2009) and subsequent extraction of the porewater with an organic solvent. However, the conventional technique can result in the overestimation of C_{free} due to artifacts from the contaminants associated with dissolved and colloidal organic matter. Another artifact is absorption and losses to glassware. Conventional methods are also labor-intensive and require sampling and extracting large volumes of water.

Another approach for estimating the C_{free} in sediment porewater is using partitioning models. Two widely used models are one-carbon and two-carbon models. The one-carbon model assumes that contaminants are associated with the natural organic carbon content of the sediment. This model is insufficient for predicting C_{free}, as it does not account for the association of the contaminants with black carbon (Maruya et al., 1996; McGroddy and Farrington, 1995). In the two-carbon model, the pollutants are associated with natural organic carbon and black carbon in the sediment. However, the latter model has been challenged due to difficulties in characterizing the different types of carbonaceous materials and partitioning contaminants to each material in a sediment sample (Ghosh et al., 2003; Hong et al., 2003). A considerable variation in the characteristics of the sediment (site-specific) leads to a vast deviation in the amounts of PAHs sorbed (3–4 orders of magnitude) (Hawthorne et al., 2006). Moreover, the unpredictable sorption behavior of some carbonaceous materials such as coal tar pitch (Khalil et al., 2006) is another challenge associated with using the two-carbon model.

1.2 PASSIVE SAMPLING METHODS FOR IN SITU MEASUREMENT OF C_{free}

1.2.1 PASSIVE SAMPLING

Challenges with direct measurement of C_{free} with the traditional methods have led to the development of passive sampling approaches using well-characterized polymeric materials. Organic polymers such as polyethylene (PE) (Adams et al., 2007;

Booij et al., 2003; Fernandez et al., 2009a; 2009b; Lohmann et al., 2004), polyoxymethylene (POM) (Cornelissen et al., 2008; Jonker and Koelmans 2001; Oen et al., 2011), and polydimethylsiloxane (PDMS) (Lampert et al., 2015; Lu et al., 2011; Smith et al., 2012) are commonly used for passive sampling. These three types of polymers have different partitioning coefficients for PAHs and PCBs that are related to the octanol-water partitioning coefficient of these chemicals (Ghosh et al., 2014). Unlike the conventional methods, passive sampling techniques are less labor-intensive and allow for accurate measurement of the contaminants that are freely dissolved in the water phase. Detection limits of contaminants are also much lower compared to the conventional methods.

When a passive sampler is placed in water/sediment with contaminants, the freely dissolved contaminants in the water phase will partition into the polymer. Over time, the contaminants will accumulate in the sampler until there is no change in the concentration of contaminants in the passive sampler. At this point, the contaminants are considered to be at equilibrium between the passive sampler and the C_{free} in the water. Once a sampler has achieved equilibrium, it can be retrieved and analyzed for contaminants. C_{free} can be estimated using the equilibrium concentration in the polymer (C_p) and the polymer-water partitioning coefficient of the contaminant (K_{pw}) as follows:

$$C_p = K_{pw}C_w \qquad (1.1)$$

As shown in Figure 1.1, passive samplers operate in three regimes: kinetic (white), intermediate (light blue), and near equilibrium (blue) (Mayer et al., 2003). Sampling under the kinetic scenario applies to relatively short sampling times and targets a time-specific concentration that must be corrected to the equilibrium condition (Lydy et al., 2014). Sampling under the equilibrium regime applies at relatively long sampling times and requires polymer-water partition constants to deduce C_{medium} from $C_{sampler}$ (Mayer et al., 2000; 2014).

Sampling under the equilibrium condition is preferred. However, mass transfer rates in sediment porewaters are slow, and it takes longer for the equilibrium to occur

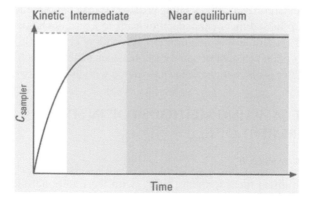

FIGURE 1.1 Generalized uptake profile for a passive sampling device.

between the samplers and aqueous phase (Booij et al., 1998; Ghosh et al., 2014). Booij et al. (2002) proposed spiking the passive samplers with known amounts of performance reference compounds (PRCs) to address this issue. The chosen PRCs, such as deuterated PAHs and specific PCB congeners, must not occur in the environment and must have similar diffusivities and partitioning properties as the target compounds. Since both adsorption and desorption, in samplers like PE, are governed by the same mass transfer processes, PRCs can be used to correct for non-equilibrium (Booij et al., 1998, Huckins et al., 2002).

Currently, passive samplers are used in surface waters (Booij et al., 2003; Fernandez et al., 2014), sediments (Booij et al., 2003; Fernandez et al., 2009b), and even for the determination of contaminant loading into wastewater treatment plants (Harman et al., 2011). Several approaches for field deployments are possible, as shown in Figure 1.2.

Deployment of passive samplers provides a time-averaged response. Additionally, passive sampling is a more cost-effective method of measuring the freely dissolved concentration of PCBs, and passive sampler-based information can better understand contaminant concentrations that result in real exposures and risks at Superfund sites (US EPA, 2012). US EPA accepts the use of passive

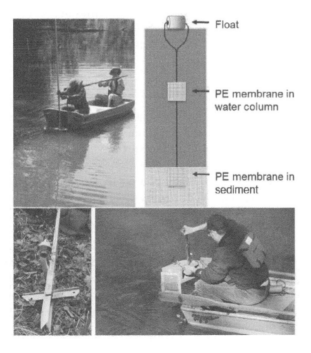

FIGURE 1.2 Passive sampler deployment approach for a tidal river. A water column sampler, enveloped in stainless steel mesh, is attached to the rope at mid-depth and tied to a buoy. Alternatively, the passive sampling frame can be inserted into the sediment using an insertion tool as shown in the bottom left picture. Porewater samplers can also be placed into a frame attached to a cinder block shown in the bottom right picture.

samplers as a reliable and accurate tool for measuring freely dissolved aqueous concentrations. In 2012 the US EPA published a set of guidelines for using passive samplers to monitor dissolved HOCs being released from sediment contamination (US EPA, 2012). In addition, later practical guidance documents have been developed to provide users of passive sampling with information on the various aspects of passive samplers, including laboratory, field, and analytical procedures involved in the application of the technology to evaluate contaminated sediments (Ghosh et al., 2014).

1.3   USE OF PASSIVE SAMPLING DATA

1.3.1   Assessing Current Site Conditions

C_{free} data obtained from passive sampling can be used to inform the current site conditions. By using passive sampling devices at several locations within a water body, contaminant "hot-spots" in sediments (that can serve as sources of pollutants to the surface water and areas with lower C_{free}) can be identified and used to delineate specific locations in the site needing immediate attention for remediation. Passive samplers can also be used to obtain information on spatial heterogeneity in the porewater concentrations, including vertical concentration profiles in sediment core samplers, which can assess the effectiveness of in situ remediation techniques such as sediment capping and sorbent amendments (Lydy et al., 2014).

1.3.2   Prediction of Pollutant Toxicity

As discussed previously, C_{free} is an excellent parameter for predicting the bioavailability of a contaminant to aquatic organisms. C_{free} can be compared with water quality criteria for performing preliminary toxicity assessment (Lydy et al., 2014). Hawthorne et al. (2007) showed that freely dissolved porewater concentrations of PAHs were better predictors for amphipod survival and for delineating toxic and non-toxic sediments than bulk sediment concentrations. In this study, no toxicity thresholds for amphipods could be identified based on bulk sediment concentration, but toxicity thresholds based on freely dissolved concentrations could correctly identify toxic and non-toxic sediments with an overall efficiency of 90%. By conducting sediment toxicity testing in conjunction with measuring the freely dissolved porewater concentration, it is possible to identify sediment "hot-spots" in need of remedial action based on polymer concentrations (Mayer et al., 2014). Furthermore, the freely dissolved pollutant concentrations in porewater can be used to develop sediment remediation goals that are protective of benthic organisms (US EPA, 2017).

1.3.3   Prediction of Pollutant Bioaccumulation

C_{free} represents the fraction of the pollutant that is thermodynamically available for uptake into aquatic organisms. This thermodynamic equilibrium is with C_{free}

in sediment porewaters for benthic organisms, whereas for pelagic invertebrates such as phytoplankton and zooplankton, the equilibrium is mostly with C_{free} in surface water. For fish, uptake can occur through surface water or porewater ventilation as well as through diet. Uptake in benthic and pelagic invertebrates can be predicted using simple equilibrium partitioning models, using biota lipid-water partitioning coefficients (Jahnke et al., 2012; Leslie et al., 2002; Tuikka et al., 2016) or through sampler-lipid partitioning coefficients (Smedes et al., 2017). For higher-level organisms such as fish where uptake through diet and gill ventilation becomes a critical pathway, bioaccumulation models such as those proposed by Gobas (1993) and Arnot and Gobas (2004) can be used. The importance of predicting bioaccumulation arises from the fact that human exposure to PCBs, PAHs, and chlorinated pesticides typically comes from fish consumption. Thus, any remedial action for an impaired waterway is ultimately geared towards reducing contaminant concentrations in fish to acceptable levels. Thus, it becomes necessary to predict changes in bioaccumulation arising from any remedial action that changes the freely dissolved concentration of contaminants in surface water and sediment porewater (US EPA, 2012).

1.3.4 PREDICTION OF POLLUTANT DEGRADATION RATES

For highly persistent pollutants such as PCBs, PAHs, and chlorinated pesticides, commonly used remediation strategies such as capping and in situ amendments are geared towards reducing these pollutants' bioavailability to aquatic organisms. However, these methods do not reduce the total mass of pollutants in the system. Thus, techniques that can degrade these pollutants in situ have long been sought after. One such technique is microbial dechlorination of PCBs in sediments, which was first observed nearly three decades ago in Hudson River sediments and is seen as a cost-effective remediation option compared to capping and dredging. However, the implementation in the field has been inhibited by a lack of understanding of the degradation kinetics, which requires accurate quantification of the freely dissolved concentration of pollutants and the degradation products in the sediment porewater. Passive sampling to measure C_{free} of parent PCBs and their dechlorination products has been used to quantify the biological dechlorination rate in laboratory experiments (Lombard et al., 2014; Needham et al., 2019), providing an opportunity to implement microbial dechlorination for remediating contaminated sediments. A pilot study using bioamended activated carbon demonstrated the effectiveness of microbial dechlorination for in situ PCB degradation (Payne et al., 2019).

1.3.5 PREDICTION OF POLLUTANT MASS TRANSFER

In environmental systems, diffusive mass transfer of pollutants across the air-water and sediment-water interfaces, based on the relative chemical activities of the contaminant in the respective phases, is an important component of the fate and transport of the pollutant. C_{free} is a suitable proxy for chemical activity and has been used in numerous studies to quantify the mass transfer rates of contaminants across environmental compartments (Apell and Gschwend, 2017; Fernandez et al., 2014).

Passive sampler data from air and water phases from the same sampling period can be used to determine the direction of mass transfer using fugacity ratios and quantify the mass of pollutants being volatilized from the water surface or being absorbed into the water body. This approach has been used to understand the role of the atmosphere as a sink or a source for pollutants (Apell and Gschwend, 2017; Ghosh et al., 2020; Liu et al., 2016) and is increasingly being incorporated in remedial investigations. Similarly, the role of sediments as a sink or source can be delineated using this approach by identifying sediments "hot spots" with dissolved porewater concentrations higher than those in the surface water. An integrated approach taking into account the spatial heterogeneity in pollutant concentrations in sediments can be also be used to derive sediment remediation goals, as has been done for the Anacostia River in Washington, DC (Ghosh et al., 2020).

1.4   ONGOING RESEARCH AND CHALLENGES

1.4.1   Passive Sampling for Emerging Contaminants

The advantages of passive sampling over conventional grab sampling have led to increasing interest in applying passive sampling techniques for emerging hydrophobic contaminants, such as polybrominated diphenyl ethers (PBDEs) and triclosan (Perron et al., 2013), fragrances, endocrine disruptors, UV filters, organophosphate flame retardants (Pintado-Herrera et al., 2016), and phthalates and alkylphenols (Posada-Ureta et al., 2016). Passive samplers are being studied to monitor polyfluorinated alky substances (PFAS) (Dixon-Anderson and Lohmann, 2018) and phenols and triazine herbicides (Gao et al., 2019). There have also been studies on developing polymeric passive-samplers for monitoring nonorganic, bioaccumulative pollutants such as mercury and methylmercury in sediment porewaters (Sanders et al. 2020; Taylor et al. 2019a).

1.4.2   Challenges and Research Needs

For typical polymeric samplers deployed under field conditions, achieving equilibrium between the polymer and water for highly hydrophobic compounds is not feasible due to impractically long deployment times. As discussed earlier, although there are methods to correct for nonequilibrium conditions, there are errors associated with the correction technique, especially for short deployment times wherein the correction factors are larger (Jalalizadeh and Ghosh 2016). Shorter deployment times can also provide increased time-resolution for detecting variation in pollutant concentrations in environmental systems. As a result, a critical area of research is improving polymeric passive sampling techniques so that the highly hydrophobic compounds in surface water or porewater are closer to or at equilibrium with the polymer in shorter deployment times. Methods for reducing equilibration times can be narrowed down to optimizing the polymer's physical aspects, such as area/volume ratio and sampler material (Booij et al., 2016).

Recent laboratory studies by Jalalizadeh and Ghosh (2016) and Jalalizadeh and Ghosh (2017) have shown that equilibration times can be reduced by inducing periodic vibrations in passive samplers deployed in sediments. Further studies are currently underway to demonstrate the application of these vibration-enhanced passive samplers under field conditions.

Another significant challenge that has been identified in several studies (Booij et al., 2016; Lydy et al., 2014; Mayer et al., 2014) is the shortage of high quality and standardized polymer-media partitioning values for target analytes (in case of equilibrium sampling) and for performance reference compounds (in case of kinetic sampling). As noted by Booij et al. (2016), the primary source of error in the measurement of C_{free} is the polymer-media partitioning coefficients and the uncertainties associated with these coefficients. Related to this aspect, insufficient quality control and lack of standardized methods for deployment, chemical analyses, and data reporting were identified by Booij et al. (2016) as critical research needs.

Taylor et al. (2019b) identified additional criteria that need to be met for increasing the acceptance and use of passive sampling in regulatory monitoring of HOCs. These include commercial availability of passive sampling products, improved lipid-water partitioning coefficients, and availability of certified reference materials such as standard spiked polymers.

REFERENCES

Adams, R. G., Lohmann, R., Fernandez, L. A., MacFarlane, J. K., & Gschwend, P. M. (2007). Polyethylene devices: Passive samplers for measuring dissolved hydrophobic organic compounds in aquatic environments. Environmental Science & Technology, 41(4), 1317–1323. https://doi.org/10.1021/es0621593

Apell, J. N., & Gschwend, P. M. (2017). The atmosphere as a source/sink of polychlorinated biphenyls to/from the Lower Duwamish Waterway Superfund site. Environmental Pollution, 227, 263–270. https://doi.org/10.1016/j.envpol.2017.04.070

Apitz, S. E., Davis, J. W., Finkelstein, K., Hohreiter, D. W., Hoke, R., Jensen, R. H., Jersak, J., Kirtay, V. J., Mack, E. E., Magar, V. S., Moore, D., Reible, D., & Stahl, R. G. (2005). Assessing and managing contaminated sediments: part I, developing an effective investigation and risk evaluation strategy. Integrated Environmental Assessment and Management, 1(1), 2–8. https://doi.org/10.1897/IEAM_2004a-002.1

Arnot, J. A., & Gobas, F. A. P. C. (2004). A food web bioaccumulation model for organic chemicals in aquatic ecosystems. Environmental Toxicology and Chemistry, 23(10), 2343. https://doi.org/10.1897/03-438

Booij, K., Hofmans, H. E., Fischer, C. V., & Van Weerlee, E. M. (2003). Temperature-dependent uptake rates of nonpolar organic compounds by semipermeable membrane devices and low-density polyethylene membranes. Environmental Science & Technology, 37(2), 361–366. https://doi.org/10.1021/es025739i

Booij, K., Robinson, C. D., Burgess, R. M., Mayer, P., Roberts, C. A., Ahrens, L., Allan, I. J., Brant, J., Jones, L., Kraus, U. R., Larsen, M. M., Lepom, P., Petersen, J., Pröfrock, D., Roose, P., Schäfer, S., Smedes, F., Tixier, C., Vorkamp, K., & Whitehouse, P. (2016). Passive Sampling in Regulatory Chemical Monitoring of Nonpolar Organic Compounds in the Aquatic Environment. Environmental Science and Technology, 50(1), 3–17. https://doi.org/10.1021/acs.est.5b04050

Booij, K., Sleiderink, H. M., & Smedes, F. (1998). Calibrating the uptake kinetics of semipermeable membrane devices using exposure standards. Environmental Toxicology and Chemistry, 17(7), 1236–1245.

Booij, K., Smedes, F., & Van Weerlee, E. M. (2002). Spiking of performance reference compounds in low density polyethylene and silicone passive water samplers. Chemosphere, 46(8), 1157–1161. https://doi.org/10.1016/S0045-6535(01)00200-4

Cho, Y.-M., Ghosh, U., Kennedy, A. J., Grossman, A., Tay, G., Tomaszewski, J. E., Smithenry, D.W., Bridges, T. S., Luthy, R. G. (2009). Field application of activated carbon amendment for in-situ stabilization of polychlorinated biphenyls in marine sediment. Environmental Science & Technology, 43, 10, 3815–3823.

Cornelissen, G., Pettersen, A., Broman, D., Mayer, P., & Breedveld, G. D. (2008). Field testing of equilibrium passive samplers to determine freely dissolved native polycyclic aromatic hydrocarbon concentrations. Environmental Toxicology and Chemistry, 27(3), 499–508. https://doi.org/10.1897/07-253.1

Dixon-Anderson, E., & Lohmann, R. (2018). Field-testing polyethylene passive samplers for the detection of neutral polyfluorinated alkyl substances in air and water. Environmental Toxicology and Chemistry, 37(12), 3002–3010. https://doi.org/10.1002/etc.4264

Eek, E., Cornelissen, G., & Breedveld, G. D. (2010). Field measurement of diffusional mass transfer of HOCs at the sediment-water interface. Environmental Science & Technology, 44(17), 6752–6759. https://doi.org/10.1021/es100818w

Fernandez, L. A., Harvey, C. F., & Gschwend, P. M. (2009a). Using perfprmance reference compounds in polyethylene passive samplers to chemicals. Environmental Science & Technology, 43, 8888–8894. https://doi.org/10.1021/es901877a

Fernandez, L. A., Lao, W., Maruya, K. A., & Burgess, R. M. (2014). Calculating the diffusive flux of persistent organic pollutants between sediments and the water column on the palos verdes shelf superfund site using polymeric passive samplers. Environmental Science & Technology, 48(7), 3925–3934. https://doi.org/10.1021/es404475c

Fernandez, L. A., Macfarlane, J. K., Tcaciuc, A. P., & Gschwend, P. M. (2009b). Measurement of freely dissolved PAH concentrations in sediment beds using passive sampling with low-density polyethylene strips. Environmental Science & Technology, 43(5), 1430–1436. https://doi.org/10.1021/es802288w

Gao, X., Xu, Y., Ma, M., Rao, K., & Wang, Z. (2019). Simultaneous passive sampling of hydrophilic and hydrophobic emerging organic contaminants in water. Ecotoxicology and Environmental Safety, 178 (January), 25–32. https://doi.org/10.1016/j.ecoenv.2019.04.014

Ghosh, U., Lombard, N., Bokare, M., Pinkney A., Yonkos, L and Harrison, R. (2020). Passive samplers and mussel deployment, monitoring, and sampling for organic constituents in Anacostia River tributaries: Final Report 2016–2018. Washington, DC. Retrieved from https://www.dropbox.com/sh/mcllipdkkr5dkhq/AADA0x1DA_ds922mtK6Fl0Yha/ 5.0 Supporting Technical Documents/5.4 Passive Sampler and Mussel Report?dl=0& preview=Passive+samplers+and+Mussel+Deployment+Year+2+Final+Report++2016-2018.pdf&subfoldcr_nav_tracking=1

Ghosh, U., Kane Driscoll, S., Burgess, R. M., Jonker, M. T. O., Reible, D., Gobas, F., Choi, Y., Apitz, S. E., Maruya, K. A., Gala, W. R., Mortimer, M., & Beegan, C. (2014). Passive sampling methods for contaminated sediments: practical guidance for selection, calibration, and implementation. Integrated Environmental Assessment and Management, 10(2), 210–223. https://doi.org/10.1002/ieam.1507

Ghosh, U., Zimmerman, J. R., & Luthy, R. G. (2003). PCB and PAH speciation among particle types in contaminated harbor sediments and effects on PAH bioavailability. Environmental Science & Technology, 37(10), 2209–2217. https://doi.org/10.1021/es020833k

Gobas, F. A. P. C. (1993). A model for predicting the bioaccumulation of hydrophobic organic chemicals in aquatic food-webs: application to Lake Ontario. Ecological Modelling, 69, 1–17. https://doi.org/10.1016/0304-3800(93)90045-T

Harman, C., Reid, M., & Thomas, K. V. (2011). In situ calibration of a passive sampling device for selected illicit drugs and their metabolites in wastewater, and subsequent year-long assessment of community drug usage. Environmental Science & Technology, 45(13), 5676–5682. https://doi.org/10.1021/es201124j

Hawthorne, S. B., Azzolina, N. A., Neuhauser, E. F., & Kreitinger, J. P. (2007). Predicting bioavailability of sediment polycyclic aromatic hydrocarbons to Hyalella azteca using equilibrium partitioning, supercritical fluid extraction, and pore water concentrations. Environmental Science & Technology, 41(17), 6297–6304. https://doi.org/10.1021/es0702162

Hawthorne, S. B., Grabanski, C. B., & Miller, D. J. (2006). Measured partitioning coefficients for parent and alkyl polycyclic aromatic hydrocarbons in 114 historically contaminated sediments: Part 1. KOC values. Environmental Toxicology and Chemistry, 25(11), 2901–2911. https://doi.org/10.1897/06-115R.1

Hong, L., Ghosh, U., Mahajan, T., Zare, R. N., & Luthy, R. G. (2003). PAH sorption mechanism and partitioning behavior in lampblack-impacted soils from former oil-gas plant sites. Environmental Science & Technology, 37(16), 3625–3634. https://doi.org/10.1021/es0262683

Huckins, J. N., Petty, J. D., Lebo, J. A., Almeida, F. V., Booij, K., Alvarez, D. A., Cranor, W. L., Clark, R. C., & Mogensen, B. B. (2002). Development of the permeability/performance reference compound approach for in situ calibration of semipermeable membrane devices. Environmental Science and Technology, 36(1), 85–91. https://doi.org/10.1021/es010991w

Jahnke, A., Mayer, P., & McLachlan, M. S. (2012). Sensitive equilibrium sampling to study polychlorinated biphenyl disposition in baltic sea sediment. Environmental Science & Technology, 46(18), 10114–10122. https://doi.org/10.1021/es302330v

Jalalizadeh, M., & Ghosh, U. (2016). In situ passive sampling of sediment porewater enhanced by periodic vibration. Environmental Science & Technology, 50(16), 8741–8749. https://doi.org/10.1021/acs.est.6b00531

Jalalizadeh, M., & Ghosh, U. (2017). Analysis of measurement errors in passive sampling of Porewater PCB concentrations under static and periodically vibrated conditions. Environmental Science & Technology, 51(12), 7018–7027. https://doi.org/10.1021/acs.est.7b01020

Jonker, M. T. O., & Koelmans, A. A. (2001). Polyoxymethylene solid phase extraction as a partitioning method for hydrophobic organic chemicals in sediment and soot. Environmental Science & Technology, 35(18), 3742–3748. https://doi.org/10.1021/es0100470

Khalil, M. F., Ghosh, U., & Kreitinger, J. P. (2006). Role of weathered coal tar pitch in the partitioning of polycyclic aromatic hydrocarbons in manufactured gas plant site sediments. Environmental Science & Technology, 40(18), 5681–5687. https://doi.org/10.1021/es0607032

Kreitinger, J. P., Neuhauser, E. F., Doherty, F. G., & Hawthorne, S. B. (2007). Greatly reduced bioavailability and toxicity of polycyclic aromatic hydrocarbons to Hyalella azteca in sediments from manufactured-gas plant sites. Environmental Toxicology and Chemistry, 26(6), 1146–1157. https://doi.org/10.1897/06-207R.1

Lampert, D. J., Thomas, C., & Reible, D. D. (2015). Internal and external transport significance for predicting contaminant uptake rates in passive samplers. Chemosphere, 119, 910–916. https://doi.org/10.1016/j.chemosphere.2014.08.063

Leslie, H. A., Ter Laak, T. L., Busser, F. J. M., Kraak, M. H. S., & Hermens, J. L. M. (2002). Bioconcentration of organic chemicals: Is a solid-phase microextraction fiber a good surrogate for biota? Environmental Science & Technology, 36(24), 5399–5404. https://doi.org/10.1021/es0257016

Liu, Y., Wang, S., McDonough, C. A., Khairy, M., Muir, D. C. G., Helm, P. A., & Lohmann, R. (2016). Gaseous and freely-dissolved PCBs in the lower great lakes based on passive sampling: spatial trends and air-water exchange. Environmental Science & Technology, 50(10), 4932–4939. https://doi.org/10.1021/acs.est.5b04586

Lohmann, R., Burgess, R. M., Cantwell, M. G., Ryba, S. A., MacFarlane, J. K., & Gschwend, P. M. (2004). Dependency of polychlorinated biphenyl and polycyclic aromatic hydrocarbon bioaccumulation in Mya arenaria on both water column and sediment bed chemical activities. Environmental Toxicology and Chemistry, 23(11), 2551–2562. https://doi.org/10.1897/03-400

Lombard, N. J., Ghosh, U., Kjellerup, B. V., & Sowers, K. R. (2014). Kinetics and threshold level of 2,3,4,5-tetrachlorobiphenyl de-chlorination by an organohalide respiring bacterium. Environmental Science & Technology, 48(8), 4353–4360. https://doi.org/10.1021/es404265d

Lu, X., Skwarski, A., Drake, B., & Reible, D. D. (2011). Predicting bioavailability of PAHs and PCBs with porewater concentrations measured by solid-phase microextraction fibers. Environmental Toxicology and Chemistry, 30(5), 1109–1116. https://doi.org/10.1002/etc.495

Lydy, M. J., Landrum, P. F., Oen, A. M., Allinson, M., Smedes, F., Harwood, A. D., Li, H., Maruya, K. A., & Liu, J. (2014). Passive sampling methods for contaminated sediments: state of the science for organic contaminants. Integrated Environmental Assessment and Management, 10(2), 167–178. https://doi.org/10.1002/ieam.1503

Maruya, K. A., Risebrough, R. W., & Horne, A. J. (1996). Partitioning of polynuclear aromatic hydrocarbons between sediments from San Francisco Bay and their porewaters. Environmental Science & Technology, 30(10), 2942–2947. https://doi.org/10.1021/es950909v

Mayer, P., Parkerton, T. F., Adams, R. G., Cargill, J. G., Gan, J., Gouin, T., Gschwend, P. M., Hawthorne, S. B., Helm, P., Witt, G., You, J., & Escher, B. I. (2014). Passive sampling methods for contaminated sediments: scientific rationale supporting use of freely dissolved concentrations. Integrated Environmental Assessment and Management, 10(2), 197–209. https://doi.org/10.1002/ieam.1508

Mayer, P., Tolls, J., Hermens, J. L. M., & Mackay, D. (2003). Equilibrium Sampling Devices. Environmental Science & Technology, 37(9), 184A–191A. https://doi.org/10.1021/es032433i

Mayer, P., Vaes, W. H. J., Wijnker, F., Legierse, K. C. H. M., Kraaij, R., Tolls, J., & Hermens, J. L. M. (2000). Sensing dissolved sediment porewater concentrations of persistent and bioaccumulative pollutants using disposable solid-phase microextraction fibers. Environmental Science & Technology, 34(24), 5177–5183. https://doi.org/10.1021/es001179g

Mcgroddy, S. E., & Farrington, J. W. (1995). Sediment porewater partitioning of polycyclic aromatic hydrocarbons in three cores from Boston Haihor, Massachusetts. Environmental Science & Technology, 29(6), 1542–1550. https://doi.org/10.1021/es00006a016

Needham, T. P., Payne, R. B., Sowers, K. R., & Ghosh, U. (2019). Kinetics of PCB microbial de-chlorination explained by freely dissolved concentration in sediment microcosms. Environmental Science & Technology, 53(13), 7432–7441. https://doi.org/10.1021/acs.est.9b01088

Oen, A. M. P., Janssen, E. M. L., Cornelissen, G., Breedveld, G. D., Eek, E., & Luthy, R. G. (2011). In situ measurement of PCB pore water concentration profiles in activated carbon-amended sediment using passive samplers. Environmental Science & Technology, 45(9), 4053–4059. https://doi.org/10.1021/es200174v

Payne, R. B., Ghosh, U., May, H. D., Marshall, C. W., & Sowers, K. R. (2019). A pilot-scale field study: In situ treatment of PCB-impacted sediments with bioamended activated carbon. Environmental Science & Technology, 53(5), 2626–2634. research-article. https://doi.org/10.1021/acs.est.8b05019

Perron, M. M., Burgess, R. M., Suuberg, E. M., Cantwell, M. G., & Pennell, K. G. (2013). Performance of passive samplers for monitoring estuarine water column concentrations: 2. Emerging contaminants. Environmental Toxicology and Chemistry, 32(10), 2190–2196. https://doi.org/10.1002/etc.2248

Pintado-Herrera, M. G., Lara-Martín, P. A., González-Mazo, E., & Allan, I. J. (2016). Determination of silicone rubber and low-density polyethylene diffusion and polymer/water partition coefficients for emerging contaminants. Environmental Toxicology and Chemistry, 35(9), 2162–2172. https://doi.org/10.1002/etc.3390

Posada-Ureta, O., Olivares, M., Zatón, L., Delgado, A., Prieto, A., Vallejo, A., Paschke, A., & Etxebarria, N. (2016). Uptake calibration of polymer-based passive samplers for monitoring priority and emerging organic non-polar pollutants in WWTP effluents. Analytical and Bioanalytical Chemistry, 408(12), 3165–3175. https://doi.org/10.1007/s00216-016-9381-7

Sanders, J. P., McBurney, A., Gilmour, C. C., Schwartz, G. E., Washburn, S., Kane Driscoll, S. B., Brown, S. S., & Ghosh, U. (2020). Development of a Novel Equilibrium Passive Sampling Device for Methylmercury in Sediment and Soil Porewaters. Environmental Toxicology and Chemistry, 39(2), 323–334. https://doi.org/10.1002/etc.4631

Smedes, F., Rusina, T. P., Beeltje, H., & Mayer, P. (2017). Partitioning of hydrophobic organic contaminants between polymer and lipids for two silicones and low density polyethylene. Chemosphere, 186, 948–957. https://doi.org/10.1016/j.chemosphere.2017.08.044

Smith, K. E. C., Rein, A., Trapp, S., Mayer, P., & Karlson, U. G. (2012). Dynamic passive dosing for studying the biotransformation of hydrophobic organic chemicals: Microbial degradation as an example. Environmental Science & Technology, 46(9), 4852–4860. https://doi.org/10.1021/es204050u

Taylor, V. F., Buckman, K. L., & Burgess, R. M. (2019a). Preliminary investigation of polymer-based in situ passive samplers for mercury and methylmercury. Chemosphere, 234, 806–814. https://doi.org/10.1016/j.chemosphere.2019.06.093

Taylor, A. C., Fones, G. R., Vrana, B., & Mills, G. A. (2019b). Applications for passive sampling of hydrophobic organic contaminants in water—A review. Critical Reviews in Analytical Chemistry, 0(0), 1–35. https://doi.org/10.1080/10408347.2019.1675043

Tuikka, A. I., Leppänen, M. T., Akkanen, J., Sormunen, A. J., Leonards, P. E. G., van Hattum, B., van Vliet, L. A., Brack, W., Smedes, F., & Kukkonen, J. V. K. (2016). Predicting the bioaccumulation of polyaromatic hydrocarbons and polychlorinated biphenyls in benthic animals in sediments. Science of the Total Environment, 563–564, 396–404. https://doi.org/10.1016/j.scitotenv.2016.04.110

US EPA. (2012). Guidelines for Using Passive Samplers to Monitor Organic Contaminants at Superfund Sediment Sites. Retrieved from https://cfpub.epa.gov/si/si_public_record_report.cfm?Lab=NHEERL&dirEntryId=238596

US EPA. (2019). Grasse River Superfund Site, Massena, NY | EPA in New York | US EPA. Retrieved September 30, 2020, from https://archive.epa.gov/epa/ny/grasse-river-superfund-site-massena-ny.html

US EPA. (2020a). Hudson River PCBs Superfund Site | US EPA. Retrieved September 30, 2020, from https://www.epa.gov/hudsonriverpcbs

US EPA. (2020b). National Recommended Water Quality Criteria - Human Health Criteria Table. Retrieved from https://www.epa.gov/wqc/national-recommended-water-quality-criteria-human-health-criteria-table

Werner, D., Hale, S. E., Ghosh, U., & Luthy, R. G. (2010). Polychlorinated biphenyl sorption and availability in field-contaminated sediments. Environmental Science & Technology, 44(8), 2809–2815. https://doi.org/10.1021/es902325t

2 Occurrence and Removal of Endocrine Disruptor Chemicals in Wastewater

V. Ponnusami* and M. Saravanan
Biomass Conversion and Bioproducts Laboratory,
Center for Bioenergy, School of Chemical & Biotechnology,
SASTRA Deemed University, Thanjavur, India

CONTENTS

2.1 INTRODUCTION

Clean and safe water for human consumption is a basic need, and meeting the per capita demand is still a significant challenge in many parts of the world. The range of contaminants found in natural water reservoirs is increasing due to anthropogenic and biological activities. Recently, there is an emerging concern over endocrine disruptor compounds (EDCs) in surface and groundwater sources and wastewater as EDCs pose a severe threat to aquatic life and human health. The evidence of EDCs related to negative health outcomes is on the rise. Even at very low concentrations in the order of a few ng/L, EDCs can have serious health effects in aquatic species [1]. The adverse health effects caused by EDCs include neurological and behavioral changes (Alzheimer's, Parkinson's, etc.), thyroid problems, infertility, early puberty, obesity, asthma, heart disease, hypertension, type 2 diabetes, low-birth weight, immune deficiency, breast/testicular/prostate cancer, etc. [2, 3].

As defined by the World Health Organization (WHO), an endocrine disruptor is "an exogenous substance or mixture that alters function(s) of the endocrine system and consequently causes adverse health effects in an intact organism, or its progeny, or (sub) populations" [4, 5]. Endocrine systems are one of the two central regulatory systems in the body. It consists of glands that secrete hormones that are carried in the bloodstream around the body.

2.2   SOURCES AND HEALTH EFFECTS OF EDCs

Wide varieties of chemicals are included in the list of established/potential EDCs. According to a recent report, over 70,000 chemicals are suspected of causing endocrine disruption [6]. EDCs include a broad range of chemicals that are structurally different. Despite their structural diversity, all these compounds have one common characteristic: they all interfere with reproduction, growth, and development of organisms. EDCs are believed to interfere with natural hormones in many different mechanisms. They can mimic natural hormones, alter their synthesis and metabolic rate, and even block them [7, 8]. How EDCs affect health are still not completely understood.

EDCs in wastewater come from a broad range of anthropogenic and naturogenic sources. Phytoestrogens, mycoestrogens, and heavy metals found on the earth's crusts are a few examples of commonly encountered natural EDCs [1]. Examples of anthropogenic sources of EDCs include processed food, pesticides, fungicides, herbicides, estrogens, phytoestrogens, antibiotics, nonsteroidal anti-inflammatory drugs, plasticizers, phenolic compounds, phthalates, fire retardants, surfactants, consumer products (toys, PVC pipes, plastic bottles) and personal care products (PPCPs) (triclosan, cosmetics, toothpaste), heavy metals, etc. [2, 8–14]. A list of representative EDCs is given in Table 2.1. Surface and groundwater resources are frequently contaminated with these compounds discharged from hospitals, industries, domestic sewage, agricultural runoffs, and animal breeding [11, 15]. As conventional effluent/sewage treatment systems cannot eradicate EDCs; they find their way to the water reservoirs and affect the ecosystem [2, 16]. These persistent compounds are transferred through the food chain and accumulate in human tissues causing various health disorders, as mentioned above.

Phytoestrogens and mycoestorgens are capable of mimicking human estrogens and affect the normal functioning of natural hormones. E1 and EE2 cause feminization in aquatic life [2]. Exposure to the heavy metals cadmium, arsenic, lead, and mercury causes various health impacts such as breast/estrogen-dependent cancers, sperm regeneration, suppression of gonadotropins, and testicular androgen, interstitial cell hyperplasia, and cystic hyperplasia of the uterus, etc. [5]. Arsenic and cadmium mimic estrogen at low concentrations and exhibit cytotoxicity at high concentrations [17]. Although these disorders are linked to hormonal imbalance, direct evidence suggesting the link to endocrine-disrupting mechanisms are only limited. Reports indicate that these diseases' etiology may be chemical toxicity of heavy metals rather than endocrine disruption. A thorough and detailed survey on clinical evidence and published reports on in vitro/in vivo studies concludes that heavy metal exposure certainly increases the risk of endocrine-related health disorders [5]. The heavy metals having oestrogenic properties are called metalloestrogens [18].

Bisphenol A is known to possess endocrine properties and can cause diabetes, hypertension, prostate cancer, breast cancer, early puberty, reproductive disorders, etc. [19, 20]. The estimated global average intake of BPA is 30.76 ng/kg body weight per day [20]. Exposure to phthalates can result in decreased fecundity, adverse obstetrical outcomes, and miscarriage [21]. Surfactants are another class of EDCs that are

TABLE 2.1
Different classes of EDCs

Class	EDC	Reference
Heavy metals	Arsenic, Cadmium, Lead	[18, 54–56]
Natural hormones (estrogens)	Estron (E1), 17β-estradiol (E2), Estriol (E3), Estetrol (E4), 17α-Ethinylestradiol (EE2)	[22, 57]
Pharmaceutical drugs	Aminopyrine, Aspirin (NSAIDs), Carbamazepine, Ciprofloxacin, Clofibrate, Diclofenac (NSAIDs), Erythromycin, Fenofibrate, Fenoprofen, Gemfibrozil, Ibuprofen (NSAIDs), Ketoprofen, Naproxen, Phenazone, Penicillin, salicylic acid, streptomycin, sulphonamides, tetracycline	[2, 24]
Industrial and domestic products, plasticizer	Alkylphenol, bis-2-ethylhexylphthalate, Bisphenol A, butyl benzyl phthalate, chlorophenol, di-cyclohexyl phthalate, di-ethylhexyl phthalate, di-ethyl phthalate, di-n-butyl phthalate (phthalates), nitrophenol, dioxins, nonylphenols, polyaromatic hydrocarbons	[22]
Pesticides, herbicides	Atrazine, benomyl, carbaryl, chlordane, chlorpyrifos, dichlorodiphenyltrichloroethane, dicofol, dieldrin, endosulfan, ethiozine, linden, linuron, molinate, paraquat, pyrethroids, petntachlorophenol, parathion, methoxychlor, malathion, toxaphene, transnonachlor,	[22, 31]
Pytoestrogen	Coumestans, isoflavones, lignans	

frequently used in daily life. The majority of the domestic and industrial surfactants constitute alkylphenols like nonylphenol or octylphenol. Nonylphenol ethoxylates are used in several industrial applications, including paints, pesticides, plastics, textiles, paper and pulp, etc. Nanophenol compounds are capable of mimicking B & C rings of 17β-estradiol. Impact on endocrine systems had been reported in freshwater and marine animals and humans due to the exposure of nonylphenol [12]. Metabolites of polyethoxylates are known to be recalcitrant and possess endocrine disruption properties [22].

Several antibiotics, anti-inflammatory, antipsychotic, antiviral, and steroidal drugs are found in sewage in many countries, including China, England, India, the United States, Canada, and Germany [11, 23, 24]. Concentrations of these compounds were typically found in the range of a few ng/L to mg/L [23]. Exposure to antibiotic compounds increase drug resistance in microbial population and alters microbial community in nature. Antiepileptic drugs are well known to produce side effects affecting reproductive endocrine systems [11]. Diclofenac, a nonsteroidal anti-inflammatory drug (NSAID), causes estrogenic effects and teratogenicity. When *Oryzias latipes* fish were exposed to 1 µg/L diclofenac, CYP1A gene expression was increased by 5.7, 9.3, and 18.4 times in gills, liver, and intestines, respectively [25]. At a concentration of 4 mg/L, it causes teratogenicity *X. laevis* [26]. Altered thyroid-stimulating hormone levels are reported in fish due to exposure of NSAIDs such as diclofenac [27].

2.3 TREATMENT TECHNOLOGIES

Most of the EDCs are recalcitrant and are not removed in conventional sewage/effluent treatment plants. Owing to the structural diversity of EDCs, a single technology may not be sufficient to remove all compounds below detectable/permissible limits. Adsorption, advanced oxidation, membrane filtration, and biological treatments are in use [28]. The choice of the best technology for a given EDC depends on its structure. As the contaminated water contains many different EDCs originated from various sources, a combination of the above treatment technologies needs to be adopted to produce desired results to lower the concentrations of the EDCs below permissible/detection limits.

2.3.1 BIOLOGICAL TREATMENTS

Biological treatments make use of microorganisms or enzymes to degrade pollutants. Microbial degradation is classified as an aerobic, anaerobic, or anoxic process based on the electron acceptor type. The structure of EDCs strongly affects the degradation rate and efficiency in biological treatment processes. It can take up to four months to completely degrade estrogens such as E1, E2, and E3 in an aerobic biological oxidation process. The use of alternate electron acceptors such as nitrates and sulphates only decreases the degradation efficiency for E1, E2, and E3. Among the various electron acceptors, in the presence of nitrate, the pollutant BPA could not be degraded [29]. Up to 90% of polychlorinated biphenyl (PCB) could be removed in biological treatment methods like ASP and biofiltration. Compounds like EE2 are reported to be highly stable and not biodegradable. Under anaerobic conditions, EE2 is found to be persistent for nearly three years [22, 30]. Nonylphenol is another example of EDCs that cannot be removed efficiently in conventional secondary treatment. Under nitrifying conditions, up to 77% of nonylphenol can be removed in the biological treatment method when the organic loading is sufficiently low [31]. Pholchan et al. (2008) studied microbial degradation of E1, E2, and EE2 in nitrate – accumulating sequential batch reactors. The first reactor was operated under aerobic conditions with a sludge retention time ranging from 1.7 to 17.1 days, and the second reactor under aerobic/anaerobic/anoxic conditions. More than 98% of E1 and E2 removal was achieved in this sequential process under optimized conditions. When SRT was less than 5.7 days, the removal efficiency was abysmal, particularly for E1. The reduction of EE2 was not significant and required alternating redox conditions [32].

One primary concern over the advanced biological treatments is that the intermediates generated are often found to be more toxic than the original compounds [33]. Therefore, the performance of any given biological treatment should not be evaluated only based on the removal efficiency. The toxicity of the final effluent must be considered to assess the treatment efficacy. A consortium of five fungi isolates, *A. Niger*, *M. Circinelloides*, *R. micorsporous*, *T. longgibrachiatum*, and *T. polyzona*, could oxidize a synthetic solution containing 1 mg each of carbamazepine, diclofenac, and ibuprofen per liter of effluent in a sequencing batch reactor. Three different amounts removed (89.77%, 95.8%, and 94.1%) of carbamazepine, diclofenac, and ibuprofen, respectively, were achieved after 24 hours of incubation. At this stage, concentrations of estrogenic intermediates, measured as estradiol equivalents, were below detection and quantification limits [33].

With recent advances in enzyme technology, enzymatic degradation of many EDCs appears to be a feasible and economical option for the removal of EDCs from wastewater. Enzymes are biocatalysts and require only mild reaction conditions. From an economic point of view, loss of enzymes is one of the major issues hampering large-scale enzymatic oxidation implementation. This necessitates recovery and reuse of enzymes, and it becomes very critical to make the enzymatic processes economically attractive [20]. Immobilization of enzyme facilitates continuous operation and improves enzyme stability and reusability [34]. Researchers had investigated packed bed and fluidized bed reactors. The packed bed approximates plug flow and provides better contact between the substrate and enzyme. In packed bed operation, additional pumping energy may be required to overcome the pressure drop across the bed. Mass transfer resistances are high in packed bed and aeration efficiency is low. Fluidized bed operation ensures better mass transfer and aeration efficiency. However, the contact between the enzyme and substrate is poor in fluidized bed operation.

Peroxidase and laccase enzymes can produce highly reactive free radicals that can completely oxidize various substrates such as bisphenol A. Optimum pH and temperature for laccase and peroxidase enzymes are 5–7 and 25–30°C, respectively. Organisms such as *Ascomycetes, Basidiomycetes, Deuteromycetes, Myceliophthora thermophile, Phanerochaete sordida, Pyncoporus sanguineus, Trametes versicolor,* and *T. pubescens* express laccase enzyme. A trinuclear copper site present in the laccase enzyme structure is identified as responsible for its redox activity [35]. Fungal laccase can oxidize various EDCs, including estrogens E1, E2, E3, and EE2. When degradation of E1 and E2 (initial concentration 5 mg/L each) was studied in a fed-batch reactor with pulse addition of laccase (500 U/l every two hours for 8 hours), 94.1% and 95.5% removal were achieved for estrogen and estradiol, respectively [36]. In the same work, degradation of E1 and E2 was studied using a continuous flow setup consisting of two slurry reactors and an ultrafiltration membrane. The enzyme was added at a 1 mg/L rate, and four hours hydraulic retention time was provided. The authors found that percentage removal of E1 and E2 increased to 95.6 and >98%, respectively. By using immobilization, enzymes could be quickly recovered and reused continuously. Up to 97% of toxicity reduction was achieved in the continuous operation studied in this work [36]. Laccase immobilized on various supports had been reviewed to remove EDCs such as bisphenol A, estrogens, tetracycline, diclofenac, etc. by different research groups. Immobilization of laccase on polytetrafluoroethylene microtubes was demonstrated by Lloret *et al.* [37] with Glutaraldehyde/paraformaldehyde as crosslinkers. At a flow rate lower than 1 µL/min (corresponding HRT 26 min), more than 95% removal of the estrogens E1, E2, and E3 (initial concentration 18 µM each) were achieved in this microreactor setup [37]. In another study, 90% removal of bisphenol A from a 10 mg/L bisphenol solution was achieved with a multi-walled carbon nanotube modified

TABLE 2.2
Fugal/enzyme treatment of selected EDCs

EDC	Funus/enzyme	Remarks	Reference
E1 and E2	Commercial laccase	95% E1 was removed and E2 was not detectable after treatment in fed batch reactor	[36]
Bisphenol A	Immobilized laccase	~100% removal	[58]
Estrogens and polyphenols in hospital waste	Immobilized laccase	>82% removal	[59]
E1 and E2	Fed batch	94.1% E1 & 95.5% E2 removal	[36]
E1 and E2	Continuous	95.6% E1 & >98% E2 removal	[36]
Bisphenol A	Batch	99% removal from 20 mg/L solution in 2 hours	[60]
Tetracycline	Free laccase Immobilized laccase	30% removal (free enzyme) 56% removal (immobilized enzyme)	[39]
Bisphenol A	Laccase in enzymatic membrane reactor	>95%	[40]
Diclofenac	Laccase in enzymatic membrane reactor	>80%	[40]
BPA, p-t-octylphenol, estrogens	Lignin peroxidase from *Phanerochaete sordida* YK-624	23.9–45% removal	[61]
BPA	Horseradish peroxidase (HRP) immobilized on poly (methyl methacrylate-co-ethyl acrylate) microfibrous membranes	93% removal efficiency obtained with immobilized enzyme; for free enzyme and bare membrane removal efficiencies were 61% and 42%, respectively	[62]

laccase immobilized on electrospun fiberous membrane [38]. The removal of tetracycline was studied using a free and immobilized enzyme from a solution containing 20 mg/L of tetracycline. Tetracycline removal efficiencies were 30% and 56%, respectively, for free enzymes and immobilized enzymes [39]. Nguyen *et al.* (2014) suspended an ultrafiltration membrane inside the enzyme reactor to facilitate the separation of enzymes and prevent wash out enzymes. Such a reactor setup used to remove diclofenac and bisphenol A from wastewater using laccase catalyst with a pollutant loading rate of 570 ± 70 µg/L d of bisphenol A and 480 ± 40 µg/L d of diclofenac. Removal efficiency for bisphenol A and diclofenac were more than 95% and 80%, respectively [40]. A brief summary of recent works on fungal degradation of EDCs is shown in Table 2.2.

2.3.2 ADSORPTION

Adsorption is widely used to remove various contaminants from aqueous streams. It is a surface phenomenon. In adsorption, the pollutant (adsorbate) is transferred from an aqueous solution (wastewater, raw water, etc.) to a solid material known as an adsorbent. The binding of the sorbate to adsorbent can be either due to physical forces or chemical binding. Adsorption has been reported as the most preferred method for removing EDCs from heavy metals to pesticides and pharmaceuticals in the recent past. There are three main advantages of adsorption over other processes: it is (i) simple, (ii) less expensive compared to other techniques, and (iii) does not produce any by-products. Much literature is available on the design and synthesis of reusable adsorbents with superior characteristics like high surface area and desired surface charges [16]. Adsorption of EDCs such as arsenic, bisphenol A (BPA), biphenyls, cadmium, dibenzofuran, dibenzo-p-dioxin, diethyl phthalate, natural and synthetic estrogens, pesticides, etc., on various adsorbents, has been reported in the literature [13, 41].

Adsorbent properties (such as specific surface area, active sites for adsorption, surface charge, hydrophobicity, particle size, pore size distribution, adsorption capacity, etc.), adsorbate properties (molecular size, solubility, ionic charge, etc.), and process variable (such as pH, temperature) determine the choice of the adsorbent for wastewater treatment processes.

Adsorption is carried out either in batch mode or in a continuous manner (Figures 2.1 and 2.2). In a batch process, a definite quantity of adsorbent is added to the aqueous solution, and the mixture is allowed to attain equilibrium. Then the slurry is passed through a filter to separate the adsorbent. The separated adsorbent may be regenerated and reused as long as the adsorbent retains its adsorption capacity, and the regeneration is viable. In continuous mode, a bed made of granular adsorbent is used. The contaminated solution is passed through it typically from the bottom to ensure proper distribution of the liquid across the bed and ensure sufficient contact between the adsorbent and adsorbate. The outlet is collected from the top, and the contaminant's concentration in the outlet stream is monitored. The bed is taken out of service once the outlet concentration reaches a breakthrough limit (typically 10% of the initial concentration in the feed), and regeneration of the bed is carried out. It is common practice to have two parallel streams of beds to ensure continuous operation. When one stream is in service, the other one is regenerated and kept ready for service. A packed bed of granular activated carbon is typically used as a polishing step in conventional sewage/effluent treatment plants.

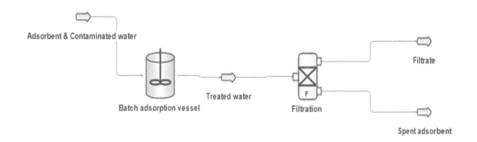

FIGURE 2.1   Schematic diagram for batch adsorption process.

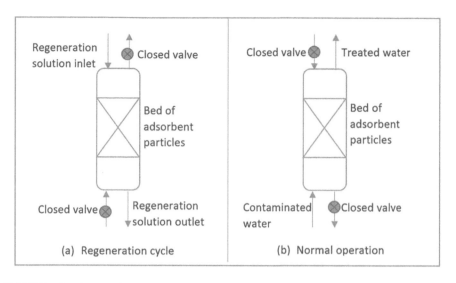

FIGURE 2.2 Schematic diagram for packed bed adsorption.

As a polishing step, granular activated carbon adsorption is highly effective even at very low concentrations of pollutants, in the range of a few ng/L, and reduces pollutant concentration below permissible limits.

As adsorption is a surface phenomenon, a higher specific surface area of the adsorbent is favorable for adsorption. Smaller particles possess higher surface area and hence high adsorption potential. However, fine particles pose operational challenges both in slurry operation as well as packed bed adsorption. The filtration of the fine particles from the slurry after adsorption is a challenge. In a bed filled with fine particles, the void fraction is very small, and it offers enormous resistance to flow and, hence, increases the pressure drop across the bed. The contaminant is adsorbed both on the external and internal surface of the adsorbents. Pores increase the specific surface area of the adsorbent. For adsorption within the pores, the solute molecule has to diffuse into the pores, which involves internal mass transfer resistance. Adsorbate molecules smaller than the pore size quickly diffuse into the pore and get adsorbed. Oppositely charged adsorbent and adsorbate have better affinity due to electrostatic attraction. The pH of the contaminated water affects the net surface charge of the adsorbent and hence strongly influences the adsorption. Temperature increases the mobility of the molecules and favors adsorption. However, at sufficiently high temperature (>60°C), desorption might occur, and hence optimum temperature must be identified for each adsorbent–adsorbate system. Availability and cost of adsorbents are other factors to be considered while choosing an adsorbent.

Activated carbon is the most commonly used adsorbent in commercial applications due to its high surface area and adsorption potential. However, as AC is considered expensive, researchers have taken various efforts in the recent past to develop alternate adsorbents that are inexpensive or possess improved adsorption potential. Bamboo biochar, $CuZnFe_2O_4$ bamboo–biochar, graphene-based monolith, metal-organic framework, modified multi-walled carbon nanotubes [42], polymer beads, and polysulfone beads are some adsorbents recently reported for removal of bisphenol A [13]. Porous polysulfone (PSf) beads were used to remove bisphenol A, biphenyls, dibenzofuran, and dibenzo-p-dioxin. The hydrophobic interaction between the pollutants and PSf was identified as an influencing factor. It has been reported that the removal efficiency was higher for the compounds having a higher octanol-water distribution coefficient [43]. Adsorption capacities of various adsorbents reported in recent literature for selected EDCs are summarized in Table 2.3.

2.3.3 ADVANCED OXIDATION

In the advanced oxidation process (AOP), water contaminated with EDCs is treated using strong oxidizing agents. Permanganate, hydroxyl and sulfate radicals are frequently used oxidizing agents [20]. The formation of free radicals and subsequent oxidation of contaminant molecules are the two main steps of any advanced oxidation process. One advantage of AOPs is they can completely mineralize pollutants without leaving residual compounds behind. However, complete mineralization by AOPs is often not cost-effective on a large scale compared to biological treatment processes. Thus, AOPs are generally recommended as polishing steps. Electrochemical oxidation, Fenton reactions, ozonation, photocatalysis, photo-Fenton, and sonolysis are some AOPs used for wastewater treatment [44]. This chapter is limited to photocatalysis, as it is recognized as the most efficient advanced oxidation process. Many reviews are available in the literature on oxidation on EDCs by different AOPs [44].

Semiconductors such as TiO_2, ZnO, and Fe_2O_3 are the most commonly used photocatalysts. These photocatalysts generate electron-hole pairs upon irradiation with photons of a suitable wavelength. The wavelength required to promote electrons in the valence band of the catalyst to its conduction band depends on the catalyst's band gap energy. Typically TiO_2 has the bandgap energy in the range of 3.0–3.2 eV. Thus, UV light is required to excite the electrons from the valence band to the conduction band in TiO_2. These charge species (e-h pair) either recombine to attain the original state or separate and move to the catalyst surface to initiate redox reactions. Pollutants that are adsorbed on the catalysts' surface are oxidized by the highly oxidative free radicals generated at the catalyst surface. Commercial TiO_2 and ZnO have been used to oxidize various EDCs such as amoxicillin, ampicillin, carbofuran, ciprofloxacin, cloxacillin, estrogen, ibuprofen, metronidazole, tetracycline, etc. [45–47]. In order to reduce the process's energy requirement, the catalyst needs to be modified suitably to absorb visible light. Making the catalysts visible light-responsive and preventing e-h pair recombination will improve the photocatalytic efficiency. Doping with Ag, Bi, C, Cl, Co, Mn, N, Ni, etc. has been tried to shift the catalyst's bandgap energy to the visible regime. Doped semiconductor catalysts were widely used to degrade bisphenol A, estrogen, methyl parathion, phthalate esters, etc. [44, 48].

Recently hydrogel-based photocatalysts had been tested for degradation of EDCs [49, 50]. The three-dimensional hydrogel of TiO_2/rGH has been shown to completely oxidize bisphenol from a 20 mg/L initial solution [51]. A hydrogel prepared by incorporating reduced graphene oxide–AgBr composite into reduced graphene had demonstrated superior performance in removing bisphenol A from wastewater compared to pristine AgBr. Electron-hole pairs generated by Ag and AgBr nanoparticles, upon

TABLE 2.3
Various adsorbents for the removal of EDCs

EDC	Adsorbent	Max. adsorption capacity (mg/g)	Remarks	Reference
Bisphenol A	Magnetic, MWCNT modified with chitosan	46.2		[42]
	Functionalized cellulose obtained from Spanish brooms	4.86	Simple and easy regeneration	[63]
	β-cyclodextrin capped graphene–magnetite nanocomposite	59.6	Physisorption; endothermic; post adsorption recovery and reuse by applying magnetic field and methanol regeneration; adsorbent is highly specific for bisphenol A and its structural analogs	[64]
	Phenolic resin based activated carbon fiber	185.19 – 277.78	Quick adsorption; 98–99.9% adsorption achieved within first 2 minutes; BET surface area of the fiber – 2342 m^2/g	[65]
	Nitrogen-doped multiwall carbon nanotubes (N-MCNT)	100.4	Adsorption through π – π electron – donor – acceptor interaction; specific surface area – 162 m^2/g; adsorption capacity 3.7–4 times higher than the undoped MCNT	[66]
	Hexadecyl–imidazole functionalized silica adsorbents	89.6	Hexadecyl–Imidazole synergistically favored BPA adsorption; optimum pH was in the range of 4–9; surface area – 169.52 m^2/g; π – π electron – donor – acceptor interaction is the major mechanism	[67]
	Bituminous coal-based activated carbon	328.3	Follows pseudo second order kinetics; Freundlich isotherm model; surface area 1060 m^2/g	[68]
	Coconut shell-based activated carbon	263.2	Follows pseudo second order kinetics; Freundlich isotherm model; surface area 916 m^2/g	[68]
	Westrvaco activated carbon (WV A1100) modified with nitric acid	432.34	Nitric acid treatment increased monolayer adsorption capacity from 382.12 mg/g to 432.34 mg/g surface area of modified AC was 1760 m^2/g	[69]
	$CoFe_2O_4$-powdered activated carbon (Sinopharm Group Chemical Reagent)	727.2	Because of magnetic property, recovery of adsorbent is after use is simple; exothermic, spontaneous process; follows Langmuir isotherm model and pseudo second order kinetics	[70]

Compound	Adsorbent	Value	Remarks	Ref.
E2	Zr/CM-β-CD	182.15	Follows Langmuir isotherm; homogeneous distribution of adsorption sites	[71]
	Zr/CM-β-CD	210.53	Follows Langmuir isotherm; homogeneous distribution of adsorption sites	[71]
Ciprofloxacin	Palygorskite – montmorillonite	107	Spontaneous, irreversible and endothermic process; strongly influenced by pH	[72]
Nofloxacin, sulfamerazine and oxytetracycline	KOH-modified cassava biochar	1.97, 0.67, and 10	BET surface area: pristine biochar– 75.3 m²/g, KOH modified biochar– 128.42m²/g; endothermic and spontaneous	[73]
Dibutyl phthalate	Red soil and black soil	4.26 and 2.252	Soil samples were blended with extraneous dissolved organic matter obtained from pig manure compost	[74]
	Calcium silicate grafted non-woven polypropylene (PP-g-CaSiO₃@SiO₂)	54.8	Lower temperature and lower pH favorable; fits Freundlich isotherm; can be recovered and reused	[75]
Phthalic acid	Zeolitic imidazolate framework (ZIF-8)	654	BET surface area – 1501 m²/g; binding forces between phthalic acid and ZIF-8 is favorable for adsorption	[76]
Phthalate esters	Magnetic zeolite/Fe₃O₄ composite	26–214	Followed internal diffusion kinetic model and Langmuir isotherm	[77]
	Zeolite	12–71	Followed internal diffusion kinetic model and Langmuir isotherm	[77]
Atrazine	P doped corn straw biochar	26.46	BET surface area 638.1 m²/g; physical adsorption; exothermic; spontaneous	[78]
Pesticides: Chlorpyrifos (Ch), Parathion (Pa), Malathion (Ma)	2-phenylethylamine functionalized graphene oxide based silica coated magnetic nanoparticles	Ch – 25.6; Pa – 135; Ma – 25.6	BET surface area 133 m²/g; follows second order kinetics; reusable	[79]
Wastewater sample containing eight different pesticides	Phenyl-modified magnetic graphene/mesoporous silica	–	BET surface area 446.5 m²/g; capacity varies with pesticide	[80]
4-chlorophenoxyacetic acid (4-CPA) and 2-(4-chlorophenoxy)-2-methyl propionic acid (CFA)	Modified activated carbon	4-CPA – 322 to 422; CFA – 298 to 442	Initial concentration 0.17–2.15 mmol L⁻¹ (4-CPA) and 0.15–1.87 mmol L⁻¹ (CFA); BET surface area 740 – 800 m²/g;	[81]

TABLE 2.4

Photocatalysts for the treatment of EDCs

EDCs	Catalysts	Remarks	Reference
Bisphenol A	MnO_2@nano hollow carbon sphere	95.3% removal within 10 minutes and complete degradation in 30 minutes	[82]
	TiO_2/wood charcoal composites	BET surface area: 45–270.89 m^2/g; removal efficiency 53.4% (1.85 times higher than pure TiO_2)	[83]
	SnO_2 quantum dots encapsulated carbon nanoflake	98% removal; synergistic adsorption and photocatalysis	[84]
	Zero-valent iron/iron carbide nanoparticles encapsulated in N-doped carbon matrix	Complete degradation of bisphenol A (initial concentration 10 mg/L); adsorption capacity 138 mg/g	[85]
	TiO_2–MoS_2–reduced graphene oxide composite	62.40% removal was achieved by combined adsorption and photocatalysis. Reduced Graphene Oxide favors and adsorption while MoS_2 acts as cocatalyst	[86]
Chlorpyrifos	rGO/ZnO	95.4% removal was achieved in 75 minutes at an initial concentration of 50 mg/L	[87]

irradiation of visible light, are quickly transferred to the surface mediated by rGO and rGH. This prevents the recombination of e-h pair and enhances oxidation-reduction reactions at the catalysts' surface, resulting in better performance. [52].

One major limitation of photocatalytic degradation as a remedial solution for removing EDCs from wastewater is the treated effluent's residual toxicity. Minimal reports are found in the literature on the assessment of the toxicity of the treated effluent. These studies indicate that though the original EDCs are degraded by photocatalysis, the treated effluent contains toxic intermediate products if the treatment is incomplete [46]. Therefore, it is essential to study the toxicity while deciding the extent of reaction required for complete treatment. However, treatment for extended duration adds up to the cost of treatment and increases the process's overall cost. There have been many attempts to synthesize photocatalysts of different structure and morphology to fine-tune their properties to achieve better performance. A few recent studies on EDCs removal by different catalysts are listed in Table 2.4.

2.4 SUSTAINABILITY

Though several technologies can be used to remove EDCs, considering the diversity of EDCs and their structure and concentrations in the effluent, an integrated treatment process is the most preferable for reducing the concentrations of EDCs below permissible levels in the treated effluent. A combination of physical, chemical, and biological methods is often required to bring down the concentration of EDCs below detectable/permissible limits without generating objectionable/toxic residues [1]. A process involving primary clarifier, ASP, membrane filtration, granular media filtration, and GAC adsorption followed by ozonation is used at Gwinnett County, Georgia, in the United States; it is efficient in removing various pharmaceutical and personal care products from the wastewater [53]. Despite various stages and types of treatments employed in wastewater treatment plants, surveys conducted on the performance of the existing plants operating in many countries had shown that EDCs are present in the treated effluent and downstream of the plant. Thus, to protect water resources and sustain the supply of adequate potable water to all, complete elimination of EDCs from wastewater must be achieved. This warrants the development of feasible and, at the same time, affordable technologies, and it is the need of the hour. It is also equally important to develop affordable and straightforward analytical techniques to detect EDCs at low concentrations in the order of ng/L.

REFERENCES

[1] J. O. Tijani, O. O. Fatoba, and L. F. Petrik, "A review of pharmaceuticals and endocrine-disrupting compounds: Sources, effects, removal, and detections," *Water, Air, & Soil Pollution*, vol. 224, no. 11, 2013.

[2] J. O. Tijani, O. O. Fatoba, O. O. Babajide, and L. F. Petrik, "Pharmaceuticals, endocrine disruptors, personal care products, nanomaterials and perfluorinated pollutants: a review," *Environmental Chemistry Letters*, vol. 14, no. 1, pp. 27–49, 2016.

[3] Ying, G.-G. "Endocrine Disrupting Chemicals. What? Where?". In Analysis of Endocrine Disrupting Compounds in Food, L.M.L. Nollet (Ed.). Chapter 1. https://doi.org/10.1002/9781118346747.

[4] W. H. Organization, "Global assessment of the state-of-the-science of endocrine disruptors. Geneva, Switzerland," in *International Programme on Chemical Safety*, 2002.

[5] A. C. Gore, *"Endocrine-disrupting chemicals: from basic research to clinical practice"*. Springer Science & Business Media, 2007.

[6] J. E. Drewes, J. Hemming, S. J. Ladenburger, J. Schauer, and W. Sonzogni, "An assessment of endocrine disrupting activity changes in water reclamation systems through the use of bioassays and chemical measurements," *Proc. Water Environ. Fed.*, vol. 2004, no. 16, pp. 47–65, 2012.

[7] P. Balaguer, V. Delfosse, and W. Bourguet, "Mechanisms of endocrine disruption through nuclear receptors and related pathways," *Curr. Opin. Endocr. Metab. Res.*, vol. 7, pp. 1–8, 2019.

[8] Cheldy Tizaoui, Olajumoke Olalade Odejimi & Ayman Abdelaziz, "Occurrence, Effects, and Treatment of Endocrine-Disrupting Chemicals in Water," Chapter 3.4 *The Water-Food-Energy Nexus*, pp 157–179, CRC Press 2018.

[9] H. B. Patisaul and S. M. Belcher, *Endocrine disruptors, brain, and behavior*, Oxford University Press, 2017.

[10] World Health Organization, "Endocrine disrupting chemicals (EDCs)," *Children's environmental health*, 2020.

[11] E. Archer, G. M. Wolfaardt, and J. H. van Wyk, "Pharmaceutical and personal care products (PPCPs) as endocrine disrupting contaminants (EDCs) in South African surface waters," *Water SA*, vol. 43, no. 4, pp. 684–706, 2017.

[12] M. Sharma, P. Chadha, and C. Madhu Sharma, "Toxicity of non-ionic surfactant 4-nonylphenol an endocrine disruptor: A review," *Int. J. Fish. Aquat. Stud.*, vol. 6, no. 2, pp. 190–197, 2018.

[13] W. T. Vieira, M. B. De Farias, M. P. Spaolonzi, M. G. C. da Silva, and M. G. A. Vieira, "Removal of endocrine disruptors in waters by adsorption, membrane filtration and biodegradation. A review," *Environ. Chem. Lett.*, vol. 18, no. 4, pp. 1113–1143, 2020.

[14] P. Dillon and D. Ellis, "Australian water conservation and reuse research program," *AWA Journal of Water*, vol. 8, no. [August 2004], pp 36–37, 2004.

[15] T. Heberer, "Occurrence, fate, and removal of pharmaceutical residues in the aquatic environment: a review of recent research data," *Toxicol. Lett.*, vol. 131, pp. 5–17, 2002.

[16] M. K. Kim and K. D. Zoh, "Occurrence and removals of micropollutants in water environment Moon-Kyung," *Environ. Eng. Res.*, vol. 21, no. 4, pp. 319–332, 2016.

[17] S. A. Ronchetti, G. V. Novack, M. S. Bianchi, M. C. Crocco, B. H. Duvilanski, and J. P. Cabilla, "In vivo xenoestrogenic actions of cadmium and arsenic in anterior pituitary and uterus," *Reproduction*, vol. 152, no. 1, pp. 1–10, 2016.

[18] N. Silva, R. Peiris-John, R. Wickremasinghe, H. Senanayake, and N. Sathiakumar, "Cadmium a metalloestrogen: Are we convinced?" *J. Appl. Toxicol.*, vol. 32, no. 5, pp. 318–332, 2012.

[19] C. L. S. Vilela, J. P. Bassin, and R. S. Peixoto, "Water contamination by endocrine disruptors: Impacts, microbiological aspects and trends for environmental protection," *Environ. Pollut.*, vol. 235, pp. 546–559, 2018.

[20] O. E. Ohore and Z. Songhe, "Endocrine disrupting effects of bisphenol A exposure and recent advances on its removal by water treatment systems. A review," *Sci. African*, vol. 5, 2019.

[21] N. M. Grindler *et al.*, "Exposure to phthalate, an endocrine disrupting chemical, alters the first trimester placental methylome and transcriptome in women," *Sci. Rep.*, vol. 8, no. 1, pp. 1–9, 2018.

[22] P. Burkhardt-Holm, "Endocrine disruptors and water quality: A state-of-the-art review," *Int. J. Water Resour. Dev.*, vol. 26, no. 3, pp. 477–493, 2010.

[23] V. Phonsiri, S. Choi, C. Nguyen, Y.-L. Tsai, R. Coss, and S. Kurwadkar, "Monitoring occurrence and removal of selected pharmaceuticals in two different wastewater treatment plants," *SN Appl. Sci.*, vol. 1, no. 7, pp. 1–11, 2019.

[24] R. Scheumann and M. Kraume, "Influence of different HRT for the operation of a Submerged Membrane Sequencing Batch Reactor (SM-SBR) for the treatment of greywater," *Desalination*, vol. 248 (September 2006), pp. 123–130, 2009.

[25] H. N. Hong, N. H. Kim, K. S. Park, S. Lee, and M. B. Gu, "Analysis of the effects diclofenac has on Japanese medaka (Oryzias latipes) using real-time PCR," *Chemosphere*, vol. 67, pp. 2115–2121, 2007.

[26] J. Chae *et al.*, "Chemosphere evaluation of developmental toxicity and teratogenicity of diclofenac using Xenopus embryos," *Chemosphere*, vol. 120, pp. 52–58, 2015.

[27] M. Saravanan, J. Hur, N. Arul, and M. Ramesh, "Toxicological effects of clofibric acid and diclofenac on plasma thyroid hormones of an Indian major carp, Cirrhinus mrigala during short and long-term," *Environ. Toxicol. Pharmacol.*, vol. 38, no. 3, pp. 948–958, 2014.

[28] T. Basile *et al.*, "Review of endocrine-disrupting-compound removal technologies in water and wastewater treatment plants: An EU perspective," *Ind. Eng. Chem. Res.*, vol. 50, pp. 8389–8401, 2011.

[29] N. Schmidt, D. Page, and A. Tiehm, "Biodegradation of pharmaceuticals and endocrine disruptors with oxygen, nitrate, manganese (IV), iron (III) and sulfate as electron acceptors," *J. Contam. Hydrol.*, vol. 203 (June), pp. 62–69, 2017.

[30] C. P. Czajka and K. L. Londry, "Anaerobic biotransformation of estrogens," *Sci. Total Environ.*, vol. 367, no. 2–3, pp. 932–941, 2006.

[31] C. K. Gadupudi, L. Rice, L. Xiao, and K. Kantamaneni, "Endocrine disrupting compounds removal methods from wastewater in the United Kingdom: A review," *Sci*, vol. 1, no. 15, pp. 1–10, 2019.

[32] P. Pholchan, M. Jones, T. Donnelly, and P. J. Sallis, "Fate of estrogens during the biological treatment of synthetic wastewater in a batch reactor," *Environ. Sci. Technol.*, vol. 42, no. 16, pp. 6141–6147, 2008.

[33] T. K. Kasonga, M. A. A. Coetzee, C. Van Zijl, and M. N. B. Momba, "Removal of pharmaceutical' estrogenic activity of sequencing batch reactor effluents assessed in the T47D-KBluc reporter gene assay," *J. Environ. Manage.*, vol. 240 (March), pp. 209–218, 2019.

[34] M. Bilal and M. N. Iqbal, "Persistence and impact of steroidal estrogens on the environment and their laccase-assisted removal," *Sci. Total Environ.*, vol. 690, pp. 447–459, 2019.

[35] M. Chen, M. Gatheru, S. Li, K. Sun, and Y. Si, "Fungal laccase-mediated humification of estrogens in aquatic ecosystems," *Water Res.*, vol. 166, p. 115040, 2019.

[36] L. Lloret, G. Eibes, G. Feijoo, M. T. Moreira, and J. M. Lema, "Degradation of estrogens by laccase from Myceliophthora thermophila in fed-batch and enzymatic membrane reactors," *J. Hazard. Mater.*, vol. 213–214, pp. 175–183, 2012.

[37] L. Lloret, G. Eibes, M. T. Moreira, G. Feijoo, J. M. Lema, and M. Miyazaki, "Improving the catalytic performance of laccase using a novel continuous-flow microreactor," *Chem. Eng. J.*, vol. 223, pp. 497–506, 2013.

[38] Y. Dai, J. Yao, Y. Song, X. Liu, S. Wang, and Y. Yuan, "Enhanced performance of immobilized laccase in electrospun fibrous membranes by carbon nanotubes modification and its application for bisphenol A removal from water," *J. Hazard. Mater.*, vol. 317, pp. 485–493, 2016.

[39] M. De Cazes *et al.*, "Design and optimization of an enzymatic membrane reactor for tetracycline degradation," *Catal. Today*, vol. 236, no. PART A, pp. 146–152, 2014.

[40] L. N. Nguyen *et al.*, "Continuous biotransformation of bisphenol a and diclofenac bylaccase in an enzymatic membrane reactor," *Int. Biodeterior. Biodegrad.*, vol. 95 (PA), pp. 25–32, 2014.

[41] A. Mojiri *et al.*, "Pesticides in aquatic environments and their removal by adsorption methods," *Chemosphere*, vol. 253, p. 126646, 2020.

[42] A. A. Mohammadi, M. H. Dehghani, A. Mesdaghinia, K. Yaghmaian, and Z. Es'haghi, "Adsorptive removal of endocrine disrupting compounds from aqueous solutions using magnetic multi-wall carbon nanotubes modified with chitosan biopolymer based on response surface methodology: Functionalization, kinetics, and isotherms studies," *Int. J. Biol. Macromol.*, vol. 155, pp. 1019–1029, 2020.

[43] C. Zhao, Q. Wei, K. Yang, X. Liu, M. Nomizu, and N. Nishi, "Preparation of porous polysulfone beads for selective removal of endocrine disruptors," *Sep. Purif. Technol.*, vol. 40, pp. 297–302, 2004.

[44] A. Cesaro and V. Belgiorno, "Removal of endocrine disruptors from urban wastewater by advanced oxidation processes (AOPs): A review," *Open Biotechnol. J.*, vol. 10, no. 1, pp. 151–172, 2016.

[45] A. Mirzaei, Z. Chen, F. Haghighat, and L. Yerushalmi, "Removal of pharmaceuticals and endocrine disrupting compounds from water by zinc oxide-based photocatalytic degradation: A review," *Sustain. Cities Soc.*, vol. 27, pp. 407–418, 2016.

[46] O. K. Dalrymple, D. H. Yeh, and M. A. Trotz, "Removing pharmaceuticals and endocrine-disrupting compounds from wastewater by photocatalysis," *J. Chem. Technol. Biotechnol.*, vol. 121, pp. 121–134, 2007.

[47] J. C. Sin, S. M. Lam, A. R. Mohamed, and K. T. Lee, "Degrading endocrine disrupting chemicals from wastewater by TiO$_2$ photocatalysis: A review," *Int. J. Photoenergy*, vol. 2012, 2012.

[48] X. Pang, N. Skillen, N. Gunaratne, D. W. Rooney, and P. K. J. Robertson, "Removal of phthalates from aqueous solution by semiconductor photocatalysis: A review," *J. Hazard. Mater.*, vol. 402 (May), p. 123461, 2020.

[49] H. Zhu, Z. Li, and J. Yang, "A novel composite hydrogel for adsorption and photocatalytic degradation of bisphenol A by visible light irradiation," *Chem. Eng. J.*, vol. 334 (August), pp. 1679–1690, 2017.

[50] F. Chen, W. An, L. Liu, Y. Liang, and W. Cui, "Highly efficient removal of bisphenol A by a three-dimensional graphene hydrogel-AgBr @ rGO exhibiting adsorption/photocatalysis synergy Highly efficient removal of bisphenol A by a three-dimensional graphene hydrogel-AgBr @ rGO exhibiting adsorption/," *Appl. Catal. B, Environ.*, vol. 217, pp. 65–80, 2017.

[51] Y. Zhang, W. Cui, W. An, L. Liu, Y. Liang, and Y. Zhu, "Combination of photoelectrocatalysis and adsorption for removal of bisphenol A over TiO_2-graphene hydrogel with 3D network structure," *Appl. Catal. B Environ.*, vol. 221 (September), pp. 36–46, 2017.

[52] F. Chen, W. An, L. Liu, Y. Liang, and W. Cui, "Highly efficient removal of bisphenol A by a three-dimensional graphene hydrogel-AgBr @ rGO exhibiting adsorption/photocatalysis synergy," *Appl. Catal. B Environ. Environ.*, vol. 217, pp. 65–80, 2017.

[53] X. Yang, R. C. Flowers, H. S. Weinberg, and P. C. Singer, "Occurrence and removal of pharmaceuticals and personal care products (PPCPs) in an advanced wastewater reclamation plant," *Water Res.*, vol. 45, no. 16, pp. 5218–5228, 2011.

[54] D. R. Wallace, "Metals as endocrine disruptors in the environment," *EC Pharmacol. Toxicol.*, vol. 2, pp. 12–14, 2019.

[55] J. J. Wirth and R. S. Mijal, "Adverse effects of low level heavy metal exposure on male reproductive function," *Syst. Biol. Reprod. Med.*, vol. 56, no. 2, pp. 147–167, 2010.

[56] C. Chen *et al.*, "Blood cadmium level associates with lower testosterone and sex hormone-binding globulin in Chinese men: from SPECT-China Study, 2014," *Biol. Trace Elem. Res.*, vol. 171, no. 1, pp. 71–78, 2016.

[57] A. J. Jafari, R. Pourkabireh Abasabad, and A. Salehzadeh, "Endocrine disrupting contaminants in water resources and sewage in Hamadan city of Iran," *Iran. J. Environ. Heal. Sci. Eng.*, vol. 6, no. 2, pp. 89–96, 2009.

[58] J. Zdarta, K. Antecka, R. Frankowski, A. Zgoła-Grześkowiak, H. Ehrlich, and T. Jesionowski, "The effect of operational parameters on the biodegradation of bisphenols by Trametes versicolor laccase immobilized on Hippospongia communis spongin scaffolds," *Sci. Total Environ.*, vol. 615, pp. 784–795, 2018.

[59] D. Becker *et al.*, "Removal of endocrine disrupting chemicals in wastewater by enzymatic treatment with fungal laccases," *Org. Process Res. Dev.*, vol. 21, pp. 480–491, 2017.

[60] F. Lassouane, H. Aït-Amar, S. Amrani, and S. Rodriguez-Couto, "A promising laccase immobilization approach for Bisphenol A removal from aqueous solutions," *Bioresour. Technol.*, vol. 271 (August), pp. 360–367, 2018.

[61] J. Wang, N. Majima, H. Hirai, and H. Kawagishi, "Effective removal of endocrine-disrupting compounds by lignin peroxidase from the white-rot fungus phanerochaete sordida YK-624," *Curr. Microbiol.*, vol. 64, no. 3, pp. 300–303, 2012.

[62] R. Xu, C. Chi, F. Li, and B. Zhang, "Immobilization of horseradish peroxidase on electrospun microfibrous membranes for biodegradation and adsorption of bisphenol A," *Bioresour. Technol.*, vol. 149, pp. 111–116, 2013.

[63] A. Tursi, E. Chatzisymeon, F. Chidichimo, A. Beneduci, and G. Chidichimo, "Removal of endocrine disrupting chemicals from water: Adsorption of bisphenol-a by biobased hydrophobic functionalized cellulose," *Int. J. Environ. Res. Public Health*, vol. 15, no. 11, 2018.

[64] K. V Ragavan and N. K. Rastogi, "Beta-Cyclodextrin capped graphene-magnetite nanocomposite for selective adsorption of Bisphenol-A," *Carbohydr. Polym.*, vol. 168, pp. 129–137, 2017.

[65] A. Srivastava *et al.*, "Effective elimination of endocrine disrupting bisphenol A and S from drinking water using phenolic resin-based activated carbon fiber: Adsorption, thermodynamic and kinetic studies," *Environ. Nanotechnology, Monit. Manag.*, vol. 14 (May), p. 100316, 2020.

[66] L. Yi *et al.*, "Enhanced adsorption of bisphenol A, tylosin, and tetracycline from aqueous solution to nitrogen-doped multiwall carbon nanotubes via cation- π and π - π electron-donor-acceptor (EDA) interactions," *Sci. Total Environ.*, vol. 719, p. 137389, 2020.

[67] Z. Wang, Y. Zhu, H. Chen, H. Wu, and C. Ye, "Fabrication of three functionalized silica adsorbents: Impact of co-immobilization of imidazole, phenyl and long-chain alkyl groups on bisphenol A adsorption from high salt aqueous solutions," *J. Taiwan Inst. Chem. Eng.*, vol. 86, pp. 120–132, 2018.

[68] W. Tsai, C. Lai, and T. Su, "Adsorption of bisphenol-A from aqueous solution onto minerals and carbon adsorbents," *J. Hazard. Mater.*, vol. 134, pp. 169–175, 2006.

[69] G. Liu, J. Ma, X. Li, and Q. Qin, "Adsorption of bisphenol A from aqueous solution onto activated carbons with different modification treatments," *J. Hazard. Mater.*, vol. 164, pp. 1275–1280, 2009.

[70] Z. Li, M. A. Gondal, and Z. H. Yamani, "Preparation of magnetic separable CoFe2O4/PAC composite and the adsorption of bisphenol A from aqueous solution," *J. Saudi Chem. Soc.*, vol. 18, no. 3, pp. 208–213, 2014.

[71] P. Tang *et al.*, "A simple and green method to construct cyclodextrin polymer for the effective and simultaneous estrogen pollutant and metal removal," *Chem. Eng. J.*, vol. 366 (November), pp. 598–607, 2019.

[72] T. M. Berhane, J. Levy, M. P. S. Krekeler, and N. D. Danielson, "Adsorption of bisphenol A and ciprofloxacin by palygorskite-montmorillonite: Effect of granule size, solution chemistry and temperature," *Appl. Clay Sci.*, vol. 132–133, pp. 518–527, 2016.

[73] J. Luo *et al.*, "Sorption of norfloxacin, sulfamerazine and oxytetracycline by KOH-modified biochar under single and ternary systems," *Bioresour. Technol.*, vol. 263 (May), pp. 385–392, 2018.

[74] W. Wu *et al.*, "Extraneous dissolved organic matter enhanced adsorption of dibutyl phthalate in soils: Insights from kinetics and isotherms," *Sci. Total Environ.*, vol. 631–632, no. 71, pp. 1495–1503, 2018.

[75] X. Wang *et al.*, "Adsorption of dibutyl phthalate in aqueous solution by mesoporous calcium silicate grafted non-woven polypropylene," *Chem. Eng. J.*, vol. 306, pp. 452–459, 2016.

[76] N. A. Khan, B. K. Jung, Z. Hasan, and S. H. Jhung, "Adsorption and removal of phthalic acid and diethyl phthalate from water with zeolitic imidazolate and metal – organic frameworks," *J. Hazard. Mater.*, vol. 282, pp. 194–200, 2015.

[77] A. Mesdaghinia *et al.*, "Removal of phthalate esters (PAEs) by zeolite/Fe3O4: Investigation on the magnetic adsorption separation, catalytic degradation and toxicity bioassay," *J. Mol. Liq.*, vol. 233, pp. 378–390, 2017.

[78] F. Suo, X. You, Y. Ma, and Y. Li, "Rapid removal of triazine pesticides by P doped biochar and the adsorption mechanism," *Chemosphere*, vol. 235, pp. 918–925, 2019.

[79] V. W. O. Wanjeri, C. J. Sheppard, A. R. E. Prinsloo, J. C. Ngila, and P. G. Ndungu, "Isotherm and kinetic investigations on the adsorption of organophosphorus pesticides on graphene oxide based silica coated magnetic nanoparticles functionalized with 2-phenylethylamine," *J. Environ. Chem. Eng.*, vol. 6, no. 1, pp. 1333–1346, 2018.

[80] X. Wang, H. Wang, M. Lu, R. Teng, and X. Du, "Facile synthesis of phenyl-modified magnetic graphene/mesoporous silica with hierarchical bridge-pore structure for efficient adsorption of pesticides," *Mater. Chem. Phys.*, vol. 198, pp. 393–400, 2017.

[81] A. Derylo-marczewska, M. Blachnio, A. W. Marczewski, and A. Swiatkowski, "Adsorption of chlorophenoxy pesticides on activated carbon with gradually removed external particle layers," *Chem. Eng. J.*, vol. 308, pp. 408–418, 2017.

[82] Y. Zhang *et al.*, "High efficiency and rapid degradation of bisphenol A by the synergy between adsorption and oxidization on the MnO2 @ nano hollow carbon sphere," *J. Hazard. Mater.*, vol. 360 (May), pp. 223–232, 2018.

[83] L. Luo *et al.*, "A novel biotemplated synthesis of TiO2/wood charcoal composites for synergistic removal of bisphenol A by adsorption and photocatalytic degradation," *Chem. Eng. J.*, vol. 262, pp. 1275–1283, 2015.

[84] D. Mohanta, "Biogenic synthesis of SnO_2 quantum dots encapsulated carbon nano flakes: An efficient integrated photocatalytic adsorbent for the removal of bisphenol A from aqueous solution," *J. Alloys Compd.*, vol. 828, p. 154093, 2020.

[85] Q. Jin *et al.*, "Simultaneous adsorption and oxidative degradation of Bisphenol A by zero-valent iron/iron carbide nanoparticles encapsulated in N-doped carbon matrix *," *Environ. Pollut.*, vol. 243, pp. 218–227, 2018.

[86] L. Luo *et al.*, "Bisphenol A removal on TiO_2 – MoS_2 – reduced graphene oxide composite by adsorption and photocatalysis," *Process Saf. Environ. Prot.*, vol. 2, pp. 274–279, 2017.

[87] A. Gulati, J. Malik, and R. Kakkar, "Mesoporous rGO @ ZnO composite: Facile synthesis and excellent water treatment performance by pesticide adsorption and catalytic oxidative dye degradation," *Chem. Eng. Res. Des.*, vol. 160, pp. 254–263, 2020.

Part II

Membrane-Based Technologies

3 Thin-Film Composites and Multi-Layered Membranes for Wastewater Treatment

Paresh Kumar Samantaray
International Institute for Nanocomposites Manufacturing
(IINM), WMG, University of Warwick, U.K.

*Suryasarathi Bose**
Department of Materials Engineering, Indian Institute
of Science, Bangalore, India

CONTENTS

3.1 INTRODUCTION

Wastewater can be defined as the "used water" by-product of domestic usage, commercial and agricultural practices, or industrial applications. It can also result from surface runoffs, stormwater, or any sewer infiltration into the water stream.[1] Due to the diverse origin of contamination, wastewater constituents can vary diversely. While the domestic sources predominantly include pathogenic contaminants and domestic waste, industrial wastewater may have oil and chemical residues, biocides, pesticides, dyes, pharmaceutical waste, alkali/acid by-products, etc. Urban runoffs containing oils, food waste, tire residues, and herbicides from soils may also infiltrate the freshwater system and lead to wastewater generation. Since freshwater remains a challenge this century, it is critical to devise strategies for sustainable re-utilization of wastewater. To do so, the wastewater must be decontaminated first.

37

Conventional processes like coagulation-flocculation, adsorption, chemical precipitation, advanced oxidation processes, and ion-exchange are some of the possible methods to remove specific contaminants selectively. However, due to the specificity of the processes towards specific contaminant removal, multi-step decontamination may be required.[2-4] Further, these processes sometimes fail to meet the stringent standards of output desired.[5-7] Membrane separation, particularly nanofiltration, reverse osmosis (RO), and forward osmosis (FO) membranes have demonstrated capabilities in remediating these challenging contaminants while complying with water standards.[8-10] These membranes are generally polyamide derivatives and are called TFC because of the dense packing of polyamide derivatives on micro/ultrafiltration support membranes with pore sizes in Å limits. The separation in such membranes occurs by the mobility of water molecules by applying high application pressure. Due to this, it can efficiently reject all the contaminants found in wastewater.[11-12] However, the current TFCs face critical challenges like scaling, cake formation, and biofouling, diminishing the membranes' performance significantly. Hence, multi-layered membranes emerged as a key counterpart to conventional TFCs. They not only can remediate all the critical contaminants in the wastewater stream but also tackle scaling, cake formation, and bacterial biofouling.[3,6]

This chapter will discuss the fabrication of thin-film nanocomposites and multi-layered membranes using state-of-the-art techniques and then highlight the recent advances of these membranes in wastewater treatment. Subsequently, key challenges in the field with the sustainability aspect of these membranes will be highlighted, and methods that can be adopted to tackle these challenges will also be formulated.

3.2 FABRICATION OF THIN-FILM COMPOSITES AND MULTI-LAYERED MEMBRANES

Long before the current technique to manufacture TFCs took over the reverse osmosis market, several other processes were predominately used in salt and other contaminant rejection. The most popular and revolutionary approach was the Loeb-Sourirajan approach in 1960, wherein the desalination membrane with low thickness using cellulose acetate as membrane material was reported.[13] Using the work of Carnell and Cassidy, in the mid-1960s, Riley et al. used a dip coating method to devise very thin films.[14] They observed that when glass plates were dipped in dilute polymer solutions and subsequently withdrawn and dried, thin polymer films were left behind, which could be withdrawn using water. These thin films could be made into a "thin-film composite" by lifting off the floating film using finely porous membrane support. Thus, using this approach, TFC based on cellulose acetate with a few hundred Å of thickness was prepared.[14-15] Cadotte et al. first demonstrated the use of commercial polyamide-based TFCs in the 1980s wherein a microporous polyethersulfone (PES) membrane was immersed in an aqueous solution of a diamine and then impregnated in a di-acyl chloride mixture dissolved in a solvent like hexane, which was cross-linked by heat-treatment at 110°C.[16-17] This process was termed interfacial polymerization (IP) because of the condensation reaction between the diamine and the di-isocyanate to form polyamide at the polymer support interface. A different combination of support surfaces (micro/ultrafiltration) and polyamide derivatives is currently available in the market.[18-20] The most common precursors for TFC fabrication is phenylenediamine (MPD), and trimesoylchloride.[21] Apart from IP, densely packed TFCs are also prepared by phase inversion.[22]

Similar to RO membranes, cellulose acetate membranes were being used in early forward osmosis desalination.[23-25] However, unlike the reverse osmosis process, in which a membrane is pressurized at high transmembrane pressure in a cross-flow configuration to yield low solute containing permeate and high solute concentrated rejection, in the forward osmosis process, natural osmotic pressure gradient due to difference in concentration in feed and draw solution drives the purification. Further, since the osmotic imbalance is the key driver in the FO process, this is a cost-effective and energy-efficient strategy and hence can be used in wastewater treatment processes over RO.[10,26,27] However, FO is a relatively slow process; hence, a small application pressure across the membrane surface can accelerate the separation process. This is called pressure assisted osmosis (PAO).[28] Currently, all FO and PAO processes use polyamide TFCs for desalination and wastewater treatment.

Before current multi-layered membranes were envisioned, in 1997, Gero Decher was among the first to report nano assemblies of polymers as an alternative to the Langmuir-Blodgett technique chemisorption, which was selective to certain classes of macromolecules only. By using consecutive adsorption of polyanions and polycations made the surface deposited surface flexible.[29] This was later utilized as one of the popular strategies to derive multilayer membranes. Multi-layered membranes are prepared either by (a) layer-by-layer (LbL) assembly: sequential deposition of a polycation or polyanion on an existing membrane surface which binds the layers via electrostatic interactions or, (b) dynamic composites: single or multiple sequential depositions of nanoparticles, or polymers onto the existing membrane surfaces. Although dynamic composites are quite similar to LbL assembly, it involves the deposition of any polymer and nanoparticles rather than just polyelectrolytes with/without nanoparticles.[3] Although, in the literature, the terms are used interchangeably, herein, we distinguish the same by appropriate nomenclature to remove ambiguity. Figure 3.1 shows the schematic representation for the preparation of TFC using IP and multi-layered membrane fabrication by LbL assembly and dynamic composite.

3.3 RECENT ADVANCES IN THE USE OF TFCS AND MULTI-LAYERED MEMBRANES IN WASTEWATER TREATMENT

Table 3.1 discusses the recent advances in dynamic composite membranes, LbL assembly, thin film composites and modified thin film composites in wastewater treatment applications.

3.4 SUSTAINABILITY TOWARDS INTENDED WATER REUSE

It is estimated that around 40% of the global population resides in urban coasts, where they have multiple water sources with variable water qualities (e.g., rivers, lakes, and seawater).[9] These dwellings generate wastewater, which in many cases gets subsequently treated in wastewater treatment plants. Further, it is estimated that there will

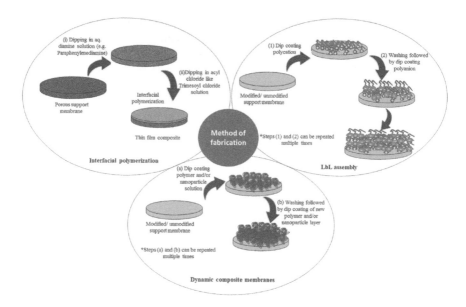

FIGURE 3.1 Schematics showing method of preparation of TFCs, LbL assembly, and dynamic composite membranes.

be a shortfall of 40% of freshwater resources by 2030, which will be coupled with a global population increase and water crisis.[53] However, the key challenge is dumping waste discharge from the critical consumption sectors and the treated water directly into the freshwater streams without appropriate water treatment. (Refer to Figure 3.2 for the scheme.) This directly contributes to the gradual conversion of these conventional freshwater sources to unusable contaminated sources.[54] If these sources are directly used for consumption, they lead to waterborne communicable diseases.[54–55] It is critical to reducing the level of contaminants in the discharge from wastewater treatment plants and the key consumption sectors. In this aspect, membranes, particularly TFC and multi-layered based membrane reactors, can play a critical role. Coupling these membrane reactors with wastewater treatment plants has resulted in membrane bioreactors (MBR) in the current scenario. Owing to the versatile design of these MBRs, these reactors can enable capturing high biomass concentrations, good effluent quality, and lower sludge production.[56–57]

Interestingly, MBR technology is expected to grow at a compounded growth rate of 10.5% per annum.[58] It's estimated that by 2019, MBR treatment plants could treat more than 5 million cubic meters of wastewater.[59] This makes the process sustainable and provides an alternative to conventional wastewater treatment processes.

3.5 KEY CHALLENGES

The first challenge in the membrane-based treatment processes is biofouling. Fouling critically declines the flux output, retards the membrane performance, and in the long run, makes it unsuitable for further use.[4,60,61] In brief, antifouling surfaces with

TABLE 3.1

Current reports of TFCs and multi-layered membranes in wastewater treatment

Membrane Type	Modifications involved	Contaminants	Key features
Dynamic composite membrane	Sequential deposition of carrageenan (CGN) and laponite on Hydrolysed polyacrylonitrile (h-PAN).[50]	• Li$^+$, Na$^+$, K$^+$, and Mg^{2+} ions. • Brilliant Blue (BB) and Rhodamine-B (RB) dyes. • Oil (motor oil)	• 99% oil and dye rejection and 99.6% salt rejection.
Dynamic composite membrane	On a double-layer PAN membrane, polydopamine and MPD were poured, and metal-organic framework (MOF)-801 dispersed in Trimesoyl chloride was added and dried.[31]	• Heavy metal ions Cd^{2+}, Ni^{2+}, and Pb^{2+}	• >94% rejection of heavy metal ions with a 30% flux increase.
Dynamic composite membrane	Mesoporous ultrafiltration membrane using Titania smectite clay over zeolite by using sequential LbL depositions.[32]	• Cr and Co heavy metals	• >80% rejection of heavy metal ions at turbidity < 1 NTU and chemical oxygen demand of >96%.
Dynamic composite membrane	Commercial TFC membranes were deposited with polyacrylic acid using in situ polymerization followed by grafting Au-Ag based photocatalyst to form multi-layered membrane.[33]	• E. coli • Ofloxacin and Methylene blue (MB) dyes.	• 80–100% bactericidal action to E. coli. • Efficiently rejected dye molecules from water.
Dynamic composite membrane	TFCs were incorporated with MgFe$_2$O$_4$ and ZnFe$_2$O$_4$ by soaking polysulfone (PSf) substrate in nanoparticle dispersion followed by IP.[34]	• Glucose based model to mimic formation and sugar industrial wastewater	• 96% of glucose rejection.
Dynamic composite membrane	On a TFC fabricated using PAN as support, polyethyleneimine (PEI) sulfobetaine methacrylate-based second TFC layer was created.[35]	• E. coli • Lysozyme and Bovine serum albumin proteins	• Lowest bacterial attachment and close to 90% flux retention were observed for both lysozyme and serum albumin-based models.

(Continued)

TABLE 3.1

Current reports of TFCs and multi-layered membranes in wastewater treatment (Cont.)

Membrane Type	Modifications involved	Contaminants	Key features
Dynamic composite membrane	Graphene oxide and PAN were dissolved in NMP to fabricate membranes. These membranes were modified by polydopamine followed by vacuum filtration of UiO-66 MOF.[36]	• Dyes and antibiotic models like methyl orange (MO), MB, Congo red (CR), RB, Oxytetracycline, Tetracycline, Ciprofloxacin, and Sulfamethoxazole. • Protein and polysaccharide models (Humic acid (HA) and Bovine Serum Albumin (BSA) respectively)	• >94% of dye and antibiotic separation, high antifouling of >95% to HA and BSA models
Dynamic composite membrane	A multi-layered membrane was designed with antibacterial copolymers as an active layer, a metal organic framework of Zinc and Fe_3O_4 as interlayer and RO as support layer.[4]	• E. coli and S. aureus as bacterial models. • BSA as protein foulant. • As and Pb as heavy metal ions.	• 100% bactericidal action and >95% of removal of heavy metal ions.
Dynamic composite membrane	A PP based film was modified with maleic anhydride using gamma irradiation followed by in situ growth of Zirconium based MOF for 4–19 cycles.[37]	• Soybean oil as model oil foulant.	• High-performance separation of oil from water.
Dynamic composite membrane	A multi-layered membrane was designed with antibacterial copolymer as an active layer, 2D MoS_2 tethered with Dithi-magnetospheres as interlayer and RO as support layer.[38]	• E. coli and S. aureus as bacterial models. • BSA as protein foulant. • As and Pb as heavy metal ions.	• 100% bactericidal action. • 95.3% of Pb removal and 96% of As removal.
Dynamic composite membrane	h-PAN was coated with PEI and borates were used to cross-link graphene oxide to the membrane.[39]	• MO and MB based dye models.	• 74% of MO rejection and 88.56% of MB rejection.
LbL assembly	A PSf support membrane was dipped in PEI solution and lignosulfonate solutions twice and cross-linked with glutaraldehyde.[40]	• Cr, Cd, Zn, Mn, Cu, and Ni heavy metals.	• >80% rejection of all the heavy metal ions by the multi-layered membrane.

LBL assembly	Polyvinylidene fluoride (PVDF) was modified with polyelectrolyte and carbon nanotubes, graphene oxide and graphitic carbon nitride was deposited LbL.[41]	Tetracycline and RB as recalcitrant model.	• 98.31% of RB removal and 84.81% of tetracycline removal under visible light irradiation.
LbL assembly	On a PAN ultrafiltration support plant polyphenol tannic acid with abundant catechol functionality and hydrophilic Jeffamine containing amine groups were deposited layer by layer to form nanofiltration membranes.[42]	MR, MO, MB, CR, RB, and Rose Bengal, based dye models, were used.	• >90% dye rejection at pH 8 was observed for all the dye models.
LBL assembly	Hydrophobic polyvinylidene fluoride (PVDF) was pre-treated with alkali and then sequentially dipped in TiO_2 and polystyrene sulfonate to form multilayer membrare.[43]	Lanasol Blue 3R as dye pollutant. BSA as protein pollutant.	• 91.42% of removal rate of Lanasol Blue 3R. • After protein fouling and 8 h of UV irradiation, 95.41% of recovery was observed.
LBL assembly	Two different types of clays; montmorillonite and kaolin were deposited on PES support membrane with layer (LbL) assembly with poly (allylamine hydrochloride) (PAH) and poly (acrylic acid) (PAA).[44]	Wastewater with total solids level ~22 wt. %	• Relatively low fouling of modified LbL membranes.
LbL assembly	TFC membranes combined with electrostatically coupled SiO_2 (silica dioxide) nanoparticles/poly (L-DOPA) (3-(3,4-dihydroxyphenyl)-1-alanine).[45]	Unchlorinated pond water.	• 50% lower microbial growth on modified TFC membranes.
LbL assembly	Polyelectrolyte multilayers of PAH and PAA were deposited on PAN-based ultrafiltration membranes to obtain multi-layered nanofiltration membranes.[46]	Synthetic secondary-treated municipal wastewater	• Versatile in removing micropollutants like Diclofenac, Ibuprofen, Naproxen, and 4n-Nonylphenol.
TFC	A TFC was formed on PEI based hollow fiber support followed by IP.[47]	Dye pollutants Safranin O and Aniline Blue	• >90% of dye rejection. • Membrane performed the best at pH 11.
TFC	TFC was formed on a PAN membrane fabricated on polyethylene terephthalate fabric.[48]	Synthetic wastewater with chemical oxygen demand of 300 mg/L.	• Forward osmosis was demonstrated for anaerobic fluidized bed bioreactor. • FO membrane was able to control ammonium and nitrogen effectively.

(Continued)

TABLE 3.1

Current reports of TFCs and multi-layered membranes in wastewater treatment (Cont.)

Membrane Type	Modifications involved	Contaminants	Key features
Modified TFC	PSf was mixed with dopamine modified Ag nanoparticles and TiO_2 and electrospun into mats. IP was carried to form dense polyamide on the surface.[49]	• *E. coli* • Tetracycline resistant bacteria and genes in secondary effluent of sewage wastewater.	• FO membranes on electrospun support for resistant bacterial remediation was demonstrated. • Permeability of bacteria in the effluent as low as 0.75%.
Modified TFC	Melamine was added to the TFC precursors while IP was performed on a PES support.[50]	• Trace organic contaminants Diazinon and Atrazine.	• 99.3% removal of Diazinon and 93.9% of Atrazine removal using melamine modified TFC membrane.
Modified TFC	Zwitterionic amide (N-aminoethyl piperazine propane sulfonate, (AEPPS) was used to modify the TFC FO membrane-active surface.[51]	• Synthetic greywater with humic acid, and casein.	• Membrane with AEPPS zwitterionic modification had better separation and antifouling performance.
Modified TFC	Polyhedral oligomeric silsesquioxane (POSS) was added to either organic or aqueous phase while performing IP on a PES support.[52]	• As and Se based heavy metal ions	• High rejections of SeO_3^{2-}, SeO_4^{2-}, and $HAsO_4^{2-}$ toward 93.9%, 96.5%, and 97.4%, respectively
Modified TFC	A FO hierarchical membrane was designed by gradient stitching of PVDF based etched membranes with an active graphene quantum dot-based TFC composite.[28]	• CR and MB based dye models and BSA as protein foulant.	• >95% of MB rejection and >85% CR rejection. • >80% of flux retention after protein fouling.

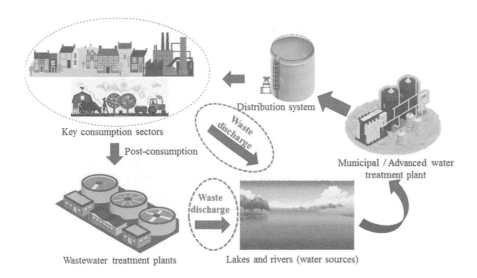

FIGURE 3.2 Schematic representation of water usage in a general urban setting. The red color indicates the unmonitored discharge of waste into freshwater sources.

biocidal active surfaces can impede microorganisms' growth while retaining the desired output.

The other key challenge is understanding the appropriate costs (capital costs and operation and maintenance costs) associated with the technology for appropriate scale-up and critical decision making. Roccaro et al. did a cost analysis comparing the MBR technology to the conventional activated sludge (CAS).[59] According to their estimation, 35–50% of MBR's capital cost is towards the purchase of membranes. However, due to market competition, this cost has reduced from 50 €/m² to nearly 30 €/m².[59] Further, the research in new TFCs and multi-layered membranes can increase the growth of this technology and reduce costs even further. The energy consumption of CAS ranges from 0.4 to 0.7 kWh/m³. On the other hand, in some cases, MBR's energy requirements are twice as high as CAS. This is due to the inherent heat production in the membrane processes wherein nearly 80% of energy input is converted into heat.[56] Additionally, energy for aeration in MBR also adds to the operational costs.

Although MBR's energy cost is on the higher end, the cost of sludge disposal is relatively low compared to that of CAS (21% for MBR and 32% for CAS).[59] The degree of instrumentation and automation in MBR technology manifests a higher water quality effluent.

Other challenges include the scalability of MOFs and 2D materials based on TFCs and multi-layered membranes. The consistency and industrial-scale up-scaling are key determinants of using these materials in bulk. However, with technological advances, sooner or later, this challenge could be possibly overcome. Apart from this, government authorities' enforcement of safe water disposal and handling practices are also pivotal. Stringent norms for disposal of waste from industries has to be enacted and monitored for violations so that the water discharge quality remains under specified regulations. Only when these are strictly obeyed and complied with will the burden on water treatment processes decrease and a sustainable circular water reuse system will be possible.

3.6 CONCLUSIONS

In this chapter, we overviewed the progress of the development of TFCs and their method synthesis towards wastewater treatment. We also discussed the different multi-layered membranes fabrication techniques, i.e., LbL assembly and dynamic composite membranes, towards wastewater treatment applications. The recent advances in this field were discussed alongside the sustainability of using these membrane based MBRs in water reuse application over the conventional CAS system. The key challenges in using these membranes from a technological and operational point of view are highlighted. It is envisaged that technological advances in TFCs and multi-layered membrane development will further enable its use in wastewater reclamation processes.

REFERENCES

(1) Tilley, E. *Compendium of sanitation systems and technologies*, Eawag: 2014.
(2) Boruah, B.; Samantaray, P. K.; Madras, G.; Modak, J. M.; Bose, S. Sustainable photocatalytic water remediation via dual active strongly coupled AgBiO3 on PVDF/PBSA membranes. *Chemical Engineering Journal* **2020**, 124777.
(3) Samantaray, P. K.; Madras, G.; Bose, S. The key role of modifications in biointerfaces toward rendering antibacterial and antifouling properties in polymeric membranes for water remediation: A critical assessment. *Advanced Sustainable Systems* **2019**, 1900017.
(4) Samantaray, P. K.; Baloda, S.; Madras, G.; Bose, S. A designer membrane tool-box with a mixed metal organic framework and RAFT-synthesized antibacterial polymer perform in tandem towards desalination, antifouling and heavy metal exclusion. *Journal of Materials Chemistry A* **2018**, *6* (34), 16664–16679.
(5) Fu, F.; Wang, Q. Removal of heavy metal ions from wastewaters: a review. *Journal of Environmental Management* **2011**, *92* (3), 407–418, DOI: 10.1016/j.jenvman.2010.11.011.
(6) Remanan, S.; Sharma, M.; Bose, S.; Das, N. C. Recent advances in preparation of porous polymeric membranes by unique techniques and mitigation of fouling through surface modification. *ChemistrySelect* **2018**, *3* (2), 609–633.
(7) Patel, P.; Biedermann, L. Will next-generation membranes rise to the water challenge? *MRS Bulletin* **2018**, *43* (6), 406–407.
(8) Samantaray, P. K.; Madras, G.; Bose, S. Water remediation aided by a graphene-oxide-anchored metal organic framework through pore-and charge-based sieving of ions. *ACS Sustainable Chemistry & Engineering* **2018**, *7* (1), 1580–1590.
(9) Blandin, G.; Verliefde, A. R.; Comas, J.; Rodriguez-Roda, I.; Le-Clech, P. Efficiently combining water reuse and desalination through forward osmosis—reverse osmosis (FO-RO) hybrids: a critical review. *Membranes* **2016**, *6* (3), 37.
(10) Lutchmiah, K.; Verliefde, A.; Roest, K.; Rietveld, L. C.; Cornelissen, E. Forward osmosis for application in wastewater treatment: a review. *Water Research* **2014**, *58*, 179–197.
(11) Kumar, S.; Ahlawat, W.; Bhanjana, G.; Heydarifard, S.; Nazhad, M. M.; Dilbaghi, N. Nanotechnology-based water treatment strategies. *Journal of Nanoscience and Nanotechnology* **2014**, *14* (2), 1838–1858, DOI: 10.1166/jnn.2014.9050.

(12) Sablani, S.; Goosen, M.; Al-Belushi, R.; Wilf, M. Concentration polarization in ultra-filtration and reverse osmosis: a critical review. *Desalination* **2001**, *141* (3), 269–289.

(13) Loeb, S.; Sourirajan, S. *Sea water demineralization by means of an osmotic membrane*, University of California, Department of Engineering: 1960.

(14) Lonsdale, H. K. The evolution of ultrathin synthetic membranes. *Journal of Membrane Science* **1987**, *33* (2), 121–136.

(15) Lonsdale, H. The growth of membrane technology. *Journal of Membrane Science* **1982**, *10* (2–3), 81–181.

(16) Cadotte, J.; Forester, R.; Kim, M.; Petersen, R.; Stocker, T. Nanofiltration membranes broaden the use of membrane separation technology. *Desalination* **1988**, *70* (1–3), 77–88.

(17) Cadotte, J.; Petersen, R.; Larson, R.; Erickson, E. A new thin-film composite seawater reverse osmosis membrane. *Desalination* **1980**, *32*, 25–31.

(18) Kadhom, M.; Deng, B. Synthesis of high-performance thin film composite (TFC) membranes by controlling the preparation conditions: Technical notes. *Journal of Water Process Engineering* **2018**.

(19) Yam-Cervantes, M.; Pérez-Padilla, Y.; Aguilar-Vega, M. TFC reverse osmosis polyamide membranes—Effect of increasing sulfonic group concentration on water flux and salt rejection performance. *Journal of Applied Polymer Science* **2018**, *135* (29), 46500.

(20) Idarraga-Mora, J. A.; Childress, A. S.; Friedel, P. S.; Ladner, D. A.; Rao, A. M.; Husson, S. M. Role of nanocomposite support stiffness on TFC membrane water permeance. *Membranes (Basel)* **2018**, *8* (4), 111, DOI: 10.3390/membranes8040111.

(21) Lalia, B. S.; Kochkodan, V.; Hashaikeh, R.; Hilal, N. A review on membrane fabrication: Structure, properties and performance relationship. *Desalination* **2013**, *326*, 77–95.

(22) Werber, J. R.; Osuji, C. O.; Elimelech, M. Materials for next-generation desalination and water purification membranes. *Nature Reviews Materials* **2016**, *1* (5), 16018.

(23) Moody, C. D. Forward osmosis extractors: theory, feasibility and design optimization **1977**.

(24) Kessler, J.; Moody, C. Drinking water from sea water by forward osmosis. *Desalination* **1976**, *18* (3), 297–306.

(25) Moody, C.; Kessler, J. Forward osmosis extractors. *Desalination* **1976**, *18* (3), 283–295.

(26) Zhao, S.; Zou, L.; Tang, C. Y.; Mulcahy, D. Recent developments in forward osmosis: opportunities and challenges. *Journal of Membrane Science* **2012**, *396*, 1–21.

(27) Cath, T. Y.; Childress, A. E.; Elimelech, M. Forward osmosis: principles, applications, and recent developments. *Journal of Membrane Science* **2006**, *281* (1–2), 70–87.

(28) Maiti, S.; Samantaray, P. K.; Bose, S. In situ assembly of a graphene oxide quantum dot-based thin-film nanocomposite supported on de-mixed blends for desalination through forward osmosis. *Nanoscale Advances* **2020**, *2* (5), 1993–2003, DOI: 10.1039/c9na00688e.

(29) Decher, G. Fuzzy nanoassemblies: toward layered polymeric multicomposites. *Science* **1997**, *277* (5330), 1232–1237.

(30) Prasannan, A.; Udomsin, J.; Tsai, H.-C.; Wang, C.-F.; Lai, J.-Y. Robust underwater superoleophobic membranes with bio-inspired carrageenan/laponite multilayers for the effective removal of emulsions, metal ions, and organic dyes from wastewater. *Chemical Engineering Journal* **2020**, *391*, 123585.

(31) He, M.; Wang, L.; Lv, Y.; Wang, X.; Zhu, J.; Zhang, Y.; Liu, T. Novel polydopamine/metal organic framework thin film nanocomposite forward osmosis membrane for salt rejection and heavy metal removal. *Chemical Engineering Journal* **2020**, *389*, 124452.

(32) Aloulou, W.; Aloulou, H.; Khemakhem, M.; Duplay, J.; Daramola, M. O.; Amar, R. B. Synthesis and characterization of clay-based ultrafiltration membranes supported on natural zeolite for removal of heavy metals from wastewater. *Environmental Technology & Innovation* **2020**, 100794.

(33) Amoli-Diva, M.; Irani, E.; Pourghazi, K. Photocatalytic filtration reactors equipped with bi-plasmonic nanocomposite/poly acrylic acid-modified polyamide membranes for industrial wastewater treatment. *Separation and Purification Technology* **2020**, *236*, 116257.

(34) Nambikkattu, J.; Kaleekkal, N. J.; Jacob, J. P. Metal ferrite incorporated polysulfone thin-film nanocomposite membranes for wastewater treatment. *Environmental Science and Pollution Research* **2020**, 1–13.

(35) Chiao, Y.-H.; Patra, T.; Ang, M. B. M. Y.; Chen, S.-T.; Almodovar, J.; Qian, X.; Wickramasinghe, R.; Hung, W.-S.; Huang, S.-H.; Chang, Y. Zwitterion Co-Polymer PEI-SBMA nanofiltration membrane modified by fast second interfacial polymerization. *Polymers* **2020**, *12* (2), 269.

(36) Fang, S.-Y.; Zhang, P.; Gong, J.-L.; Tang, L.; Zeng, G.-M.; Song, B.; Cao, W.-C.; Li, J.; Ye, J. Construction of highly water-stable metal-organic framework UiO-66 thin-film composite membrane for dyes and antibiotics separation. *Chemical Engineering Journal* **2020**, *385*, 123400.

(37) Gao, J.; Wei, W.; Yin, Y.; Liu, M.; Zheng, C.; Zhang, Y.; Deng, P. Continuous ultra-thin UiO-66-NH 2 coatings on a polymeric substrate synthesized by a layer-by-layer method: a kind of promising membrane for oil–water separation. *Nanoscale* **2020**, *12* (12), 6658–6663.

(38) Samantaray, P. K.; Baloda, S.; Madras, G.; Bose, S. Interlocked dithi-magnetospheres–decorated MoS2 nanosheets as molecular sieves and traps for heavy metal ions. *Advanced Sustainable Systems* **2019**, 1800153.

(39) Yan, X.; Tao, W.; Cheng, S.; Ma, C.; Zhang, Y.; Sun, Y.; Kong, X. Layer-by-layer assembly of bio-inspired borate/graphene oxide membranes for dye removal. *Chemosphere* **2020**, 127118.

(40) Xie, M.-Y.; Wang, J.; Wu, Q.-Y. Nanofiltration membranes via layer-by-layer assembly and cross-linking of polyethyleneimine/sodium lignosulfonate for heavy metal removal. *Chinese Journal of Polymer Science* **2020**, 1–8.

(41) Shi, Y.; Wan, D.; Huang, J.; Liu, Y.; Li, J. Stable LBL self-assembly coating porous membrane with 3D heterostructure for enhanced water treatment under visible light irradiation. *Chemosphere* **2020**, 126581.

(42) Guo, D.; Xiao, Y.; Li, T.; Zhou, Q.; Shen, L.; Li, R.; Xu, Y.; Lin, H. Fabrication of high-performance composite nanofiltration membranes for dye wastewater treatment: mussel-inspired layer-by-layer self-assembly. *Journal of Colloid and Interface Science* **2020**, *560*, 273–283.

(43) Luo, J.; Chen, W.; Song, H.; Liu, J. Fabrication of hierarchical layer-by-layer membrane as the photocatalytic degradation of foulants and effective mitigation of membrane fouling for wastewater treatment. *Science of the Total Environment* **2020**, *699*, 134398.

(44) Hong, J. S.; Yu, J.; Lee, I. Role of clays in fouling-resistant clay-embedded polyelectrolyte multilayer membranes for wastewater effluent treatment. *Separation Science and Technology* **2017**, *52* (13), 2108–2119.

(45) Alhumaidi, M. S.; Arshad, F.; Aubry, C.; Ravaux, F.; McElhinney, J.; Hasan, A.; Zou, L. Electrostatically coupled SiO2 nanoparticles/poly (L-DOPA) antifouling coating on a nanofiltration membrane. *Nanotechnology* **2020**, *31* (27), 275602.

(46) Abtahi, S. M.; Marbelia, L.; Gebreyohannes, A. Y.; Ahmadiannamini, P.; Joannis-Cassan, C.; Albasi, C.; de Vos, W. M.; Vankelecom, I. F. Micropollutant rejection of annealed polyelectrolyte multilayer based nanofiltration membranes for treatment of conventionally-treated municipal wastewater. *Separation and Purification Technology* **2019**, *209*, 470–481.

(47) Shao, L.; Cheng, X. Q.; Liu, Y.; Quan, S.; Ma, J.; Zhao, S. Z.; Wang, K. Y. Newly developed nanofiltration (NF) composite membranes by interfacial polymerization for Safranin O and Aniline blue removal. *Journal of Membrane Science* **2013**, *430*, 96–105.

(48) Kwon, D.; Kwon, S. J.; Kim, J.; Lee, J.-H. Feasibility of the highly-permselective forward osmosis membrane process for the post-treatment of the anaerobic fluidized bed bioreactor effluent. *Desalination* **2020**, *485*, 114451.

(49) Chen, H.; Zheng, S.; Meng, L.; Chen, G.; Luo, X.; Huang, M. Comparison of novel functionalized nanofiber forward osmosis membranes for application in antibacterial activity and TRGs rejection. *Journal of Hazardous Materials* **2020**, *392*, 122250.

(50) Rastgar, M.; Shakeri, A.; Karkooti, A.; Asad, A.; Razavi, R.; Sadrzadeh, M. Removal of trace organic contaminants by melamine-tuned highly cross-linked polyamide TFC membranes. *Chemosphere* **2020**, *238*, 124691.

(51) Wang, J.; Xiao, T.; Bao, R.; Li, T.; Wang, Y.; Li, D.; Li, X.; He, T. Zwitterionic surface modification of forward osmosis membranes using N-aminoethyl piperazine propane sulfonate for grey water treatment. *Process Safety and Environmental Protection* **2018**, *116*, 632–639.

(52) He, Y.; Tang, Y. P.; Chung, T. S. Concurrent removal of selenium and arsenic from water using polyhedral oligomeric silsesquioxane (POSS)–polyamide thin-film nanocomposite nanofiltration membranes. *Industrial & Engineering Chemistry Research* **2016**, *55* (50), 12929–12938.

(53) UN Goal 6: Ensure access to water and sanitation for all. https://www.un.org/sustainabledevelopment/water-and-sanitation/.

(54) Montgomery, M. A.; Elimelech, M., *Water and Sanitation in Developing Countries: Including Health in the Equation.* ACS Publications: 2007.

(55) Shannon, M. A.; Bohn, P. W.; Elimelech, M.; Georgiadis, J. G.; Marinas, B. J.; Mayes, A. M. Science and technology for water purification in the coming decades. *Nature* **2008**, *452* (7185), 301–10, DOI: 10.1038/nature06599.

(56) Van Dijk, L.; Roncken, G. Membrane bioreactors for wastewater treatment: the state of the art and new developments. *Water Science and Technology* **1997**, *35* (10), 35–41.

(57) Stephenson, T.; Brindle, K.; Judd, S.; Jefferson, B. *Membrane Bioreactors for Wastewater Treatment*, IWA publishing: 2000.

(58) Kraume, M.; Drews, A. Membrane bioreactors in waste water treatment–status and trends. *Chemical Engineering & Technology* **2010**, *33* (8), 1251–1259.

(59) Roccaro, P.; Vagliasindi, F. G. Membrane bioreactors for wastewater reclamation: Cost analysis. In *Current Developments in Biotechnology and Bioengineering*, Elsevier: 2020; pp. 311–322.

(60) Samantaray, P. K.; Bose, S. Cationic biocide anchored graphene oxide-based membranes for water purification. *Proceedings of the Indian National Science Academy* **2018**, *84* (3), 669–679.

(61) Padmavathy, N.; Samantaray, P. K.; Ghosh, L. D.; Madras, G.; Bose, S. Selective cleavage of the polyphosphoester in cross-linked copper based nanogels: enhanced antibacterial performance through controlled release of copper. *Nanoscale* **2017**, *9* (34), 12664–12676.

4 Membrane Distillation: An Efficient Technology for Desalination

Sanjay Remanan and Narayan Chandra Das*
Rubber Technology Centre, Indian Institute of
Technology-Kharagpur, 721302, West Bengal, India

CONTENTS

4.1 INTRODUCTION

Historical development in MD dates from a patent filed in 1963 by Bodell. Some publications also appeared in late 1960 by Findley. He used different membrane materials for direct contact membrane distillation such as paper, cellophane, and nylon. Improvement in the membrane manufacturing technology helps to produce higher porosity membranes with increased water permeability. Development in membrane module architecture and clear understanding of the concepts such as vapor pressure, temperature, and membrane fouling increased the competitiveness of the membrane distillation based separation process.

The process involves the phase change of feed liquid to vapor due to the thermal gradient across the membrane followed by the entry of vapors into the pores and then diffuses to the permeate side. Vapor pressure gradient across the membrane causes the transport of vapor molecules to the permeate side. Advantages of the membrane distillation includes

- 100% theoretical rejection of the pollutants such as salts, macromolecules, microorganisms and colloids.
- Compared to RO, the operating pressure required is negligible to a few kPa. MD is more efficient in removing ions and nonvolatile pollutants compared to RO and can achieve the water flux similar to the RO processes.
- Requirement of a lower operating temperature and reuse of waste heat energy available from the various industrial operations. Cost efficiency can be achieved through coupling the MD system to solar and geothermal energy.
- Less demanding of membrane mechanical properties.
- Reduced interaction between the feed solution and membrane, and hence less fouling related issues.
- Lower equipment cost and increased process safety.

Apart from the several advantages, a major disadvantage of the MD is the wetting of the membrane during the operation. The feed solution should not wet the membrane and only allow the permeation of the vapor molecules; otherwise, the separation efficiency diminishes. Lawson and Lloyd stated that a membrane is called 'MD membrane' if it fulfills the following criterias:

- The membrane should be a porous substrate
- The membrane should not be wetted by liquid under consideration
- No capillary condensation take place inside the membrane pores
- The membrane must not alter the vapor–liquid equilibrium of the different components present in the process liquids

What differs in the case of conventional distillation and MD is the requirement of a large vapor space by micrometer thickness membrane. In the conventional processes, high vapor pressure facilitates the close contact between the vapor and liquid while, as in MD, a hydrophobic microporous membrane establishes the liquid-vapor interface. This indicates the MD plants would consume less physical space than the traditional distillation methods. As mentioned, the required operating temperature for the MD is lower because there is no need to heat the feed solution to a very high temperature.

In MD processes, the hot feed is allowed to flow through the one side of a hydrophobic membrane. The feed would not be able to enter into the membrane pores due to the solvent repelling, and high LEP, hence there will be a formation of the liquid-vapor interface at the pore entrance.

Four basic MD configurations are widely employed for MD based applications:

1. Direct contact membrane distillation (DCMD)
2. Air-gap membrane distillation (AGMD)
3. Sweeping gas membrane distillation (SGMD)
4. Vacuum membrane distillation (VMD)

Depending upon the permeate composition, flux, and volatility, a suitable type of MD process can be employed. DCMD is simple, straightforward and a commonly used MD process for applications such as desalination and concentration of juices. When the volatile components or dissolved gas are to be removed, VMD or AGMD can be employed. During the MD processes, liquid molecules vaporize at the feed side and form a liquid-vapor interface, and vapor molecules diffuse through the membrane pores and are either condensed in a medium or eliminated as vapors from the permeate side. An overview of MD membrane characteristics is shown in Table 4.1.

Herein, various membrane distillation types and their property requirements and desalination applications are described. The selection of a membrane material, fouling and membrane property requirement to outperform in the separation processes are discussed. Practical and sustainable approaches for the membrane separation using MD are reviewed from the recent literature. This chapter provides a clear and concise understanding on fundamentals of MD, various types, and its desalination application.

TABLE 4.1

Overview of the optimal membrane properties and characterization methods for membrane distillation (Adapted from Eykens et al. 2016)

Parameter	Symbol	Recommended	Characterization method
Contact angle	θ	>90°	Static sessile drop method
			Dynamic sessile drop
Liquid entry pressure	LEP	>2.5 bar	Liquid entry pressure measurement
Porosity	$\in$	80–90%	Gas permeation test (effective porosity)
			Electron microscopy (surface porosity)
			Liquid pycnometer (bulk porosity)
Pore diameter	d_{av}, d_{max}	0.1–1 μm	Gas permeation test (d_{av})
			Wet/dry flow method (pore distribution)
			Mercury porosimetry (pore distribution)
			Electron microscopy (pore distribution)
Thickness	δ	30–60 μm	Digital micrometer

4.2 PROPERTIES REQUIRED FOR MEMBRANES

4.2.1 MEMBRANE

Membrane material must have high chemical and thermal stability. Higher chemical stability allows the membrane to work with various feed solutions and cleaning agents. High thermal stability is necessary for the increased service life and uninterrupted operation during the separation process. Membranes with a micropore architecture are suitable for MD processes and can be prepared by the conventional phase inversion (nonsolvent-induced phase separation and thermally induced phase separation) method. Stretching is another method for microporous membrane preparation. Membrane thickness must be maintained to achieve a compromise between mass transfer and heat transfer. For higher mass transfer in MD, thin membranes are desirable, but at the same time, there must not be any heat energy lost. As in the case of a multilayered membrane, mass-transfer is facilitated by selecting the thin hydrophobic membrane, and heat loss is optimized by the following thick layer. Generally, the lower thermal conductive membranes are desirable for the MD processes. Polymers such as polyethylene (PE), polypropylene (PP), and fluorinated polymers are widely exploited for various MD applications and modified ceramic materials, and graphene-derived materials are also used for preparing the MD membranes (Larbot et al. 2004, Goh and Ismail 2019).

4.2.2 WETTING

Since the membrane operations are based on the formation of the vapor-liquid interface, the membrane should not allow the liquid entry, and it must have sufficient antiwetting property. That means the membrane should have high hydrophobicity (Yao et al., 2020). Hydrophobic polymers such as PVDF, PTFE, and PP are widely used for the preparation of the MD membrane. Membranes can be single or multilayer, but at least one layer should satisfy the hydrophobic characteristics (Khayet 2011). Generally, the membrane surface with a water contact angle of more than 90° is used for the MD application.

LEP decides the wetting of the membrane. It is defined as the minimum transmembrane pressure required for water or other solutions to enter into the membrane pores by overcoming the membrane hydrophobicity. Hence, membrane hydrophobicity must be high enough to prevent the liquid entry. A membrane must have high LEP, or it deteriorates the permeate quality. Higher liquid entry pressure can be achieved by selecting the membrane with a high contact angle (or low surface energy or high hydrophobicity) and small pore size. LEP is mathematically expressed as

$$LEP = P_{liquid} - P_{vapor} = \frac{-2\gamma B Cos\theta}{r_{max}} \qquad (4.1)$$

where γ represents the surface tension of the liquid, θ is the contact angle at the solid-liquid interface, B is the geometric factor that depends on the porous structure, and r_{max} is the largest pore radius (Zhang et al. 2020).

4.2.3 POROSITY

Membranes with pore size varying from submicrometer to several microns are typically used for the MD application. Two factors must be taken into consideration while selecting the membrane: (1) micron-size pores must allow the vapors across the membrane, and (2) pore size of the membrane must not be very large such that liquid penetration is prevented. A narrow pore size distribution prevents the liquid entry into the pores. As mentioned, a decrease in pore size can influence higher LEP, but a greater reduction in membrane pore size reduces efficiency. Therefore, a better tradeoff between the membrane permeability and LEP is required for the efficient operation of MD processes. It is recommended to have a high porosity membrane that is useful for providing a large room for evaporation. Also, higher porosity of the membrane yields higher water flux, which improves the MD performance.

4.2.4 MEMBRANE FOULING RESISTANCE

Fouling is the aggregation of the pollutants on the membrane surface and in pores of the membrane that deteriorates the permeability and separation efficiency. Fouling is an Achilles' heel in the separation processes and is caused by the pollutants, which can be grouped as colloidal, organic, inorganic, and biofoulant (Tijing et al. 2014). In a desalination process, the deposition of salts on the membrane surface and in pores due to the evaporation of the feed solution can compromise membrane wetting properties. Increased fouling leads to inefficient separation and reduces permeate quality. MD membranes tend to undergo less fouling as compared to the pressure-driven membranes. The antifouling property of the hydrophobic membrane is usually improved by treatments such as chemical modification, which inhibits the foulant growth over the membrane surface.

In summary, a MD process is facilitated by the use of low surface energy materials or hydrophobic surfaces. MD membranes should offer higher thermal stability and low thermal conductivity at the same time, and they should offer the least mass transfer resistance. An increase in feed solution temperature and the feed solution's velocity has a positive effect on the membrane flux (Santoro et al. 2019). Also, membranes must have sufficient porosity to get high water permeability, and micropores offer sufficiently high LEP, which increases the rejection percentage.

4.3 MATERIALS FOR MEMBRANE DISTILLATION

The primary requirement for the MD membrane is the hydrophobicity or low surface energy. This means the membrane surface is nonwetted by the process liquids and only permits vapors and gases through the pores. Hence, polymer materials are commonly selected for the MD membrane preparation. Polymers such as PP, PVDF, and PTFE are usually used, and among these, PTFE has the highest hydrophobicity, high chemical, and thermal stability. However, a major disadvantage is difficulty in processing. Other polymers such as PP microporous membranes prepared by stretching the extruded films or phase inversion from the solvents, which dissolves PP at high temperatures. PVDF is another example and is commonly prepared by the NIPS method. The selection criteria of a membrane employed for the membrane distillation application involves a balance between low thermal conductivity, high water permeability, optimum pore size, high porosity, and high separation performance (Khayet 2011). Studies showed that maintaining a better temperature gradient and membrane with low thermal conductivity and high porosity yields high permeability (Al-Obaidani et al. 2008).

The membrane can be prepared by several methods such as stretching, phase inversion, and vacuum filtration. Among this, phase inversion is the easiest and most widely used method. In this method, the polymer is solubilized in a suitable solvent, phase-separated from a nonsolvent, which is usually water. Phase inversion involves liquid-liquid phase separation in which a polymer is initially dissolved in a solvent, cast on a substrate, and phase-separated from a nonsolvent (Remanan et al. 2018). The nonsolvent is called a gelation medium, and phase separation is achieved in two steps. Initially, a thin layer of membrane is formed, followed by further solvent/nonsolvent dissolution from the layer beneath with finger-like structure formation. When nonsolvent is in contact with the solvent, thermodynamic instability begins, and the polymer starts to phase-separate. More solvent is in contact with the nonsolvent, the demixing increases, and porous polymer structure precipitates. Membrane porous structure depends upon the solvent/nonsolvent miscibility. The membrane can be classified into three major groups as follows:

   i. Flat sheet thin membrane
  ii. Hollow fiber membrane
 iii. Multilayered flat sheet membrane

4.3.1 Flat Sheet Thin Membrane

Generally, flat sheet thin membranes are prepared by the phase inversion method, and PVDF homopolymer and vinylidene copolymers are mainly used (Tijing et al. 2014). Incorporation of various hydrophobic nanofillers such as graphene can further improve the membrane hydrophobicity. For instance, Woo et al. recently prepared an AGMD using the PVDF-HFP polymer incorporated with graphene filler (Woo et al. 2016). At an optimal concentration of the filler, they observed an enhanced hydrophobicity (>160°), higher LEP, and high porosity. These conditions favored the MD to yield greater salt rejection and high membrane flux. Figure 4.1 shows the flux and salt rejection of the commercial membrane and optimized graphene filler–added PVDF-HFP membrane. The modified membrane exhibited improved water permeability and high salt rejection consistently for 60 hours compared to the commercial membranes. Poor salt rejection observed for commercial PVDF membrane at the end of 60 hours while as graphene/PVDF-HFP membrane maintaining 100% salt rejection. Other hydrophobic coating or modification on membrane surface has profound increment in antiwetting property of the membranes; few such examples are the fabrication of MD membrane with nanoparticles such as TiO_2, and SiO_2 based water repellent hydrophobic coatings (Razmjou et al. 2012, Efome et al. 2015). Similar approaches were adopted for the surface modification to enhance

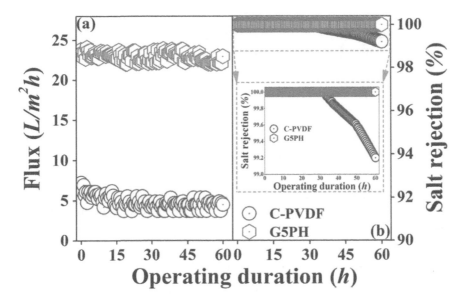

FIGURE 4.1 (a) Flux and (b) salt rejection of the 5 wt% graphene polyvinylidene fluoride-co-hexafluoropropylene (G5PH) electrospun nanofiber membrane and commercial PVDF membrane (Adapted from Woo et al. 2016).

the hydrophobicity of MD membrane surfaces were reported in various literature (Yang et al. 2014, Lee et al. 2016, Nthunya et al. 2019). Modified membranes show higher salt rejection, consistent water flux, improved mechanical properties, and low wetting for a prolonged time (Grasso et al. 2020). Incorporation of hydrophobic additives can increase the LEP and alter the pore size and porosities of the membrane. Very large pore size and porosities are not favored in the MD as it might deteriorate the permeate quality. Hence, optimized membrane characteristics such as pore size and porosities increase the LEP, which reduces the membrane fouling. It is observed that the addition of SiO_2 nanoparticles and modification of membrane by fluorosilanization increased the hydrophobicity, decrease the pore size, and increase the LEP of the electrospun PVDF membrane. The modified membrane maintained a constant water flux and higher salt rejection in a VMD process (Dong et al. 2015).

4.3.2 Hollow Fiber Membrane

Commercially available hollow fiber membranes are also modified with hydrophobic coatings to enhance the membrane hydrophobicity and reduce the fouling (Li et al. 2019). A membrane can be wet by the process liquid in three stages: surface wetting, partial wetting, and full wetting, as shown in Figure 4.2. Let us consider one example: coating of the PVDF hollow fiber membrane surface by Hyflon AD 60, and the contact angle was increased from 94° for PVDF to 139° or even higher for the composite membrane (Tong et al. 2016). The Hyflon AD 60 is a vinylidene copolymer-based material that influences the membrane characteristics: two times increase in LEP,

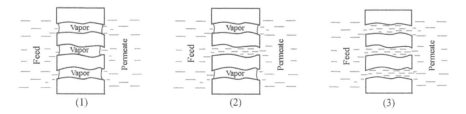

FIGURE 4.2 Mechanism of membrane wetting: (1) surface wetting, (2) partial wetting, (3) fully wetting (Adapted from Li, Zhang, et al. 2019).

increase in mechanical properties, less flux decline, and improvement in membrane antiwetting or hydrophobic properties. An increase in membrane hydrophobicity reduces the membrane fouling (Zhang et al. 2017).

4.3.3 MULTILAYERED FLAT SHEET MEMBRANE

As in the case of a multilayered flat sheet membrane, more than one layer of hydrophobic or hydrophobic/hydrophilic combinations are possible to aid the wetting resistance and efficient vapor transport across the membrane. In an example, GO-modified PTFE membrane were also used for MD applications, in which GO was coated over the commercial PTFE surface, showed a reduction in water contact angle from 110° (for PTFE) to 90°. However, GO-coated PTFE membrane exhibited maximum flux at a higher temperature, such as 80°C. Since the GO coating is present over the PTFE membrane surface, there will be less thermal conductivity losses and lower temperature polarization. Hence, the higher temperature at the feed side can be constantly maintained, which is essential for contributing to higher flux (Bhadra, Roy, and Mitra 2016). A multilayered membrane induces higher hydrophobicity over the other hydrophobic membrane such as PVDF, and one such example is the fabrication of CNT on PVDF membrane and its modification using FAS. Modified membrane showed consistent water flux, lower conductivity at the permeate side, and very high contact angle (Figure 4.3). The modified membrane has high water flux at a higher temperature and fairly low conductivity (Wang et al. 2020). Fluorosilane based modification facilitates the antifouling property through the superhydrophobic and slippery nature of the surface.

Practically, the hollow fibrous membrane has more advantages compared to the flat sheet membrane; the latter one is more prone to wetting compared to the hollow fibrous membrane due to the presence of spacers and a possible pressure drop in the MD system. Flat sheet membranes can be specifically tuned to get higher porosity, and a larger pore size can better perform compared to the hollow fibrous membrane in terms of specific thermal energy consumption. At the same time, the hollow fibrous membrane shows a lower pressure drop and electrical energy consumption in the MD module. Figure 4.4 indicates the flux of both flat sheet and hollow fibrous commercial membranes at a specific feed temperature and two different feed velocities with the varying function of contact length. At high feed velocity and lower contact lengths, maximum membrane flux is observed (Ali et al. 2019). Other recent approaches

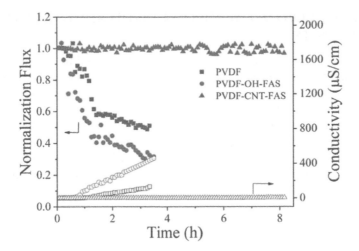

FIGURE 4.3 Normalized water vapor flux and conductivity for the permeate versus time regarding the antiwetting property for different hydrophobic membranes as PVDF, PVDF-OH-FAS and PVDF-CNT-FAS. SDS was added into the feed to give a concentration of 0.4 mM at the time "0" and before this a stable flux was already achieved using feed containing 5 g/L NaCl (Adapted from Wang et al. 2020).

for the prediction of seawater desalination by a large-scale DCMD unit (for both flat sheet and hollow fiber membranes) using experimental results developed from the laboratory-scale membrane modules (Dong et al. 2017).

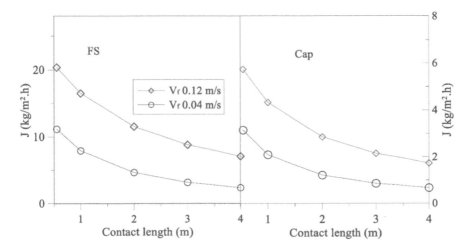

FIGURE 4.4 Dependence of flux on contact length for flat sheet (FS) and capillary (Cap) membranes at two different feed velocities and T_{fin} of 70°C. In all cases, the permeate inlet temperature and velocity were set at 20°C and 0.04 m/s, respectively (Adapted from Ali et al. 2019).

4.4　DIRECT CONTACT MEMBRANE DISTILLATION (DCMD) PROCESSES

DCMD is the simplest and one of the most widely studied membrane configurations for the desalination application. In this process, a hot feed solution is in direct contact with the membrane surface, and lower temperature liquid is circulated at the permeate side of the membrane. As a result of the temperature difference, a vapor pressure gradient establishes across the membrane, feed solution is evaporated, and vapors are transported through the pores in the membrane and condense at the permeate liquid (Figure 4.5). Heat loss due to conduction in the solution is the major disadvantage of this configuration. Membrane flux (J_w) can be calculated as

$$J_w = B_m \left(P_{mf} - P_{mp} \right) \qquad (4.2)$$

where P_{mf} and P_{mp} are the partial pressure of water at the feed and permeate side and B_m is the membrane distillation coefficient. In a DCMD process, mass transport assumes a linear relationship between the mass flux (J_w) and vapor pressure difference across the membrane. Partial pressure of the water at the feed and permeate

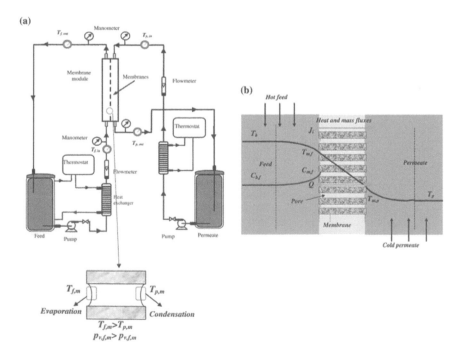

FIGURE 4.5　Schematic of (a) typical experimental DCMD system and (b) heat and mass transfer through the membrane in DCMD configuration (Adapted from Khayet 2011).

side can be determined using the Antoine equation at the corresponding feed and permeate temperatures:

$$P^v = \left(23.328 - \frac{3841}{T - 45} \right) \tag{4.3}$$

where P^v is the water vapor pressure in Pascal and T is the corresponding temperature in Kelvin (Qtaishat et al. 2008).

Saline solution has low surface tension and tends to wet the membrane surface during the MD process. Hence, the membrane needs to have antiwetting property against both water as well as low surface tension liquids like saline solution. This wetting behavior is called omniphobic. Membrane antiwetting property is improved when the surface is modified with the fluorosilane based coating that imparts superhydrophobicity and high LEP (Li et al. 2019, Zhang et al. 2020). Similar modifications were also carried out for the PVDF-HFP modified ENM with TiO_2 nanoparticles that showed very high contact angle, LEP, higher flux and salt rejection in a DCMD system (Lee et al. 2016). Hence, membrane surface coating and mixing modifier with the dope solution during the membrane preparation are two possible ways to get high contact angle, high LEP, and optimized pore properties for the desalination membrane. DCMD membranes are fouled due to inorganic and organic foulants and that can be reduced by suitably modifying the membrane surfaces. For instance, scaling or inorganic fouling was reduced by altering the membrane surface architecture by micromolding phase separation (μPS). This technology can be used to form the nanopillar like structure on the membrane surface. Micropatterned surfaces, along with superhydrophobic coating, enhance the water contact angle and allow a suspended wetting state, which predominantly reduces the scaling and organic fouling (Xiao et al. 2020) (Figure 4.6). The patterned surface contributes to higher membrane flux in the direction parallel to the

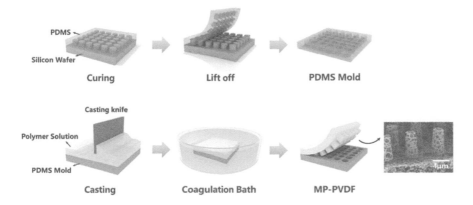

FIGURE 4.6 Schematic for the fabrication of micro-pillar PVDF (MP-PVDF) membrane: (top) preparation of PDMS mold based on silicon wafer; (bottom) casting PVDF polymer solution followed by immersion precipitation to prepare membranes with pillars (Adapted from Xiao et al. 2019).

corrugations (Liu et al. 2020). Alterations to the membrane preparation method also contribute to the formation of high contact angle isotropic membranes using the phase inversion method. One such example is the formation of the PVDF membrane by the bottom-up method in which the isotropic membrane is prepared by the phase inversion method. The top skin layer of the membrane is removed by putting a nonwoven layer of cloth on the top side during the membrane casting (Tian et al. 2020).

The influence of nanoparticles on the VEDCMD performance indicates that different particles positively increase the membrane flux. Different temperature regimes can be applied; however, different nanocomposite membranes show higher flux at different temperatures. This improvement in property is due to the uniform dispersion of filler in the membrane matrix. However, nanoparticles beyond the threshold limit may increase the polymer dope solution's viscosity and filler aggregation, which results in the flux reduction. Through this method, water can be produced at an economical rate as compared to RO with similar or even higher permeate flux (Fahmey et al. 2019). The incorporation of nanoparticles increases the membrane water permeability and desalination property compared to the control membrane.

4.5 AIR-GAP MEMBRANE DISTILLATION (AGMD) PROCESS

In AGMD, only feed solution is in direct contact with the membrane, and permeate is obtained by condensing the water vapor on a condensation plate. As shown in Figure 4.7, an air gap is provided between the permeate side of the membrane surface and condensation plate to reduce the energy due to conduction. However, this air gap offers resistance to mass transfer and is the major disadvantage of this process. A decrease in mass transfer leads to low productivity (Meindersma, Guijt, and de Haan 2006). A coolant medium is circulated behind the condensation plate (cold surface), and vapor molecules transport across the membrane and air gap, reach this condensation plate, and condense

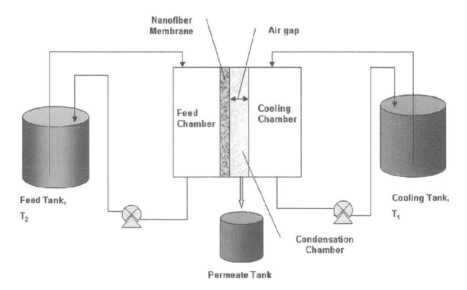

FIGURE 4.7   Diagram of AGMD system (Adapted from Feng et al. 2008).

on it. Permeate of AGMD is collected at the bottom of the membrane module (Essalhi and Khayet 2015). The advantage of AGMD processes is that less heat is lost due to conduction.

Convective heat transfer from the hot feed solution to the membrane surface allows the water vapor to reach the membrane pores. Then conductive heat transfer occurs across the membrane cross-section, and vapors finally reach the condensation plate through the stagnant air gap (diffusion). The vapor condenses here, and the condensation plate is cooled by a cold solution by convective heat transfer. In the AGMD process, chances for membrane fouling is low, and if it occurs, it can hamper the overall water flux. Salt accumulation on the membrane surface during the desalination can be cleaned by the ultrasonication technique, and nearly the same flux can be achieved after cleaning the membrane (Naji et al. 2020).

Membrane flux determined using the similar formula as explained in the DCMD process and partial pressure can be determined using the Antoine equation. If salt solution in the feed is taken into consideration, equation (4.2) can be written as (Khalifa et al. 2015),

$$J_w = B_w \left(\gamma_{wf} . x_{wf} . P_f - P_{cd} \right) \tag{4.4}$$

where γ_{wf} is the activity coefficient and x_{wf} is the water mole fraction of the feed and for the aqueous NaCl solution. The activity coefficient is given as

$$\gamma_{wf} = 1 - (0.5 x_{NaCl}) - (10 x^2{}_{NaCl}) \tag{4.5}$$

Where x_{NaCl} is the mole fraction of the NaCl in water. Overall mass transfer coefficient B_w can be determined using the following equation:

$$B_w = \frac{\varepsilon P D_{wa}}{RT (\delta \tau + b) |P_a| ln} \tag{4.6}$$

where ε is the membrane porosity, P is the total pressure inside the pore, D_{wa} is the diffusion coefficient of water vapor into the air, R is the gas constant, T is the absolute temperature, δ is the membrane thickness, τ is the membrane tortuosity, b is the airgap width, and $|P_a|ln$ is the mean air pressure calculated from the following equation:

$$|P_a| ln = \frac{(P_{a,mf} - P_{a,cd})}{\ln \left(\dfrac{P_{a,mf}}{P_{a,cd}} \right)} \tag{4.7}$$

where

$$PD_{wa} = 1.895 \times 10^{-5} \times T^{2.072} \tag{4.8}$$

And tortuosity is obtained from the porosity using the following equation:

$$\tau = \frac{1}{\varepsilon} \tag{4.9}$$

4.6   VACUUM MEMBRANE DISTILLATION (VMD) PROCESS

In this process, vacuum pressure is applied at the permeate side of the membrane, as shown in Figure 4.8. Permeate vacuum pressure is less than the pressure required for volatile solvents to be separated from the hot feed solution. The process is characterized by the lower operating temperature, lower hydrostatic pressure, and high salt rejection. VMD permits high partial pressure gradients and hence exhibits high water permeability (El-Bourawi et al. 2006, Abu-Zeid et al. 2015, Criscuoli and Carnevale 2015).

In a particular study, ECTFE membranes were prepared by the TIPS method in which ATBC was used as diluent. Different ECTFE concentration and quenching temperatures that were employed for membrane preparation showed high liquid entry pressure, improved mechanical properties, and high salt rejection (99.9%) (Xu et al. 2019). PVDF/SiO$_2$ nanocomposite membranes exhibit a higher water contact angle and LEP with an increase in SiO$_2$ concentration. Pore size increased to a point and then decreased while porosity decreased upon SiO$_2$ filler addition. High membrane flux and rejection (>99.98%) of artificial seawater and brackish water were exhibited by the nanocomposite membrane (Efome et al. 2015).

4.7   SWEEPING GAS MEMBRANE DISTILLATION (SGMD) PROCESS

In this process, an inert gas is circulated at the permeate side of the membrane, which carries the water vapor molecules and condenses in an external condenser. The advantage of SGMD is that it has low thermal polarization, a lower hindrance to the mass transfer due to the air gap (as in VMD), and no wetting at the membrane permeate side. SGMD is similar to the VMD such that in the former, permeate is collected in an external condenser. However, compared to the other three MD processes, SGMD is associated with the complexity of collecting permeate from the external condenser, requirement of an efficient condenser, and difficulty in heat recovery (González, Amigo, and Suárez 2017).

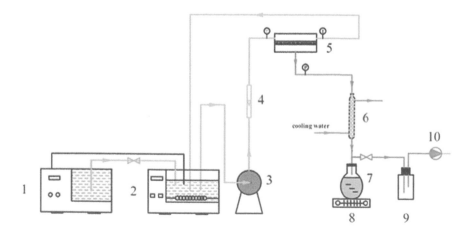

FIGURE 4.8   Experimental setup for vacuum membrane distillation (Adapted from Tang et al. 2016).

In summary, four basic configurations are presently used for MD applications. The DCMD system is the direct process and widely employed in laboratories among the four different membrane configurations. However, other approaches, such as hybrid configurations, are also developed for better performance.

4.8 SUSTAINABLE CONSIDERATIONS FOR MD SYSTEMS

Generally, the MD membrane surfaces should have low thermal conductivity to keep a constant thermal gradient across the membrane. A polymeric membrane having higher thermal conductivity has the least thermal resistance, and therefore conduction of heat across the membrane increases, which in turn reduces the vaporization heat. The reduction in the vaporization heat will eventually reduce the membrane flux and thermal efficiency (Al-Obaidani et al. 2008). Among the commonly used MD membrane, materials such as PP, PVDF, and PTFE, the last one has the highest thermal conductivity, which results in a greater reduction in water permeability and thermal efficiency, as shown in Figure 4.9. One approach that reduces thermal conductivity is employing the multilayer membrane with hydrophobic polymers having thermal conductivity in the range of 0.2 W/m K to reduce the thermal conductivity. Polymers such as cellulose acetate, PVC, PS, and PU can be employed as a substrate layer, which has less thermal conductivity (in the range of 0.13–0.18 W/m K). A thin hydrophobic top layer followed by hydrophilic membrane substrate reduces the conductive heat loss across the membrane. This combination has improved thermal

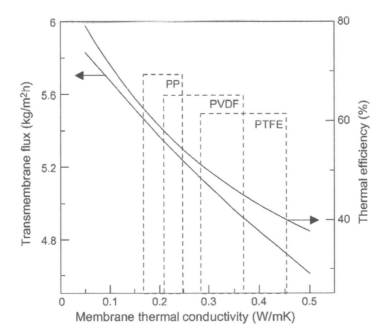

FIGURE 4.9 Simulation results of the effects of the membrane thermal conductivity on the DCMD performance (Adapted from Al-Obaidani et al. 2008).

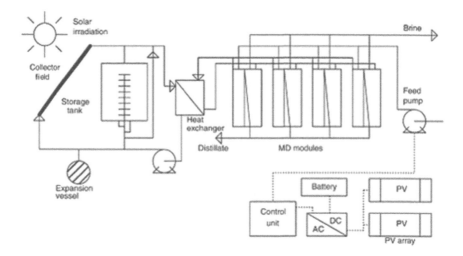

FIGURE 4.10 Diagrammatic representation of the solar-heated large MD system for producing potable water (Adapted from Qtaishat and Banat 2013).

resistance and results in higher water flux for a long time (Al-Obaidani et al. 2008). Hence, thermal energy expenditure can influence the overall operating cost of the MD process and the final permeate water cost. Hence, by using the waste heat energy and heat energy from renewable resources can reduce the operating cost of the MD processes and permeate water cost comparable to the RO.

Solar energy as a renewable energy resource is economical and environmentally friendly. Solar energy can be used for the autonomous operation of the MD plant in the semi-arid and remote accessible regions where the availability of electricity and drinking water is an issue. This technology is used for either generation of the thermal energy required for the process-liquid or producing the electricity to drive the membrane processes. Solar desalination can be classified into two types: direct and indirect collection systems. The direct type uses solar energy to produce distillate in the solar collector while two different subsystems are used: one for solar energy collection and the other for desalination (Banat et al. 2007a). The direct system is simple and can be employed for the small scale desalination application while an indirect system (Figure 4.10) uses the solar energy for heat energy required for the process and electrical energy required for the membrane operation. When adequate sunlight is available, the system can start to produce distilled water. Production of a low cost and reliable MD module can make the MD technologies compete with the existing RO based separation (Qtaishat and Banat 2013). A few studies have reported on the installation of solar-powered membrane distillation (SPMD) units for desalination and are successfully operating with the production rate of 120 L/day (Banat et al. 2007b, Wang et al. 2009).

4.9 SUMMARY

Membrane distillation is a hybrid separation process that involves thermal distillation and membrane separation. Unlike the pressure-driven membrane, MD uses the temperature difference across the feed and permeate side of the membrane that generates a vapor pressure difference at the feed and permeate side. Vapor evaporated at the feed side transports through the membrane pores and is collected at the permeate side. The concentrated salt solution at the feed side is rejected during the separation. Hence, a 100% rejection of the dissolved ions is possible by the MD process; it is a promising technology for desalination applications. However, operating conditions such as insufficient LEP and scaling (fouling) can cause a change in the wetting behavior of the membrane, which can reduce the membrane selectivity.

ABBREVIATIONS

LEP	Liquid entry pressure
PE	Polyethylene
PP	Polypropylene
PVDF	Polyvinylidene fluoride
PTFE	Polytetrafluoroethylene
GO	Graphene oxide
FAS	1H,1H,2H,2H-perfluorodecyltriethoxysilane
PVDF-HFP	Polyvinylidene fluoride-co-hexafluoropropylene
ENM	Electrospun nanofiber membrane
ECTFE	Poly(ethylene chlorotrifluoroethylene)
VEDCMD	Vacuum enhanced direct contact membrane distillation
ATBC	Acetyl tributyl citrate
PVC	Polyvinyl chloride
PS	Polystyrene
PU	Polyurethane

REFERENCES

Abu-Zeid, Mostafa Abd El-Rady, Yaqin Zhang, Hang Dong, Lin Zhang, Huan-Lin Chen, and Lian Hou. 2015. "A comprehensive review of vacuum membrane distillation technique." *Desalination* 356:1–14, doi: https://doi.org/10.1016/j.desal.2014.10.033.

Al-Obaidani, Sulaiman, Efrem Curcio, Francesca Macedonio, Gianluca Di Profio, Hilal Al-Hinai, and Enrico Drioli. 2008. "Potential of membrane distillation in seawater desalination: Thermal efficiency, sensitivity study and cost estimation." *Journal of Membrane Science* 323 (1):85–98. doi: https://doi.org/10.1016/j.memsci.2008.06.006.

Ali, Aamer, Alessandra Criscuoli, Francesca Macedonio, and Enrico Drioli. 2019. "A comparative analysis of flat sheet and capillary membranes for membrane distillation applications." *Desalination* 456:1–12. doi:https://doi.org/10.1016/j.desal.2019.01.006.

Banat, Fawzi, Nesreen Jwaied, Matthias Rommel, Joachim Koschikowski, and Marcel Wieghaus. 2007a. "Desalination by a "compact SMADES" autonomous solarpowered membrane distillation unit." *Desalination* 217 (1):29–37. doi: https://doi.org/10.1016/j.desal.2006.11.028.

Banat, Fawzi, Nesreen Jwaied, Matthias Rommel, Joachim Koschikowski, and Marcel Wieghaus. 2007b. "Performance evaluation of the "large SMADES" autonomous desalination solar-driven membrane distillation plant in Aqaba, Jordan." *Desalination* 217 (1):17–28. doi: https://doi.org/10.1016/j.desal.2006.11.027.

Bhadra, Madhuleena, Sagar Roy, and Somenath Mitra. 2016. "Desalination across a graphene oxide membrane via direct contact membrane distillation." *Desalination* 378:37–43. doi: https://doi.org/10.1016/j.desal.2015.09.026.

Criscuoli, A., and M. C. Carnevale. 2015. "Desalination by vacuum membrane distillation: The role of cleaning on the permeate conductivity." *Desalination* 365:213–219. doi: https://doi.org/10.1016/j.desal.2015.03.003.

Dong, Guangxi, Jeong F. Kim, Ji Hoon Kim, Enrico Drioli, and Young Moo Lee. 2017. "Open-source predictive simulators for scale-up of direct contact membrane distillation modules for seawater desalination." *Desalination* 402:72–87. doi: https://doi.org/10.1016/j.desal.2016.08.025.

Dong, Zhe-Qin, Xiao-Hua Ma, Zhen-Liang Xu, and Zhi-Yun Gu. 2015. "Superhydrophobic modification of PVDF–SiO2 electrospun nanofiber membranes for vacuum membrane distillation." *RSC Advances* 5 (83):67962–67970. doi: 10.1039/C5RA10575G.

Efome, Johnson E., Mohammadali Baghbanzadeh, Dipak Rana, Takeshi Matsuura, and Christopher Q. Lan. 2015. "Effects of superhydrophobic SiO2 nanoparticles on the performance of PVDF flat sheet membranes for vacuum membrane distillation." *Desalination* 373:47–57. doi: https://doi.org/10.1016/j.desal.2015.07.002.

El-Bourawi, M. S., Z. Ding, R. Ma, and M. Khayet. 2006. "A framework for better understanding membrane distillation separation process." *Journal of Membrane Science* 285 (1):4–29. doi: https://doi.org/10.1016/j.memsci.2006.08.002.

Essalhi, M., and M. Khayet. 2015. "10. Fundamentals of membrane distillation." In *Pervaporation, Vapour Permeation and Membrane Distillation*, edited by Angelo Basile, Alberto Figoli and Mohamed Khayet, 277–316. Oxford: Woodhead Publishing.

Eykens, L., K. De Sitter, C. Dotremont, L. Pinoy, and B. Van der Bruggen. 2016. "Characterization and performance evaluation of commercially available hydrophobic membranes for direct contact membrane distillation." *Desalination* 392:63–73. doi: https://doi.org/10.1016/j.desal.2016.04.006.

Fahmey, Mohamed S., Abdel-Hameed Mostafa El-Aassar, Mustafa M.Abo-Elfadel, Adel Sayed Orabi, and Rasel Das. 2019. "Comparative performance evaluations of nanomaterials mixed polysulfone: A scale-up approach through vacuum enhanced direct contact membrane distillation for water desalination." *Desalination* 451:111–116. doi: https://doi.org/10.1016/j.desal.2017.08.020.

Feng, C., K. C. Khulbe, T. Matsuura, R. Gopal, S. Kaur, S. Ramakrishna, and M. Khayet. 2008. "Production of drinking water from saline water by air-gap membrane distillation using polyvinylidene fluoride nanofiber membrane." *Journal of Membrane Science* 311 (1):1–6. doi: https://doi.org/10.1016/j.memsci.2007.12.026.

Goh, P. S., and A. F. Ismail. 2019. "Chapter 9 Graphene-based Membranes for Water Desalination Applications." In *Graphene-based Membranes for Mass Transport Applications*, 188–210. The Royal Society of Chemistry.

González, Daniel, José Amigo, and Francisco Suárez. 2017. "Membrane distillation: Perspectives for sustainable and improved desalination." *Renewable and Sustainable Energy Reviews* 80:238–259. doi: https://doi.org/10.1016/j.rser.2017.05.078.

Grasso, G., F. Galiano, M. J. Yoo, R. Mancuso, H. B. Park, B. Gabriele, A. Figoli, and E. Drioli. 2020. "Development of graphene-PVDF composite membranes for membrane distillation." *Journal of Membrane Science* 604:118017. doi: https://doi.org/10.1016/j.memsci.2020.118017.

Khalifa, A., D. Lawal, M. Antar, and M. Khayet. 2015. "Experimental and theoretical investigation on water desalination using air gap membrane distillation." *Desalination* 376:94–108. doi: https://doi.org/10.1016/j.desal.2015.08.016.

Khayet, Mohamed. 2011. "Membranes and theoretical modeling of membrane distillation: A review." *Advances in Colloid and Interface Science* 164 (1):56–88. doi: https://doi.org/10.1016/j.cis.2010.09.005.

Larbot, André, Laetitia Gazagnes, Sebastian Krajewski, Malgorzata Bukowska, and Kujawski Wojciech. 2004. "Water desalination using ceramic membrane distillation." *Desalination* 168:367–372. doi: https://doi.org/10.1016/j.desal.2004.07.021.

Lee, Eui-Jong, Alicia Kyoungjin An, Tao He, Yun Chul Woo, and Ho Kyong Shon. 2016. "Electrospun nanofiber membranes incorporating fluorosilane-coated TiO2 nano-composite for direct contact membrane distillation." *Journal of Membrane Science* 520:145–154. doi: https://doi.org/10.1016/j.memsci.2016.07.019.

Li, Xipeng, Huiting Shan, Min Cao, and Baoan Li. 2019. "Facile fabrication of omniphobic PVDF composite membrane via a waterborne coating for anti-wetting and anti-fouling membrane distillation." *Journal of Membrane Science* 589:117262. doi: https://doi.org/10.1016/j.memsci.2019.117262.

Li, Xue, Yongxing Zhang, Jingyi Cao, Xiaozu Wang, Zhaoliang Cui, Shouyong Zhou, Meisheng Li, Enrico Drioli, Zhaohui Wang, and Shuaifei Zhao. 2019. "Enhanced fouling and wetting resistance of composite Hyflon AD/poly(vinylidene fluoride) membrane in vacuum membrane distillation." *Separation and Purification Technology* 211:135–140. doi: https://doi.org/10.1016/j.seppur.2018.09.071.

Liu, Yongjie, Jin Wang, Zechun Xiao, Li Liu, Dongdong Li, Xuemei Li, Huabing Yin, and Tao He. 2020. "Anisotropic performance of a superhydrophobic polyvinyl difluoride membrane with corrugated pattern in direct contact membrane distillation." *Desalination* 481:114363. doi: https://doi.org/10.1016/j.desal.2020.114363.

Meindersma, G. W., C. M. Guijt, and A. B. de Haan. 2006. "Desalination and water recycling by air gap membrane distillation." *Desalination* 187 (1):291–301. doi: https://doi.org/10.1016/j.desal.2005.04.088.

Naji, Osamah, Raed A. Al-juboori, Les Bowtell, Alla Alpatova, and Noreddine Ghaffour. 2020. "Direct contact ultrasound for fouling control and flux enhancement in air-gap membrane distillation." *Ultrasonics Sonochemistry* 61:104816. doi: https://doi.org/10.1016/j.ultsonch.2019.104816.

Nthunya, Lebea N, Leonardo Gutierrez, Arne R Verliefde, and Sabelo D Mhlanga. 2019. "Enhanced flux in direct contact membrane distillation using superhydrophobic PVDF nanofibre membranes embedded with organically modified SiO$_2$ nanoparticles." 94 (9):2826 2837. doi: 10.1002/jctb.6104.

Qtaishat, M., T. Matsuura, B. Kruczek, and M. Khayet. 2008. "Heat and mass transfer analysis in direct contact membrane distillation." *Desalination* 219 (1):272–292. doi: https://doi.org/10.1016/j.desal.2007.05.019.

Qtaishat, Mohammed Rasool, and Fawzi Banat. 2013. "Desalination by solar powered membrane distillation systems." *Desalination* 308:186–197. doi: https://doi.org/10.1016/j.desal.2012.01.021.

Razmjou, Amir, Ellen Arifin, Guangxi Dong, Jaleh Mansouri, and Vicki Chen. 2012. "Superhydrophobic modification of TiO$_2$ nanocomposite PVDF membranes for applications in membrane distillation." *Journal of Membrane Science* 415–416:850–863. doi: https://doi.org/10.1016/j.memsci.2012.06.004.

Remanan, Sanjay, Maya Sharma, Suryasarathi Bose, and Narayan Ch. Das. 2018. "Recent Advances in Preparation of Porous Polymeric Membranes by Unique Techniques and Mitigation of Fouling through Surface Modification." *ChemistrySelect* 3 (2):609–633. doi: 10.1002/slct.201702503.

Santoro, Sergio, Ivan Vidorreta, Isabel Coelhoso, Joao Carlos Lima, Giovanni Desiderio, Giuseppe Lombardo, Enrico Drioli, Reyes Mallada, Joao Crespo, and Alessandra Criscuoli. 2019. "Experimental evaluation of the thermal polarization in direct contact membrane distillation using electrospun nanofiber membranes doped with molecular probes." *Molecules* 24 (3):638. doi: https://doi.org/10.3390/molecules24030638

Tang, Na, Chunlei Feng, Huaiyuan Han, Xinxin Hua, Lei Zhang, Jun Xiang, Penggao Cheng, Wei Du, and Xuekui Wang. 2016. "High permeation flux polypropylene/ethylene vinyl acetate co-blending membranes via thermally induced phase separation for vacuum membrane distillation desalination." *Desalination* 394:44–55. doi: https://doi.org/10.1016/j.desal.2016.04.024.

Tian, Miaomiao, Shushan Yuan, Florian Decaesstecker, Junyong Zhu, Alexander Volodine, and Bart Van der Bruggen. 2020. "One-step fabrication of isotropic poly(vinylidene fluoride) membranes for direct contact membrane distillation (DCMD)." *Desalination* 477:114265. doi: https://doi.org/10.1016/j.desal.2019.114265.

Tijing, Leonard D., June-Seok Choi, Sangho Lee, Seung-Hyun Kim, and Ho Kyong Shon. 2014. "Recent progress of membrane distillation using electrospun nanofibrous membrane." *Journal of Membrane Science* 453:435–462. doi: https://doi.org/10.1016/j.memsci.2013.11.022.

Tong, Daqing, Xiaozu Wang, Mohammad Ali, Christopher Q. Lan, Yong Wang, Enrico Drioli, Zhaohui Wang, and Zhaoliang Cui. 2016. "Preparation of Hyflon AD60/PVDF composite hollow fiber membranes for vacuum membrane distillation." *Separation and Purification Technology* 157:1–8. doi: https://doi.org/10.1016/j.seppur.2015.11.026.

Wang, Xuyun, Lin Zhang, Huajian Yang, and Huanlin Chen. 2009. "Feasibility research of potable water production via solar-heated hollow fiber membrane distillation system." *Desalination* 247 (1):403–411. doi: https://doi.org/10.1016/j.desal.2008.10.008.

Wang, Yuting, Minyuan Han, Lang Liu, Jingmei Yao, and Le Han. 2020. "Beneficial CNT Intermediate Layer for Membrane Fluorination toward Robust Superhydrophobicity and Wetting Resistance in Membrane Distillation." *ACS Applied Materials & Interfaces* 12 (18):20942–20954. doi: 10.1021/acsami.0c03577.

Woo, Yun Chul, Leonard D. Tijing, Wang-Geun Shim, June-Seok Choi, Seung-Hyun Kim, Tao He, Enrico Drioli, and Ho Kyong Shon. 2016. "Water desalination using graphene-enhanced electrospun nanofiber membrane via air gap membrane distillation." *Journal of Membrane Science* 520:99–110. doi: https://doi.org/10.1016/j.memsci.2016.07.049.

Xiao, Zechun, Hong Guo, Hailong He, Yongjie Liu, Xuemei Li, Yuebiao Zhang, Huabing Yin, Alexey V. Volkov, and Tao He. 2020. "Unprecedented scaling/fouling resistance of omniphobic polyvinylidene fluoride membrane with silica nanoparticle coated micropillars in direct contact membrane distillation." *Journal of Membrane Science* 599:117819. doi: https://doi.org/10.1016/j.memsci.2020.117819.

Xiao, Zechun, Zhansheng Li, Hong Guo, Yongjie Liu, Yanshai Wang, Huabing Yin, Xuemei Li, Jianfeng Song, Long D. Nghiem, and Tao He. 2019. "Scaling mitigation in membrane distillation: From superhydrophobic to slippery." *Desalination* 466:36–43. doi: https://doi.org/10.1016/j.desal.2019.05.006.

Xu, Ke, Yuchun Cai, Naser Tavajohi Hassankiadeh, Yangming Cheng, Xue Li, Xiaozu Wang, Zhaohui Wang, Enrico Drioli, and Zhaoliang Cui. 2019. "ECTFE membrane fabrication via TIPS method using ATBC diluent for vacuum membrane distillation." *Desalination* 456:13–22. doi: https://doi.org/10.1016/j.desal.2019.01.004.

Yang, Chi, Xue-Mei Li, Jack Gilron, Ding-feng Kong, Yong Yin, Yoram Oren, Charles Linder, and Tao He. 2014. "CF4 plasma-modified superhydrophobic PVDF membranes for direct contact membrane distillation." *Journal of Membrane Science* 456:155–161. doi: https://doi.org/10.1016/j.memsci.2014.01.013.

Yao, Minwei, Leonard D. Tijing, Gayathri Naidu, Seung-Hyun Kim, Hideto Matsuyama, Anthony G. Fane, and Ho Kyong Shon. 2020. "A review of membrane wettability for the treatment of saline water deploying membrane distillation." *Desalination* 479:114312. doi: https://doi.org/10.1016/j.desal.2020.114312.

Zhang, Wei, Yubing Lu, Jun Liu, Xipeng Li, Baoan Li, and Shichang Wang. 2020. "Preparation of re-entrant and anti-fouling PVDF composite membrane with omniphobicity for membrane distillation." *Journal of Membrane Science* 595:117563. doi: https://doi.org/10.1016/j.memsci.2019.117563.

Zhang, Yongxing, Xiaozu Wang, Zhaoliang Cui, Enrico Drioli, Zhaohui Wang, and Shuaifei Zhao. 2017. "Enhancing wetting resistance of poly(vinylidene fluoride) membranes for vacuum membrane distillation." *Desalination* 415:58–66. doi: https://doi.org/10.1016/j.desal.2017.04.011.

5 Membrane Distillation for Wastewater Treatment

Yuhe Cao and Ranil Wickramasinghe*
University of Arkansas

CONTENTS

5.1 INTRODUCTION

Human water consumption is increasing rapidly in various industries and urban life [1], and a severe shortage of water has been seen throughout the world [2]. Consequently, wastewater needs to be purified and reused, which has become a widely accepted approach to sustain our clean water supply and support economic development [3]. Reliable and efficient methods of water purification are urgently required. The acceptance of membrane technology in wastewater treatment increases since it can handle a wide range of feed water with high operational reliability [4]. Pressure-driven and osmoticly-driven membrane processes, such as reverse osmosis (RO), nanofiltration (NF), ultrafiltration (UF), microfiltration (MF), and forward osmosis (FO), have been widely used to purify different types of feed wastewater. A promising thermal-based membrane separation technology—membrane distillation (MD)—can help sustainably reduce worldwide water-energy stress [5].

In MD, volatile components in the feed solution evaporate and cross a microporous hydrophobic membrane; then, these volatile components condene in a distillate (permeate) solution [5]. Nonvolatile solutes are retained, and a highly pure distillate solution is produced [5, 6]. The vapor pressure difference between the feed and distillate is the driving force for MD, which causes evaporation at the feed side of the membrane and condensation at the distillate side. When MD is used for wastewater treatment, water vapor from a hot feed stream will pass through the hydrophobic membrane and condense in the permeate side [7]. MD can be used to enhance water recovery and decrease the amount of concentrate requiring disposal for other separation techniques such as RO, electrodialysis (ED), crystallization, and bioreactors [8].

This chapter provides an overview of the innovative and sustainable development of MD in wastewater treatment. Membrane materials and structures are first outlined. The development of conventional and new MD configurations is discussed. The applications of MD in wastewater treatment are reviewed. Membrane fouling and wetting phenomenon are also discussed. Finally, the outlook for the future is briefly highlighted, and conclusions are drawn.

5.2   MEMBRANE MATERIALS AND STRUCTURES

5.2.1   MEMBRANE MATERIALS

Hydrophobicity is the essential requirement for MD membranes for aqueous applications, intrinsic or modified; hydrophobic polymers with low surface energy are required to fabricate the MD membranes [9]. Commercial polymers commonly used for the fabrication of MD membranes are polytetrafluoroethylene (PTFE), polypropylene (PP), polyvinylidene fluoride (PVDF), poly (vinylidene fluoride-co-hexafluoropropylene) (PVDF-HFP), and Hyflon® AD (Solvay Plastics) [9–13]. Among these commercially available polymer materials, PTFE, PE, PVDF, and PP are the most commonly used material for membrane preparation [9, 14]. To enhance vapor transport and minimize heat losses, the membranes should have a low resistance to mass transfer and low thermal conductivity. Ceramic membranes are often more expensive than polymeric membranes. Meanwhile, they have high thermal conductivity and could lose much heat through the membrane matrix [15–17]. Therefore, ceramic membranes are often not considered optimal MD membranes.

5.2.2   MEMBRANE STRUCTURES

Different membrane structures have been proposed. Typically all these membranes must have a hydrophobic layer, which is not wetted by the feed liquid [14]. Four types

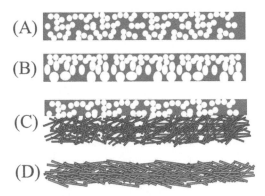

FIGURE 5.1 Schematic illustration of the different membrane structures: (A) isotropic membrane, (B) integral asymmetric membrane, (C) composite membrane, and (D) nanofibrous membrane.

of membrane morphologies are shown in Figure 5.1. Isotropic membranes have the same morphology throughout the entire cross-section, as is shown in Figure 5.1A, which requires that the membrane material is hydrophobic.

Alternatively, asymmetric membranes (Figure 5.1B) made with hydrophobic material have less mass transfer resistance than isotropic membrane. Composite membranes have attracted wide attention from many researchers [18–25], where a nonwoven or scrim support is used to provide the mechanical strength [22, 26], as is illustrated in Figure 5.1C.

Composite membranes such as a hydrophobic PVDF layer on the top of hydrophilic polyvinyl alcohol blended polyethylene glycol (PVA/PEG) sublayer [27], or polyvinylidene fluoride (PVDF) membrane supported by a polyester (PET) filament woven fabric [24], or a polytetrafluoroethylene (PTFE) layer on a scrim-backing support layer of polypropylene (PP) [25], have been reported. All the aforementioned composite membranes have an active top hydrophobic layer.

Another essential type of membrane is the nanofibrous membrane fabricated by electrospinning, as is shown in Figure 5.1D. The electrospun nanofibrous membranes have high porosity, an interconnected open pore structure, pore sizes ranging from tens of nanometers to several micrometers, tailorable membrane thickness, and high gas permeability [28]. A large surface area-to-pore volume ratio makes electrospun nanofibrous membranes most desirable as MD membranes to generate a high water vapor flux [29].

5.3 CONVENTIONAL AND NEW MD CONFIGURATIONS

Based on the methods used to recover the permeate, the MD process is divided into four traditional configurations: (1) direct contact membrane distillation (DCMD), where a cold liquid is in direct contact with the membrane at the permeate side; (2) sweeping gas membrane distillation (SGMD), where gas is pumped on the permeate side to aid the evaporation of less volatile substances of interest; (3) vacuum membrane distillation (VMD), where the permeate side is subjected to a lower pressure

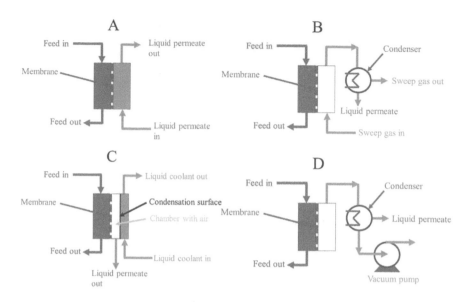

FIGURE 5.2 Schematic diagram of conventional membrane configurations: (A) direct contact membrane distillation (DCMD); (B) sweeping gas membrane distillation (SGMD); (C) air gap membrane distillation (AGMD); and (D) vacuum membrane distillation (VMD).

than the feed side; and (4) air gap membrane distillation (AGMD), where an air gap acts as the condensing surface at the permeate side [8, 30–32].

Figure 5.2 shows these configurations; each has its merit and demerits. For example, DCMD is the most straightforward MD configuration with the highest heat loss by conduction [33, 34]. AGMD has less conductive heat loss and wetting, while it creates additional resistance for the mass transfer, and the module is not easy to design [35]. SGMD is a suitable configuration to remove volatile components and dissolved gases without wetting from the permeate side, requiring a large condenser and large sweep gas volume [33]. VMD has high flux and negligible conductive heat loss; however, it has a high risk of pore wetting and fouling [35]. Therefore, the MD configuration should be chosen based on the energy and equipment requirements, and the properties of the feed and the target compound that is to be removed.

The specific energy consumption of these MD configurations can be easily higher than 1256 kWh/m^3 (estimated from gain output ratio) if there is no heat recovery [36]. Several newer MD configurations with lower energy consumption and improved permeation flux have been investigated [9, 37, 38]. The AGMD and SGMD processes can be combined in a process named thermostatic sweeping gas membrane distillation (TSGMD), as is shown in Figure 5.3A, where the sweeping gas is passed through the gap between the condensation surface and the membrane. The vapor is partially condensed on the condensation surface, and the remaining vapor is condensed over an external condenser. The external condenser can decrease the sweeping gas's temperature, which will accelerate mass transfer by generating a higher driving force. Hence, for example, wastewater recovery can be enhanced [33, 39].

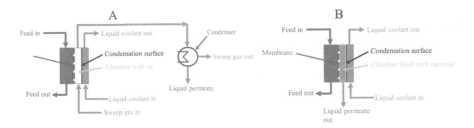

FIGURE 5.3 Schematic diagram of new membrane configurations: (A) thermostatic sweeping gas membrane distillation (TSGMD); and (B) material gap membrane distillation (MGMD).

Material-gap membrane distillation (MGMD) is another variant of MD combining both DCMD and AGMD configurations. A stagnant cold liquid could fill the air gap in the AGMD (usually the produced distilled water). This configuration is also named liquid gap membrane distillation (LGMD). Alternatively, the air gap in AGMD could be filled with conductive materials such as porous support, sand, sponge (polyurethane), or metal mesh [8]. The vapor flux improvement can be from 200 to 800% [40–43], as is shown in Figure 5.3B.

5.4 MEMBRANE MODIFICATION

As discussed in the previous section, the desired MD membrane should be hydrophobic to prevent the membrane pores from wetting. Some popular polymeric materials with intrinsic hydrophobicity, including polypropylene (PP), polytetrafluoroethylene (PTFE), and polyvinylidene fluoride (PVDF), have been broadly used to fabricate porous membranes in early work. Hydrophilic membrane surfaces could be modified to become hydrophobic. Two methods, coating and grafting, have been proposed for hydrophobic surface modifications [44]. By producing covalent bonding between the membrane surface and the graft molecules, a more permanent, thin layer on the membrane surface could be formed [45, 46]. As for the coating approach, one should be aware that the stability between the coating layer and membrane surface degrades during long-term operation, and thus the hydrophobicity can be lost.

A significant effort has been focused on constructing superhydrophobic surfaces for the MD membrane, which means the water contact angle should be greater than 150° [47]. Three main approaches have been applied to fabricate superhydrophobic surfaces: building up rough hierarchical structures on a hydrophobic surface, modifying a rough surface by low-surface-energy materials, and combining rough hierarchical structures with low-surface-energy materials. Both chemical components and geometrical structures play an essential role in the solid surface's wettability performance, as is inspired by the lotus leaf surface superhydrophobic microstructure [48].

Another important modification of membranes involves the incorporation of nanoparticles. By adding micron and nanoscale materials, hierarchical micro/nano surface morphology with multilevel surface roughness can be generated. The surface can be fluorosilanized with fluoroalkylsilane coupling agents to improve the

TABLE 5.1

Examples for membrane modification

Method	Base membrane	Specific modification	Reference
Grafting	Hydrophilic porous polyacrylonitrile	Dipping in fluorine-containing solution (1H, 1H, 2H, 2H-perfluorodecyl methacrylate) followed by plasma irradiation	[44]
Grafting	Hydrophilic polyethersulfone	CF_4 plasma modification	[49]
Grafting	Cellulose nitrate membrane	Plasma polymerization of octafluorocyclobutane and vinyltrimethylsilicon/carbon tetrafluoride	[50]
Coating	PVDF hollow fibers	Teflon® AF 2400 is coated on the membrane surface	[51]
Coating and grafting	PVDF membrane	Coating SiO_2 nanoparticles and hydrophobic modification using perfluorooctyltrichlorosilane	[52]
Spray coating	PVDF membrane	Spraying a mixture of polydimethylsiloxane and hydrophobic SiO_2 nanoparticles on base membrane	[53]
Incorporation of nanoparticles	PVDF/nonwoven composite membranes	Hydrophobic modified calcium carbonate nanoparticles in the PVDF casting solution	[54]
Incorporation of nanoparticles	PVDF membrane	Combining hydrophobic modified SiO_2 nanoparticles with polydimethylsiloxane (PDMS) and then spraying them on the membrane surface.	[55]
Incorporation of nanoparticles	PVDF membrane	depositing TiO_2 nanoparticles via a low temperature hydrothermal process and then fluorosilanized	[53]

hydrophobicity further and reduce the surface free energy. Table 5.1 lists some examples of membrane surface modification.

5.5 WASTEWATER TREATMENT

As discussed previously, membrane separation processes such as MF, UF, NF, RO and FO have been widely employed to treat various types of wastewater. The unique characteristics of MD have made this process an excellent option to treat wastewater and produce high-quality water on-site when waste heat and low-cost thermal energy are available. MD processes could be combined with other membrane processes to enhance water recovery and decrease the amount of concentrate requiring disposal, making it a suitable wastewater treatment with zero (or near zero) liquid discharge.

5.5.1 MINING INDUSTRY WASTEWATER TREATMENT

The exploitation of surface water and groundwater resources through mining activities is becoming a serious environmental issue. MF/UF-NF and RO membrane processes were integrated to treat the industrial mining wastewater [56, 57], while a limited water recovery ratio (around 30% to 60%) could be achieved [58]. MD could be employed to enhance the water recovery ratio further, and meanwhile, the minerals could also be recovered. More than 80% of water could be recovered using an integrated forward osmosis–MD process [59].

5.5.2 Textile Wastewater Treatment

Textile wastewater could cause serious problems for both underground and surface waters. The hot effluents discharged from textile industries have temperatures ranging from 353 K to 363 K. Therefore, by utilizing the energy from the effluents, MD is considered an appropriate method for textile wastewater treatment [60]. DCMD was employed to treat industrial dyeing wastewater and characteristic pollutants [61] by two commercial hydrophobic membranes made of PTFE and PVDF. Results showed excellent rejection for nonvolatile pollutants under mild temperature and limited hydraulic pressure. In terms of membrane fouling caused by dyes, it seems the PTFE membrane has enhanced resistance to dye absorption because of its strong negative charge and chemical structure [62]. A flake-like dye-dye structure formed on the membrane surface rather than in the membrane pores, which was loose, and therefore the membrane flux was quickly recovered by thoroughly flushing with water. DCMD using a PTFE membrane for treating dye mixtures showed stable flux and superior color removal during five days of operation.

5.5.3 Treatment of Radioactive Wastewater

DCMD was used to treat low-level radioactive wastewater [63]. The experimental results showed that DCMD could separate almost all Cs^+, Sr^{2+}, and Co^{2+} from wastewater. DCMD was also used to remove nuclides and boron from highly saline radioactive wastewater [64], which can help reduce the volume of radioactive brines and the costs associated with solidification and long-term storage. Membrane surface chemistry is also essential for radioactive wastewater treatment performance. Fluorinated surface modifying macromolecules were used to improve MD performance. This modification led to higher values of the liquid entry pressure of water, higher rejection factors, and lower adsorption of radioactive isotopes on the membrane surface [65]. DCMD has been shown to be a promising separation process for low-level radioactive wastewater treatment.

5.5.4 Treatment of Oily Wastewater

Oily wastewater from industry and domestic sewage is one of the main types of wastewater [66, 67]. Among these oily wastewaters, the amount of hydraulic fracturing generated oily wastewater, namely produced water, increases rapidly with oil consumption and subsequent oil exploitation. Produced water is a very complex wastewater stream with a variety of constituents: a complex mixture of organic substances (mainly dispersed and dissolved oil compounds) and inorganic substances (high concentrations of mineral salts with total dissolved solids TDS up to tens of thousands ppm) [68]. Very few reports are available on using MD solely for treating oily feeds since the organic compounds could cause severe fouling during the membrane [69]; therefore, other pretreatment processes integrated with MD have been widely investigated for produced water treatment.

A hybrid FO-MD system was employed to recover water from oily wastewater [66]; water molecules were transferred from oily water to the draw of the FO process and then to the distillate of the MD process. After that the MD process produced high purity water and regenerated the draw solution of the FO process, and high osmotic pressure of the FO process was continuously maintained to accelerate water transfer. At least 90% feed water recovery could be achieved while oil and NaCl were almost wholly rejected.

Besides treating produced water, another type of oily wastewater, a typical bilge water collected from a harbour without pretreatment, was also tested by a UF-MD integrated process [70]. After the UF pretreatment, the UF permeate contained less than 5 ppm of oil. Further purification of the UF permeates by MD resulted in complete removal of oil from wastewater and a very high reduction of the total organic carbon (TOC) (99.5%) and total dissolved solids (99.9%). To decrease the fouling of FO process, an integrated UF-FO-MD system was also investigated [71]. UF pretreated oily water was simultaneously used as a draw solution for FO and feed for MD in s FO-MD subsystem by utilizing its high salinity and temperature. Results showed UF could mitigate membrane fouling of the downstream FO-MD subsystem and high-quality water could be obtained by utilization of waste heat at low-energy cost.

5.5.5 OTHER PRETREATMENT/HYBRID MD PROCESSES FOR WASTEWATER TREATMENT

To decrease membrane fouling and scaling, other types of pretreatment/hybrid MD process were investigated. Integrated crystallization was employed to reduce scalant loading of multivalent ions, such as barium and calcium, which increased the total water recovery from 20% to 62.5% [72]. Foam fractionation was used to remove surface-active materials, causing membrane wetting during textile mill effluent treatment [73]. Ninety-one and a half percent of water recovery was achieved at the end of a 65-day trial without membrane wetting.

The integrated electrocoagulation–forward osmosis–MD process was employed to treat the produced water [74]. The electrocoagulation pretreatment for FO could significantly remove TOC and total suspended solids (TSS) and showed stable performance with minimal fouling. Results showed that electrocoagulation could significantly aid in fouling mitigation and achieve stable FO-MD performance. Besides the physical and chemical pretreatments applied for MD, biological processes were also considered with the MD process. An integrated bioreactor was used to biologically remove organics and nutrients from industrial wastewater [75]. By lowering the retentate organic and nutrient concentration, the bioreactor successfully delayed membrane wetting by 1.7–3.6 times, which reduced the frequency of membrane cleaning and drying. Biological treatment was also investigated to digest the surfactants from the textile wastewater [73], which reduced the TOC) from 100 to 26 mg/L. Therefore, organic fouling of the MD process was suppressed, and the MD operation time was significantly increased.

5.6 MEMBRANE FOULING AND WETTING PHENOMENON

5.6.1 MEMBRANE FOULING

Like any other membrane process, MD membranes also suffer from fouling during operation, which is a significant problem in membrane-based separation processes. There are mainly four factors affecting the membrane fouling [76]: (a) foulant characteristics (concentration, molecular size, solubility, diffusivity, hydrophobicity, charge, etc.); (b) membrane properties (hydrophobicity, surface roughness, pore size, surface charge, and surface functional groups); (c) operating conditions (flux, solution temperature, and flow velocity); and (d) feedwater characteristics (solution chemistry, pH, ionic strength, and presence of organic/inorganic matter) [30]. The foulants in the feed water will determine the type of membrane fouling: inorganic fouling, organic fouling, and biological fouling. Understanding the interaction between the foulant and the membrane is essential to develop a fouling resistant membrane.

5.6.2 MEMBRANE WETTING

Wastewater containing wetting compounds (e.g., oils, surfactants) is a big challenge. Membrane wetting reduces the rejection of nonvolatile solutes during MD, which must be addressed for long-term industrial implementation of MD [77]. Usually, the liquid entry pressure (LEP) is used as the primary metric for measuring membrane wettability. Membrane wetting occurs when the hydraulic pressure on the membrane surface is greater than the LEP [5]. The LEP can be determined by [30]

$$\text{LEP} = \frac{-4B\sigma\cos\theta}{d_{max}} \tag{5.1}$$

B is a pore geometric factor (in cylindrical pores $B = 1$), σ is the surface tension of the solution, θ is the contact angle between the solution and the membrane surface, and d_{max} the diameter of the largest pore. As can be seen from Equation (5.1), LEP depends on many parameters, including the maximum membrane pore size, the surface tension of the liquid, the contact angle of the liquid on the membrane surface, and the geometrical structure of the membrane.

Higher LEP means the membrane has a larger antiwetting capability. Generally, membrane wetting can be classified as follows [78]: nonwetted, surface-wetted, partially-wetted, and fully-wetted. During the nonwetted or surface-wetted stages, there are no nonvolatile solutes from the feed penetrating the membrane. The rejection rate remains almost the same as the nonwetted stage even though the surface is wetted to a significant depth, e.g., 100–200 μm, which can still provide a liquid/vapor interface; a gap between the feed stream and the permeate stream ensures high rejection. As the feed solution penetrates deeper into the membrane pores, partial wetting can occur, causing deterioration of permeate quality. Once the membrane is fully wetted, the membrane no longer acts as a barrier for liquid transport, resulting in viscous water flow through the membrane pores [79, 80].

5.6.3 Approaches to Control Fouling and Cleaning

The fouling layer on top of the membrane could lead to additional thermal and hydraulic resistances [81]. Two types of fouling could occur: external surface fouling and pore-blocking [82]. Fouling on the external surface is usually reversible and can be removed by cleaning with chemicals, while the internal fouling or pore blocking is irreversible in most cases. Therefore, to maintain high performance and extend membrane lifetime, fouling prevention and membrane cleaning approaches are needed for MD.

Several approaches could be employed to control the fouling: pretreatment, membrane flushing, gas bubbling, temperature and flow reversal, surface modification (discussed in the previous section), using a magnetic field, microwave irradiation, and use of antiscalants [30]. Among these approaches, pretreatment approaches were widely investigated for MD implementation in wastewater treatment since many contaminants should be removed before entering the MD system. We will focus on this approach in this section.

Coagulation with polyaluminum chloride was effective for significantly reducing the solids and organic loading in biologically treated coking wastewater, which significantly reduces the propensity for membrane fouling. Besides chemical coagulation, electrocoagulation was also used as a pretreatment for produced water, which removed ~90% TSS and ~60% TOC from the raw produced water and resulted in lower membrane fouling during DCMD [83].

Both flocculation-sedimentation and flocculation-sedimentation-microfiltration (FSMF) were employed to pretreat the shale gas wastewater before MD [84]. Results indicate FSMF was more efficient at suppressing flux decline and also the most effective way to suppress fouling at high temperature differences.

Chemical cleaning is a promising method to recover the permeate flux after fouling. After MD treatment of wastewater from the petrochemical industry, the deposited foulant layer was entirely removed, and the membrane surface was restored as clean as a fresh membrane by using 2 wt. % citric acid cleaning [85]. In desalination of recirculating cooling water, Wang et al. [86] utilized 2 wt. % HCl or 2 wt. % NaOH solutions to remove the $CaCO_3$ scales from fouled hollow fiber membranes, which removed the majority of the deposits and recovered membrane hydrophobicity. Physical cleaning-water flushing was employed to clean the MD membrane after the treatment of dye wastewater. The aggregated flake type dye foulants could not enter the membrane pores and were therefore easily removed by water flush [62], which enabled the long-term application of DCMD.

5.6.4 Approaches to Control Wetting

Membrane wetting is usually related to fouling. Wetting could be avoided by either fabricating membranes with low affinity for the feed components or the use of a robust pretreatment of the feed. As mentioned in the previous section, membrane surface modification could change the membrane geometrical structure and surface chemistry, which can be tuned to increase the surface nonwettability. Moreover, surface free energy can be decreased by functionalization with low surface energy materials. Surfactants are the most challenging organic foulants in treating wastewater, which could easily cause membrane wetting.

Composite membranes were also developed to suppress surfactant wetting. A thin layer of agarose hydrogel was attached to the hydrophobic porous Teflon membrane's surface to suppress surfactant wetting during MD for industrial wastewater treatment [87]. The results showed that no wetting occurred during the 24 hour period of MD against 10 mg/L of SDS or Tween20.

5.7 TECHNO-ECONOMIC ANALYSIS OF MD-BASED WASTEWATER TREATMENT PROCESS

In terms of sustainability, MD also has the advantage that it can be coupled with other separation techniques to achieve zero liquid discharge. To determine if MD is a conceptually solid and economical fit in the zero liquid discharge scheme of processing, techno-economic evaluation of the MD process for highly saline wastewaters treatment with a zero liquid discharge was investigated [88]. The results showed that MD could potentially be ~40% more cost-effective than mechanical vapor compression for a system capacity of 100 m^3/day feed water, and up to ~75% more cost-effective if the MD is driven with free waste heat. Integrated MD and other membrane processes could be successfully used with textile dye bath waste streams to achieve zero liquid discharge.

The techno-economic analysis indicated that the success of the integrated approach in practice would rely chiefly on the MD process, which constitutes 70–90% of the benefit/cost ratio [89]. Tavakkoli et al. [90] investigated a detailed techno-economic assessment to understand the cost drivers and assess the total cost of treating high salinity produced water using DCMD. The results reveal that the total cost of treating produced water using MD is $5.70/$m^3_{feed}$, which decreases significantly to $0.74/$m^3_{feed}$ when MD is integrated with a source of waste heat, which can potentially come from the natural gas compressor stations.

5.8 FUTURE PERSPECTIVES AND CONCLUSIONS

MD is an attractive technology that can sustainably reduce worldwide water-energy stress, potentially powered by renewable energies or waste heat. It has been demonstrated that MD is becoming competitive with RO if the waste heat can be utilized to power MD. Solar energy and geothermal energy can be used to drive MD, which can be a continuous source of thermal energy. Nevertheless, MD plants need to be evaluated to determine how MD performs if they are driven by solar and geothermal energy. However, a commercial breakthrough on a large-scale MD application in wastewater treatment has not been achieved so far.

There are some limitations for the state of the art RO and evaporation technology, such as high pressure with difficulty treating high salinity water and corrosion in evaporators [88]. To compete with them, MD provides a unique market opportunity and driver for commercializing the technology as a chain process to achieve zero liquid discharge [31].

MD is a promising technology for the recovery of water from wastewater streams in the future. However, experimental large-scale pilot plants with continuous monitoring of the different processes must still overcome the current challenges. Besides, the development of practical membrane antifouling modifications and antifouling material development is also needed to improve membrane antifouling characteristics and the integrated membrane system.

ACKNOWLEDGEMENTS

Funding was provided by the Arkansas Research Alliance and the Ross E. Martin Chair for Emerging Technologies.

REFERENCES

[1] D. Ghazanfari, D. Bastani, S.A. Mousavi, Preparation and characterization of poly (vinyl chloride) (PVC) based membrane for wastewater treatment, Journal of Water Process Engineering 16 (2017) 98–107.

[2] C.H. Neoh, Z.Z. Noor, N.S.A. Mutamim, C.K. Lim, Green technology in wastewater treatment technologies: Integration of membrane bioreactor with various wastewater treatment systems, Chemical Engineering Journal 283 (2016) 582–594.

[3] I. Gehrke, A. Geiser, A. Somborn-Schulz, Innovations in nanotechnology for water treatment, Nanotechnology, Science Applications 8 (2015) 1–17.

[4] P.S. Goh, T.W. Wong, J.W. Lim, A.F. Ismail, N. Hilal, Chapter 9: Innovative and sustainable membrane technology for wastewater treatment and desalination application, in C.M. Galanakis (ed.), Innovation Strategies in Environmental Science, Elsevier 2020, pp. 291–319.

[5] D. González, J. Amigo, F. Suárez, Membrane distillation: Perspectives for sustainable and improved desalination, Renewable and Sustainable Energy Reviews 80 (2017) 238–259.

[6] T.Y. Cath, V.D. Adams, A.E. Childress, Experimental study of desalination using direct contact membrane distillation: a new approach to flux enhancement, Journal of Membrane Science 228(1) (2004) 5–16.

[7] Y. Liao, R. Wang, A.G. Fane, Fabrication of bioinspired composite nanofiber membranes with robust superhydrophobicity for direct contact membrane distillation, Environmental Science & Technology 48(11) (2014) 6335–41.

[8] P. Biniaz, N. Torabi Ardekani, M. Makarem, M. Rahimpour, Water and wastewater treatment systems by novel integrated membrane distillation (MD), ChemEngineering 3(1) (2019) 8.

[9] P. Wang, T.-S. Chung, Recent advances in membrane distillation processes: Membrane development, configuration design and application exploring, Journal of Membrane Science 474 (2015) 39–56.

[10] B.S. Lalia, E. Guillen-Burrieza, H.A. Arafat, R. Hashaikeh, Fabrication and characterization of polyvinylidenefluoride-co-hexafluoropropylene (PVDF-HFP) electrospun membranes for direct contact membrane distillation, Journal of Membrane Science 428 (2013) 104–115.

[11] J. Zhang, J.-D. Li, S. Gray, Effect of applied pressure on performance of PTFE membrane in DCMD, Journal of Membrane Science 369(1–2) (2011) 514–525.

[12] M. Khayet, Membranes and theoretical modeling of membrane distillation: a review, Advances in Colloid and Interface Science 164(1–2) (2011) 56–88.

[13] L. Camacho, L. Dumée, J. Zhang, J.-d. Li, M. Duke, J. Gomez, S. Gray, Advances in membrane distillation for water desalination and purification applications, Water 5(1) (2013) 94–196.

[14] L. Eykens, K. De Sitter, C. Dotremont, L. Pinoy, B. Van der Bruggen, Membrane synthesis for membrane distillation: A review, Separation and Purification Technology 182 (2017) 36–51.

[15] S. Cerneaux, I. Strużyńska, W.M. Kujawski, M. Persin, A. Larbot, Comparison of various membrane distillation methods for desalination using hydrophobic ceramic membranes, Journal of Membrane Science 337(1–2) (2009) 55–60.

[16] Z.D. Hendren, J. Brant, M.R. Wiesner, Surface modification of nanostructured ceramic membranes for direct contact membrane distillation, Journal of Membrane Science 331(1–2) (2009) 1–10.

[17] J.-W. Wang, L. Li, J.-W. Zhang, X. Xu, C.-S. Chen, β-Sialon ceramic hollow fiber membranes with high strength and low thermal conductivity for membrane distillation, Journal of the European Ceramic Society 36(1) (2016) 59–65.

[18] M. Qtaishat, M. Khayet, T. Matsuura, Guidelines for preparation of higher flux hydrophobic/hydrophilic composite membranes for membrane distillation, Journal of Membrane Science 329(1–2) (2009) 193–200.

[19] M. Qtaishat, D. Rana, M. Khayet, T. Matsuura, Preparation and characterization of novel hydrophobic/hydrophilic polyetherimide composite membranes for desalination by direct contact membrane distillation, Journal of Membrane Science 327(1–2) (2009) 264–273.

[20] C. Feng, R. Wang, B. Shi, G. Li, Y. Wu, Factors affecting pore structure and performance of poly(vinylidene fluoride-co-hexafluoro propylene) asymmetric porous membrane, Journal of Membrane Science 277(1–2) (2006) 55–64.

[21] M. Khayet, J.I. Mengual, T. Matsuura, Porous hydrophobic/hydrophilic composite membranes, Journal of Membrane Science 252(1–2) (2005) 101–113.

[22] D. Winter, J. Koschikowski, D. Düver, P. Hertel, U. Beuscher, Evaluation of MD process performance: Effect of backing structures and membrane properties under different operating conditions, Desalination 323 (2013) 120–133.

[23] J.A. Prince, V. Anbharasi, T.S. Shanmugasundaram, G. Singh, Preparation and characterization of novel triple layer hydrophilic–hydrophobic composite membrane for desalination using air gap membrane distillation, Separation and Purification Technology 118 (2013) 598–603.

[24] R. Huo, Z. Gu, K. Zuo, G. Zhao, Preparation and properties of PVDF-fabric composite membrane for membrane distillation, Desalination 249(3) (2009) 910–913.

[25] J.-G. Lee, Y.-D. Kim, W.-S. Kim, L. Francis, G. Amy, N. Ghaffour, Performance modeling of direct contact membrane distillation (DCMD) seawater desalination process using a commercial composite membrane, Journal of Membrane Science 478 (2015) 85–95.

[26] L. Eykens, K. De Sitter, C. Dotremont, L. Pinoy, B. Van der Bruggen, Characterization and performance evaluation of commercially available hydrophobic membranes for direct contact membrane distillation, Desalination 392 (2016) 63–73.

[27] P. Peng, A.G. Fane, X. Li, Desalination by membrane distillation adopting a hydrophilic membrane, Desalination 173(1) (2005) 45–54.

[28] C. Feng, K.C. Khulbe, T. Matsuura, S. Tabe, A.F. Ismail, Preparation and characterization of electrospun nanofiber membranes and their possible applications in water treatment, Separation and Purification Technology 102 (2013) 118–135.

[29] S.S. Ray, S.-S. Chen, N.C. Nguyen, H.T. Nguyen, Chapter 9 - Electrospinning: A Versatile Fabrication Technique for Nanofibrous Membranes for Use in Desalination, in S. Thomas, D. Pasquini, S.-Y. Leu, D.A. Gopakumar (eds.), Nanoscale Materials in Water Purification, Elsevier 2019, pp. 247–273.

[30] L.D. Tijing, Y.C. Woo, J.-S. Choi, S. Lee, S.-H. Kim, H.K. Shon, Fouling and its control in membrane distillation—A review, Journal of Membrane Science 475 (2015) 215–244.

[31] A. Alkhudhiri, N. Darwish, N. Hilal, Membrane distillation: A comprehensive review, Desalination 287 (2012) 2–18.

[32] B.B. Ashoor, S. Mansour, A. Giwa, V. Dufour, S.W. Hasan, Principles and applications of direct contact membrane distillation (DCMD): A comprehensive review, Desalination 398 (2016) 222–246.

[33] A.A. Kiss, O.M. Kattan Readi, An industrial perspective on membrane distillation processes, Journal of Chemical Technology & Biotechnology 93(8) (2018) 2047–2055.

[34] H. Susanto, Towards practical implementations of membrane distillation, Chemical Engineering and Processing: Process Intensification 50(2) (2011) 139–150.

[35] E. Drioli, A. Ali, F. Macedonio, Membrane distillation: Recent developments and perspectives, Desalination 356 (2015) 56–84.

[36] A. Criscuoli, M.C. Carnevale, E. Drioli, Evaluation of energy requirements in membrane distillation, Chemical Engineering and Processing: Process Intensification 47(7) (2008) 1098–1105.

[37] K. Zhao, W. Heinzl, M. Wenzel, S. Büttner, F. Bollen, G. Lange, S. Heinzl, N. Sarda, Experimental study of the memsys vacuum-multi-effect-membrane-distillation (V-MEMD) module, Desalination 323 (2013) 150–160.

[38] Y.Z. Tan, L. Han, N.G.P. Chew, W.H. Chow, R. Wang, J.W. Chew, Membrane distillation hybridized with a thermoelectric heat pump for energy-efficient water treatment and space cooling, Applied Energy 231 (2018) 1079–1088.

[39] M. Essalhi, M. Khayet, Chapter Three - Membrane Distillation (MD), in S. Tarleton (Ed.), Progress in Filtration and Separation, Academic Press, Oxford, 2015, pp. 61–99.

[40] L. Francis, N. Ghaffour, A.A. Alsaadi, G.L. Amy, Material gap membrane distillation: A new design for water vapor flux enhancement, Journal of Membrane Science 448 (2013) 240–247.

[41] J. Swaminathan, H.W. Chung, D.M. Warsinger, F.A. AlMarzooqi, H.A. Arafat, J.H. Lienhard V, Energy efficiency of permeate gap and novel conductive gap membrane distillation, Journal of Membrane Science 502 (2016) 171–178.

[42] D. Winter, J. Koschikowski, M. Wieghaus, Desalination using membrane distillation: Experimental studies on full scale spiral wound modules, Journal of Membrane Science 375(1–2) (2011) 104–112.

[43] J. Swaminathan, H.W. Chung, D.M. Warsinger, J.H. Lienhard V, Membrane distillation model based on heat exchanger theory and configuration comparison, Applied Energy 184 (2016) 491–505.

[44] L. Liu, F. Shen, X. Chen, J. Luo, Y. Su, H. Wu, Y. Wan, A novel plasma-induced surface hydrophobization strategy for membrane distillation: Etching, dipping and grafting, Journal of Membrane Science 499 (2016) 544–554.

[45] J. Meng, J. Li, Y. Zhang, S. Ma, A novel controlled grafting chemistry fully regulated by light for membrane surface hydrophilization and functionalization, Journal of Membrane Science 455 (2014) 405–414.

[46] G. Zuo, R. Wang, Novel membrane surface modification to enhance anti-oil fouling property for membrane distillation application, Journal of Membrane Science 447 (2013) 26–35.

[47] Z. Yuan, H. Chen, J. Tang, X. Chen, D. Zhao, Z. Wang, Facile method to fabricate stable superhydrophobic polystyrene surface by adding ethanol, Surface and Coatings Technology 201(16) (2007) 7138–7142.

[48] J. Zha, N. Batisse, D. Claves, M. Dubois, L. Frezet, A.P. Kharitonov, L.N. Alekseiko, Superhydrophocity via gas-phase monomers grafting onto carbon nanotubes, Progress in Surface Science 91(2) (2016) 57–71.

[49] X. Wei, B. Zhao, X.-M. Li, Z. Wang, B.-Q. He, T. He, B. Jiang, CF4 plasma surface modification of asymmetric hydrophilic polyethersulfone membranes for direct contact membrane distillation, Journal of Membrane Science 407–408 (2012) 164–175.

[50] Y. Wu, Y. Kong, X. Lin, W. Liu, J. Xu, Surface-modified hydrophilic membranes in membrane distillation, Journal of Membrane Science 72(2) (1992) 189–196.

[51] K.-J. Lu, J. Zuo, T.-S. Chung, Tri-bore PVDF hollow fibers with a superhydrophobic coating for membrane distillation, Journal of Membrane Science 514 (2016) 165–175.

[52] W. Zhang, Y. Li, J. Liu, B. Li, S. Wang, Fabrication of hierarchical poly (vinylidene fluoride) micro/nano-composite membrane with anti-fouling property for membrane distillation, Journal of Membrane Science 535 (2017) 258–267.

[53] A. Razmjou, E. Arifin, G. Dong, J. Mansouri, V. Chen, Superhydrophobic modification of TiO2 nanocomposite PVDF membranes for applications in membrane distillation, Journal of Membrane Science 415–416 (2012) 850–863.

[54] D. Hou, G. Dai, H. Fan, J. Wang, C. Zhao, H. Huang, Effects of calcium carbonate nanoparticles on the properties of PVDF/non-woven fabric flat-sheet composite membranes for direct contact membrane distillation, Desalination 347 (2014) 25–33.

[55] J. Zhang, J.-D. Li, M. Duke, M. Hoang, Z. Xie, A. Groth, C. Tun, S. Gray, Modelling of vacuum membrane distillation, Journal of Membrane Science 434 (2013) 1–9.

[56] B.C. Ricci, C.D. Ferreira, A.O. Aguiar, M.C.S. Amaral, Integration of nanofiltration and reverse osmosis for metal separation and sulfuric acid recovery from gold mining effluent, Separation and Purification Technology 154 (2015) 11–21.

[57] M.C.S. Amaral, L.B. Grossi, R.L. Ramos, B.C. Ricci, L.H. Andrade, Integrated UF–NF–RO route for gold mining effluent treatment: From bench-scale to pilot-scale, Desalination 440 (2018) 111–121.

[58] C. Quist-Jensen, F. Macedonio, E. Drioli, Integrated Membrane Desalination Systems with Membrane Crystallization Units for Resource Recovery: A New Approach for Mining from the Sea, Crystals 6(4) (2016) 36.

[59] M. Xie, L.D. Nghiem, W.E. Price, M. Elimelech, A Forward Osmosis–Membrane Distillation Hybrid Process for Direct Sewer Mining: System Performance and Limitations, Environmental Science & Technology 47(23) (2013) 13486–13493.

[60] A. Criscuoli, J. Zhong, A. Figoli, M.C. Carnevale, R. Huang, E. Drioli, Treatment of dye solutions by vacuum membrane distillation, Water Research 42(20) (2008) 5031–7.

[61] F. Li, J. Huang, Q. Xia, M. Lou, B. Yang, Q. Tian, Y. Liu, Direct contact membrane distillation for the treatment of industrial dyeing wastewater and characteristic pollutants, Separation and Purification Technology 195 (2018) 83–91.

[62] A.K. An, J. Guo, S. Jeong, E.J. Lee, S.A.A. Tabatabai, T. Leikness, High flux and anti-fouling properties of negatively charged membrane for dyeing wastewater treatment by membrane distillation, Water Research 103 (2016) 362–371.

[63] H. Liu, J. Wang, Treatment of radioactive wastewater using direct contact membrane distillation, Journal of Hazardous Materials 261 (2013) 307–15.

[64] X. Wen, F. Li, X. Zhao, Removal of nuclides and boron from highly saline radioactive wastewater by direct contact membrane distillation, Desalination 394 (2016) 101–107.

[65] M. Khayet, Treatment of radioactive wastewater solutions by direct contact membrane distillation using surface modified membranes, Desalination 321 (2013) 60–66.

[66] S. Zhang, P. Wang, X. Fu, T.S. Chung, Sustainable water recovery from oily wastewater via forward osmosis-membrane distillation (FO-MD), Water Research 52 (2014) 112–21.

[67] U. Daiminger, W. Nitsch, P. Plucinski, S. Hoffmann, Novel techniques for iol/water separation, Journal of Membrane Science 99(2) (1995) 197–203.

[68] P. Jain, M. Sharma, P. Dureja, P.M. Sarma, B. Lal, Bioelectrochemical approaches for removal of sulfate, hydrocarbon and salinity from produced water, Chemosphere 166 (2017) 96–108.

[69] L. Han, Y.Z. Tan, T. Netke, A.G. Fane, J.W. Chew, Understanding oily wastewater treatment via membrane distillation, Journal of Membrane Science 539 (2017) 284–294.

[70] M. Gryta, K. Karakulski, A.W. Morawski, Purification of oily wastewater by hybrid UF/MD, Water Research 35(15) (2001) 3665–3669.

[71] D. Lu, Q. Liu, Y. Zhao, H. Liu, J. Ma, Treatment and energy utilization of oily water via integrated ultrafiltration-forward osmosis–membrane distillation (UF-FO-MD) system, Journal of Membrane Science 548 (2018) 275–287.

[72] J. Kim, H. Kwon, S. Lee, S. Lee, S. Hong, Membrane distillation (MD) integrated with crystallization (MDC) for shale gas produced water (SGPW) treatment, Desalination 403 (2017) 172–178.

[73] N. Dow, J. Villalobos García, L. Niadoo, N. Milne, J. Zhang, S. Gray, M. Duke, Demonstration of membrane distillation on textile waste water: assessment of long term performance, membrane cleaning and waste heat integration, Environmental Science: Water Research & Technology 3(3) (2017) 433–449.

[74] K. Sardari, P. Fyfe, S. Ranil Wickramasinghe, Integrated electrocoagulation – Forward osmosis – Membrane distillation for sustainable water recovery from hydraulic fracturing produced water, Journal of Membrane Science 574 (2019) 325–337.

[75] S. Goh, J. Zhang, Y. Liu, A.G. Fane, Fouling and wetting in membrane distillation (MD) and MD-bioreactor (MDBR) for wastewater reclamation, Desalination 323 (2013) 39–47.

[76] C.Y. Tang, T.H. Chong, A.G. Fane, Colloidal interactions and fouling of NF and RO membranes: a review, Advances in Colloid and Interface Science 164(1–2) (2011) 126–43.

[77] M. Gryta, Long-term performance of membrane distillation process, Journal of Membrane Science 265(1) (2005) 153–159.

[78] M. Gryta, Influence of polypropylene membrane surface porosity on the performance of membrane distillation process, Journal of Membrane Science 287(1) (2007) 67–78.

[79] M. Rezaei, D.M. Warsinger, V.J. Lienhard, M.C. Duke, T. Matsuura, W.M. Samhaber, Wetting phenomena in membrane distillation: Mechanisms, reversal, and prevention, Water Research 139 (2018) 329–352.

[80] M. Rezaei, D.M. Warsinger, J.H. Lienhard V, W.M. Samhaber, Wetting prevention in membrane distillation through superhydrophobicity and recharging an air layer on the membrane surface, Journal of Membrane Science 530 (2017) 42–52.

[81] E. Curcio, X. Ji, G. Di Profio, A.O. Sulaiman, E. Fontananova, E. Drioli, Membrane distillation operated at high seawater concentration factors: Role of the membrane on $CaCO_3$ scaling in presence of humic acid, Journal of Membrane Science 346(2) (2010) 263–269.

[82] T.V. Knyazkova, A.A. Maynarovich, Recognition of membrane fouling: testing of theoretical approaches with data on NF of salt solutions containing a low molecular weight surfactant as a foulant, Desalination 126(1) (1999) 163–169.

[83] K. Sardari, P. Fyfe, D. Lincicome, S. Ranil Wickramasinghe, Combined electrocoagulation and membrane distillation for treating high salinity produced waters, Journal of Membrane Science 564 (2018) 82–96.

[84] H. Cho, Y. Choi, S. Lee, Effect of pretreatment and operating conditions on the performance of membrane distillation for the treatment of shale gas wastewater, Desalination 437 (2018) 195–209.

[85] T.-H. Khaing, J. Li, Y. Li, N. Wai, F.-s. Wong, Feasibility study on petrochemical wastewater treatment and reuse using a novel submerged membrane distillation bioreactor, Separation and Purification Technology 74(1) (2010) 138–143.

[86] J. Wang, D. Qu, M. Tie, H. Ren, X. Peng, Z. Luan, Effect of coagulation pretreatment on membrane distillation process for desalination of recirculating cooling water, Separation and Purification Technology 64(1) (2008) 108–115.

[87] P.-J. Lin, M.-C. Yang, Y.-L. Li, J.-H. Chen, Prevention of surfactant wetting with aga-rose hydrogel layer for direct contact membrane distillation used in dyeing wastewater treatment, Journal of Membrane Science 475 (2015) 511–520.

[88] R. Schwantes, K. Chavan, D. Winter, C. Felsmann, J. Pfafferott, Techno-economic comparison of membrane distillation and MVC in a zero liquid discharge application, Desalination 428 (2018) 50–68.

[89] I. Vergili, Y. Kaya, U. Sen, Z.B. Gönder, C. Aydiner, Techno-economic analysis of textile dye bath wastewater treatment by integrated membrane processes under the zero liquid discharge approach, Resources, Conservation and Recycling 58 (2012) 25–35.

[90] S. Tavakkoli, O.R. Lokare, R.D. Vidic, V. Khanna, A techno-economic assessment of membrane distillation for treatment of Marcellus shale produced water, Desalination 416 (2017) 24–34.

6 Applications of Electrospun Membranes for Wastewater Treatment

Aiswarya M., Aquib Mohammed,
Akash M. Chandran, S. Varun and
Prasanna Kumar S. Mural*
Materials Chemistry and Polymer Technology Group,
Department of Chemical Engineering, National Institute
of Technology Calicut, India

CONTENTS

6.1 INTRODUCTION

Environmental pollution has been on the rise for the past few decades due to alarming population growth and increased industrialization. One of the most extensive and free resources on the planet has always been water. However, we are presently witnessing a reversal in this scenario because of water scarcity caused by pollution. The ever-changing healthcare settings, markets, and consumption have drastically increased the demand for freshwater [Shirazi, A. et al., 2017]. Ingestion of contaminated drinking water can lead to various health problems. Consequently, there is an urgent need to pave new technological advancements for water/wastewater treatment, novel desalination, and water reclamation [Ray, Saikat Sinha (2016)].

Nanotechnology advancements have opened doors to fabricate innovative, novel, and nanoengineered materials to meet the rising demand for freshwater. Nanofibrous membranes possess high porosity and can be fabricated with extensive flexibility of the pore size, surface area to weight ratio, pore volume, and many other parameters. They can be tuned to give desired permeability, selectivity, and strength with or without incorporating a wide variety of fillers [Snowdon, Monika R., and Liang, Robert L. (2018)]. There are several methods to fabricate these membranes, including phase separation, nanoparticle self-assembly, synthetic templates, drop-casting, and electrospinning. The most widespread technique is electrospinning due to its high productivity, low cost, simplicity, reproducibility, and high surface area to volume ratio. Electrospinning can be used to create durable and highly efficient membranes for required applications. In the present day, revolutionary technologies in filtration avenues with low cost and energy requirements are a must, and nanostructured membranes fabricated via electrospinning are the best solution to this.

6.1.1 History

Electrospinning is an application of the electrospraying technique in which the liquid emerging out of an emitter is subjected to a very high voltage. William Gilbert carried out the first-ever recorded observation of electrospraying in the late 16th century [Gilbert (1967)]. A cone-shaped liquid water structure was observed to be formed when an electrically charged amber piece was brought near it. It was further observed that small droplets of water were ejected from the cone [Gilbert & Wright (1628)]. The term "electrospinning" was derived from "electrostatic spinning," which came into existence in 1897 as described by Baron Rayleigh [Huang, Zheng-Ming (2003)]. J.F. Cooley was one of the first scientists who filed patents for the process of electrospinning in May 1900 [Cooley, J. F. (1900)] and February 1902 [Cooley, John F. (1902]. John Zeleny conducted an in-depth study of electrospraying, focusing on how fluid droplets behaved at the end of metal capillaries [Zeleny, John (1914)]. This paved the path to the formulation of mathematical models for such fluid particles' behavior when subjected to electrostatic forces.

A series of patents filed by Anton Formhals from 1934 to 1944 brought about further developments in the electrospinning technique for commercially manufacturing textile yarns [Formals, A. (1934), (1938), (1940), (1944)]. C.L. Norton, in 1936, patented the method of assisting fiber formation using an air blast for electrospinning from a melt instead of a solution [Norton, Charles L. (1936)]. In 1938, electrospun fiber filter materials were known as "Petryanov filters," which were generated by Nathalie D. Rozenblum and Igor V. Petryanov-Sokolov [Lushnikov, Alex (1997)]. This led to the manufacture of electrospun smoke filter elements for gas masks in subsequent years.

Geoffrey Ingram Taylor was the first to describe the formation of a polymer droplet's cone-shaped structure at the needle tip when an electrical field was applied. This conical shape came to be known as the "Taylor cone" in 1964 [Taylor, G. (1964)].

Taylor described that the potential across the conical shape's surface was the same and further stated that the shape is in steady-state equilibrium. He went on to publish more works detailing the phenomenon in the following years [Taylor, G. (1966), (1969)]. In 1988, Eric Simon reported that substrates of vitro cells could be incorporated with electrospun fibrous mats of polystyrene and polycarbonate specifically produced by solution electrospinning in the nano as well as the submicron-scale. He also reported that the polarity changes in the applied electric field resulted in slight variations on the fibers [Simon, Eric M. (1988)].

The current interest in the subject was initiated in the early 1990s. Darell H. Reneker and Greg Rutledge and many other research groups demonstrated that nanofibers of several organic compound polymers could be produced through the electrospinning technique [Jayesh & Reneker (1995)]. There has been a rise in the number of published works on the matter ever since. Hohman et al. studied the various instabilities and their relative growth rates in the polymer jet after the Taylor cone formation, and attempts were made to describe the whipping instability [Hohman (2001)]. While many companies such as Donaldson Company and Freudenberg are producing products for air filtration installed with electrospun fibers, several others like NanoTechnics, Kato Tech, and eSpin Technologies are emerging with the idea of bringing the electrospinning technique and its advantages to the market.

6.2   ELECTROSPINNING TECHNIQUE

The general idea of this processing technique is relatively simple. The setup consists of three main components: a high voltage power supply, a metal collecting screen, and a polymer solution within a capillary tube attached to a small diameter needle. High voltage is applied between the polymer solution and the grounded collector, which act as electrodes. This electrically charges the jet of polymer melt emerging out of the needle. A syringe pump is used to precisely control the flow rate of the melt. A conductive screen is installed as a stationary plate or as a rotating drum to collect the spun fibers [Al-Hazeem, Nabeel Zabar Abed (2018)]. Figure 6.1 shows a schematic representation of a simple electrospinning setup.

The surface of the polymer droplet at the needle tip gets electrified with the application of the electric field. It leads to the formation of repulsive charges on the surface of the droplet. Besides, these charges are attracted to the opposite electrode. These forces finally overcome the surface tension, and the liquid droplet thus gets strained. The droplet's hemispherical shape is stretched to form a conical shape known as the Taylor cone [Bagbi, Yana (2019)]. At a specific critical voltage, the surface tension is entirely disabled by the electrostatic forces, resulting in the ejection of a charged jet of the liquid polymer solution from the tip of the cone. As the jet propagates through air, it experiences bending instabilities caused by its surface's repulsive charges. This imparts a whipping motion (Figure 6.1), which makes the fiber oscillate and move in a chaotic manner aiding in further fiber elongation. [Homaeigohar, Shahin & Mady (2014)]. The traversal of the jet is accompanied by solvent evaporation, leading to solidification of the jet to form a charged nanofiber. Finally, the charged jet is deposited randomly on the collector, forming a nonwoven nanofiber mat. The nanofibers' dimensions and morphology can be controlled by modifying the various parameters involved [Bagbi, Yana (2018)].

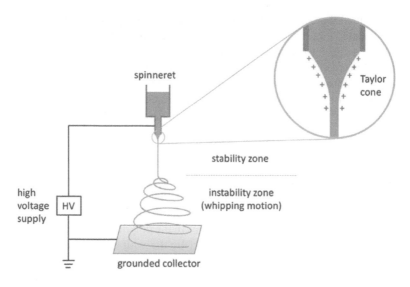

FIGURE 6.1 A typical electrospinning setup and Taylor cone formation [Ditaranto, Nicoletta (2018)].

6.2.1 ELECTROSPINNING PARAMETERS

Numerous operating parameters influence the quality, dimensions and morphology of the electrospun nanofibers. These parameters contribute significantly to the understanding of the optimum conditions for electrospinning. The meticulous control of these parameters can obtain the desired morphologies and dimensions of fibers. There are three categories of electrospinning parameters of vital importance: (a) polymer solution parameters, (b) processing parameters, and (c) ambient parameters [Al-Hazeem, Nabeel Zabar Abed (2018)]. Some of the crucial parameters in each category are listed in Table 6.1.

6.2.1.1 Polymer Solution Parameters

The different properties coming under this category are the most significant in determining the fiber morphology. These parameters are closely related to one another,

TABLE 6.1

Operating parameters affecting electrospinning process

Polymer solution parameters	Processing parameters	Ambient parameters
Concentration	Applied voltage	Humidity
Viscosity	Feed flow rate	Temperature
Molecular weight	Needle diameter	Pressure
Surface tension	Tip-to-collector distance	Type of atmosphere
Conductivity	Electric field strength	
Dielectric constant		

and changing one would simultaneously affect certain others. Hence, it is hard to control one parameter individually. The important parameters are listed in Table 6.1. The surface tension influences the bead formation along the length of the fiber. The extent to which the fiber elongation occurs is greatly affected by the solution's viscosity and electrical properties. Accordingly, alterations in fiber diameter occur.

6.2.1.1.1 Molecular Weight

The polymer's molecular weight affects the viscosity, surface tension, and conductivity of the solution, which will have significant impacts on the morphology of the electrospun nanofibers. It was observed that beads were formed at lower molecular weights of the polymer used while keeping the concentration constant [Kozki, Antti (2004)]. Very high molecular weights lead to micro ribbons forming even at low concentrations [Zhao, Y. Y. (2005)]. Akduman et al. studied the nanofibers produced by electrospinning aqueous solutions prepared with different molecular weights ranges of PVA at fixed concentrations [Akduman, C. (2014)].

6.2.1.1.2 Concentration/Viscosity

The concentration of the solution has a direct impact on its viscosity and surface tension. The electrospun nanofiber diameter increases with the viscosity of the solution [Deitzel, Joseph M., (2001)]. Hence, the concentration must be maintained at a certain range to carry out electrospinning. Below this range, chain entanglements will not occur. As the viscosity/concentration is increased beyond this range, it becomes very problematic to pump the solution through the syringe [Kameoka, Jun, (2003)]. Electrospray has been observed at low concentrations instead of electrospinning due to its low viscosity and high surface tension [Angammana, Chitral J. (2016)]. The concentration parameter affects not just the dimensions but also the morphology aspects of the nanofiber. At low concentrations, beads and beaded fibers are obtained, and as the concentration increases, the beads elongate to give spindle-shaped fibers. Fibers of enlarged uniform thicknesses are obtained once the concentration is further increased. Zeng et al. studied electrospinning PLA nanofibers with concentration varying from 1% to 5% [Zeng, Jun (2003)].

6.2.1.1.3 Surface Tension

As discussed before, the electrospinning process is initiated once the charged droplet overcomes its surface tension. As the jet travels towards the collector, a decrease in the surface area per unit mass of the fluid is observed. Due to this, at higher concentrations, there is a tendency for solvent molecules to flock together and attain a spherical shape owing to its surface tension. Thus bead formation takes place on the electrospun fibers. However, as the solution's viscosity increases, there will be increased interaction among polymer and solvent molecules. In this case, the bead formation tendency is reduced as the solvent molecules spread out on the stretched polymer fiber [Ramakrishna, Seeram. (2005)]. The surface tension parameter is also altered by the solvent used. Yang et al. carried out the electrospinning of PVP with different solvents such as dichloromethane (MC), N, N-dimethylformamide(DMF), and ethanol [Yang, Qingbiao (2004)].

6.2.1.1.4 Conductivity

As the conductivity of the solution increases, its capacity to carry charges also increases. Adding ions to the solution is one way to increase its conductivity. Salts and polyelectrolytes can increase the amount of carried charges and increase their stretching in the electric field. Altering the pH of the solution is another way to increase its conductivity. Fiber jets with high conductivity will be subject to higher elongation on the application of high voltage. This leads to the formation of uniform fibers instead of beaded ones. It is also observed that the nanofiber diameter reduces with an increase in the conductivity of the solution [Beachley, Vince, and Xuejun Wen. (2009)]. As more ions aid in increasing the solution's conductivity, the critical voltage requirement for electrospinning gets reduced [Son, Won Keun (2004)]. An increase in the number of charges present is also observed to increase bending instabilities [Choi, Jae Shin (2004)].

6.2.1.2 Processing Parameters

These sets of parameters are another category that affects the electrospinning process. Although these parameters influence the morphology of the electrospun nanofibers, it is not as crucial as the polymer solution parameters. A set of commonly employed electrospun polymers with their functions and electrospinning parameters has been listed in Table 6.2.

6.2.1.2.1 Applied Voltage

The applied voltage is the key factor in inducing charges on the polymer solution. The electric field generated initiates the jet formation once the electrostatic forces have nullified the surface tension. The typical voltage applied to start electrospinning is around 6 kV. Increasing the applied voltage can lead to a decrease in the fiber length due to further stretching. The shape of the drop at the tip of the syringe is also affected by the applied voltage, and this, in turn, causes variations in the fiber structure and morphology [Beachley, Vince, and Xuejun Wen. (2009)]. At high voltages, the high amount of induced charges facilitates the faster acceleration of the charged jet, thereby increasing the amount of polymer ejection from the needle tip. This produces fibers with larger diameters from a slightly unstable Taylor cone [Zhang, Chunxue (2005)]. It is also observed that the bead formation tendency increases at high voltages with its shape changing from spherical to spindle-like. Fibers with rough surfaces are formed when electrospun at high voltages. To decrease the bead formation tendency, they used salt particles to increase the charge density on the droplet [Zong, Xinhua (2002)]. Shao et al. reported changes in PVDF nanofibers' morphology concerning the applied voltage ranging from 9 kV to 21 kV [Shao, Hao (2015)]. The crystallinity of the electrospun nanofiber is another parameter altered by the applied voltage. The electric field orients the polymer molecules in an orderly fashion, and hence a higher crystallinity is induced in the fiber.

6.2.1.2.2 Flow Rate

The quantity of polymer solutions available for the electrospinning process depends upon the feed rate. The flow rate is usually kept low so that the droplet has enough

TABLE 6.2
Electrospun polymers with their functions and electrospinning parameters

Polymer System	Target Separation	Electrospinning Parameters	Reference
Alginate/PET	*Ultrafiltration of dye solution*	Applied Voltage: 25 kV Feed Rate: 0.7 mL/h Tip-to-collector Distance: 18 cm	Mokhena, Teboho Clement, and Adriaan Stephanus Luyt (2017)
Polysulfone/ Cellulose Acetate	*Ultrafiltration of BSA and PEG mixtures*	Applied Voltage: 17 and 25 kV Feed Rate: 2 mL/h and 3 mL/h Tip-to-collector Distance: 10 cm	Dobosz, Kerianne M., et al. (2017)
Polyamide	*Bacterial removal*	Applied Voltage: 25 kV Feed Rate: 1 mL/h Tip-to-collector Distance: 10 cm	Daels, Nele, et al. (2016)
PVDF	*Dead-end filtration*	Applied Voltage: 25 kV Feed Rate: 1 mL/h Tip-to-collector Distance: 15 cm	Li, Zongjie, et al. (2016)
Ag Fly Ash/PU	*Filtration of TiO2 nanoparticles*	Applied Voltage: 15 kV Feed Rate: 1 mL/h Tip-to-collector Distance: 18 cm	Pant, Hem Raj, et al. (2014)
Polyamide/PAN	*Nanofiltration of divalent salts and sugar*	Applied Voltage: 15 kV Feed Rate: 1.2 mL/h Tip-to-collector Distance: 18 cm	Yoon, Kyunghwan, et al. (2009)
BSA/PANGMA	*Filtration of Au nanoparticles*	Applied Voltage: 15–20 kV Feed Rate: 1.1 mL/h Tip-to-collector Distance: 25 cm	Elbahri, Mady, et al. (2012)
β-cyclodextrin/ CA	*Adsorption filtration of phenanthrene*	Applied Voltage: 15 kV Feed Rate: 1 mL/h Tip-to-collector Sistance: 10 cm	Celebioglu, Asli, Serkan Demirci, and Tamer Uyar (2014)
PVDF-co-HFP	*DCMD*	Applied Voltage: 21 kV Feed Rate: 1 mL/h Tip-to-collector Distance: 20 cm	Yao, Minwei, et al. (2016)
SiO$_2$ NPs/PVDF	*DCMD*	Applied Voltage: 25 kV Feed Rate: 0.3 mL/h Tip-to-collector Distance: 15 cm	Li, Xiong, et al. (2015)

time to get polarized. This also helps in solvent evaporation to a greater extent. When the feed rate is kept high, beaded fibers are formed as there is little time for solvent evaporation before getting deposited on the collector. Moreover, the extent of stretching of the jet decreases as the surface charge density is low. A low flow rate can result in high charge distribution. Hence, the fiber undergoes bending instabilities efficiently, which aids in forming nanofibers with small diameters [Liu, Yong (2008)]. Zargham et al. showed that when the feed rate is lowered while keeping all other parameters constant, there is a decrease in the sizes of the beads formed along with an increase in the fiber diameter. It can be further concluded that, as the feed rate is reduced, more number of bead-free nanofibers can be obtained [Zargham, Shamim (2012)].

6.2.1.2.3 Diameter of Needle

The diameter of the needle plays a pretty important role in determining the fiber diameter. It has been observed that nanofibers with smaller diameters were formed when the needle diameter was small. This is because, as the droplet size decreases, the surface tension increases, leading to slower initiation of the charged jet. This decreases the jet acceleration, allowing sufficient time for it to get stretched [Zhao, Shengli (2004)]. Moreover, clogging in the needle can be reduced by using a smaller internal diameter. The needle diameter determines the amount of polymer solution that comes out, which decides the size of the droplet formed at the needle tip. However, the needle diameter must not be too small, making it very difficult to pump the viscous solution out.

6.2.1.2.4 Distance Between Tip and Collector

The distance between the tip and the collector has heavy influences on the flight time and the strength of the electric field. As the distance is reduced, the electric field generated is increased, which increases the jet acceleration. This will reduce the time available for solvent evaporation. There are also cases where beads were observed to form when the distance was kept too low [Megelski, Silke (2002)]. This can be explained by the electric field's increased strength in the short distance between the tip and the collector. As the distance increases, the extent of elongation of the fibers increases, giving an ample amount of time for solvent evaporation. This will generate smooth fibers with low diameters [Reneker, Darrell H. (2000)].

6.2.1.3 Ambient Parameters

Interactions between the polymer solution and its environment can have effects on the morphology of the electrospun nanofibers. As the surrounding electric field heavily influences the process, changes in the surroundings will also affect it.

6.2.1.3.1 Humidity

A low value of humidity will aid in evaporating the solvent completely at a higher rate. At high humidity, there are chances for water condensation on the fiber surface, which may, in turn, affect the morphology of the fiber [Megelski, Silke (2002)]. It has also been reported that small circular pores were formed on the surface of fibers during high humidity. These pores increased in size as the humidity increased [Casper, Cheryl L. (2004)].

6.2.1.3.2 Pressure

The effect of pressure on the electrospinning process is studied by enclosing the process. Pressure lower than the atmospheric pressure will cause the polymer solution to flow out of the syringe with ease, which may lead to unstable initiation of the charged jet. The solution bubbles up rapidly at the needle tip with further reduction in pressure. The electrical charges get discharged at very low pressures, and electrospinning cannot be carried out. As a general observation, it can be stated that an increase in pressure has very little or no effects on the electrospinning process [Ramakrishna, Seeram (2005)].

6.3 WATER TREATMENT USING ELECTROSPUN NANOFIBROUS MEMBRANES

Water treatment by membrane separation technology can be carried out by extrapolating the physical, chemical, or biological nature of substances/molecules. Water treatment by physical separation is accomplished in processes such as filtration and sedimentation, among which filtration is of industrial importance. The four main types of filtrations are microfiltration (MF), ultrafiltration (UF), nanofiltration (NF) and reverse osmosis (RO). The mode of filtration is chosen based on the size of the particle to be removed or retained, and pressure requirements (Figure 6.2). Chemical water treatment will involve disinfection or redox reactions between the particle of interest and the interface. The interface here could be functional/affinity membranes. Biological water purification is carried out using aerobic and anaerobic bacteria [Snowdon, Monika R., and Robert L. Liang (2018)].

Recently, electrospun nanofibrous membranes (ENMs) are the subject of utmost interest in water treatment. These 3-D structures are an attractive alternative over conventional 2-D membranes. Some of the desirable properties of ENMs are its uniform and controllable pore size distribution, light weight, fully interconnected pores, and high surface area. Furthermore, owing to their small fiber dimensions, ENMs have high porosity per unit area. In effect, high porosity enhances permeability through the membrane and leads to reduced transport resistance. Another major positive attribute for ENM is the availability of a wide variety of industrial materials for membrane fabrication and its tunable features. In addition to that, ENMs offer us a threefold separation mechanism combining screen filtration, depth filtration, and surface adsorption on spun fibers [Shirazi, A. (2017)].

The properties of the polymer used for the fabrication of electrospun membranes are crucial parameters that govern the material's choice. These could be the chemical structure of the polymer, its durability, or thermal stability. According to the final application of the ENMs, polymers or polymer combinations are selected from main eight structures: polyvinyl alcohol (PVA), polyacrylonitrile (PAN), polyester (PE), polyvinylidene fluoride (PVDF), polyurethane (PU), cellulose acetate (CA), polyethersulfone (PES) and chitosan [Snowdon, Monika R., and Robert L. Liang (2018)].

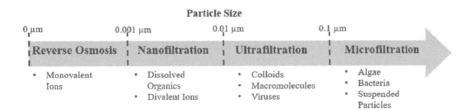

FIGURE 6.2 The four main kinds of filtration with the examples of the smallest substances that can be filtered [Snowdon, Monika R., and Robert L. Liang (2018)].

6.4 ENM APPLICATIONS FOR WATER TREATMENT

6.4.1 SIZE EXCLUSION

6.4.1.1 Microfiltration

The range of pore sizing for MF is 0.1–10 μm. The pressure requirement for the microfiltration process is between 1–5 atm [Liang, Hu (2014)]. MF is principally useful for both purification and disinfection. This is accomplished by the separation of suspended particles, bacteria, and algae from drinking water. Microfiltration is often carried out as a prefiltration step to reverse osmosis (RO) in order to maintain the flux through the osmosis membranes. When the feed consists of micron-sized particles, microfiltration membranes act as screen filters, and the flux remains high throughout the filtration process. However, if the feed contains submicron particles, membrane clogging occurs, which in turn drops down the flux and dramatically escalates the pressure drop [Snowdon, Monika R., and Robert L. Liang (2018)]. Here, the microfiltration membrane takes the role of a depth filter, requiring regular cleaning to prevent fouling. In this way, microfiltration membranes protect the RO system leading to a reduction of the overall cost.

ENMs are highly investigated for microfiltration and ultrafiltration applications because of their excellent water permeability. Compared to traditional microfiltration membranes, ENMs are far more efficient in removing microparticles from wastewater as they utilize significantly low amounts of energy.

PAN ENMs are low-cost and easy to-handle, along with possessing tailorable fiber properties, which can be changed by tuning the electrospinning parameters. Electrospun membranes made of PAN and its composites have been widely used for microfiltration of suspended particles. Satinderpal Kaur, et al. have proven the applicability of PAN ENMs for desalination of salt water with salt concentrations up to 2000 ppm. Researchers have also reported PAN ENMs for unique uses, such as removal of TiO_2 microparticles from industrial wastewaters. Microfiltration has moved a step forward by studies on coke removal from petrochemical wastewater using Nylon-6 ENMs [Mirtalebi, Elham (2014)]. In addition to this, dead-end microfiltration can also be efficiently carried out by using ENMs. For this, PES ENMs have been used to filter out bacteria and fungus [Khezli, Sakine (2016)]. Zongjie Li, et al. fabricated a novel tree-like microfiltration membrane made of PVDF with tetrabutylammonium chloride salt (TBAC) additive for dead-end filtration.

Apart from using a single polymeric material, we can also rely upon hybrid polymeric structures. Some hybrids used so far for MF are PAN/PET [Wang, Ran (2012)] and PES/PET [Homaeigohar, S.S., K. Buhr (2010)] composite membranes. These hybrids were found to have enhanced MF performance with a high rejection ratio.

PAN and PES ENMs are susceptible to natural mechanical breakdown attributed to improper orientation and electrospun polymer chain interaction. Structural durability can be enhanced by thermal treatment of the nanofiber matrix or by the addition of fillers. The thermal treatment of ENMs strengthens the membrane structure by inducing inter-fiber adhesion. In filler addition, carbon fillers such as carbon nanotubes (CNTs) and graphene oxide (GO) are considered the best fillers for polymeric ENM reinforcement. The inclination towards carbon fillers is undeniable, as the presence of low weight percentages of these inexpensive additives can offer excellent thermal and chemical resistance of composite as a supplement to required mechanical strength. The addition of fillers is not limited to the nature of the ENMs; that is, fillers can also be incorporated into hybrid matrixes such as TiO_2-PAN. Other than carbon fillers, many inorganic fillers are also widely reported. For example, novel nanofillers such as zirconia (ZrO_2) have been used for the mechanical strengthening of PES ENMs [Homaeigohar, Seyed Shahin, and Mady Elbahri (2012)]. Another simple method suggested for ameliorating the durability of the ENMs is by initially spinning the fibers on a more robust support.

6.4.1.2   Ultrafiltration

Ultrafiltration is used to segregate viruses, emulsions, colloids, and macromolecules such as proteins at a pressure range of 2–10 atm [Liang, Hu (2014)]. The membrane pore size for ultrafiltration is from 0.01 μm to 0.1 μm. At this pore size, fouling problems and flux loss problems occur. In order to surpass these, ENMs can be used as the middle layer of a thin film composite (TFC) membrane. The high permeability nature of the ENM-based middle layer will lead to improved filtration efficiency, compared to a conventional non-spun membrane as the central layer. Kyunghwan Yoon, et al. have established this by using PAN ENM, coated with a chitosan top layer for wastewater filtration. Also, using a TFC for ultrafiltration can help to withstand the required pressure for the filtration operation. In general, a TFC has three layers: the top layer is fabricated out of a "non-porous" hydrophilic coating; the middle support layer is made of an ENM; and the bottom layer is made out of a "non-woven" microfibrous substrate (Figure 6.3) [Ray, Saikat Sinha (2016)].

Among the polymers preferred for ENMs, PES and PVDF are considered useful for TFCs, owing to their excellent mechanical properties. They also have superior thermal properties and chemical stability. The enhanced chemical stability is due to the lack of active functional groups on their surfaces. Researchers have proposed

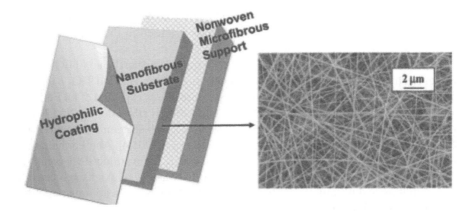

FIGURE 6.3   Schematic representation of a typical TFC membrane with the SEM-image of the middle layer made of electrospun polymer (From [Wang, Xuefen (2005)], reproduced with permission from American Chemical Society.)

that the use of PVDF as a combination with chitosan (<1%) is highly effective for ultrafiltration purposes [Zhao, Zhiguo (2013)]. Copolymers of PVDF, such as PVDF-co-HFP, are also presented to separate oil from water with 100% efficiency [Ganesh, Venkatesan Anand (2016)].

For oily wastewater filtrations, a unique blend of PVA and MWCNT as a top layer, along with a PAN electrospun mid-layer composite membrane, can be used [You, Hao (2013)]. Other alterations of PVA used for the same purpose are PVA/Nylon 6 [Islam, Md Shahidul (2017)], PVA crosslinked with glutaraldehyde [Wang, Xuefen (2005)], PVA/PVA hydrogel [Wang, Xuefen (2006)], and UV-cured PVA [Tang, Zhaohui (2009)]. Biopolymer based ENMs can also be considered in this regard. For instance, incorporating 0.5–1% chitin polysaccharide nanowhiskers with PVDF has been proved applicable for oily wastewater treatment [Gopi, Sreerag (2018)]. Other polysaccharides such as cellulose are also found useful for certain water purification operations. Cellulose/PAN composite, which can expertly filter off viruses, is a brilliant example [Sato, Anna (2011)].

Specific proteins and polymers dissolved in industrial wastewater can be filtered for their economic value. An ultrafiltration ENM helpful for this purpose is a cellulose acetate/polysulfone polymer system. It discriminates BSA and PEG from their mixtures [Dobosz, Kerianne M. (2017)].

The need for the separation of dyes from effluents rejected from textile industries is now needed more than ever. If they are rejected into the surrounding water bodies without separation, it can lead to critical environmental danger. ENMs have thus revealed their significance in the green environment world by the development of alginate/PET composite for the ultrafiltration of dye solutions [Mokhena, Teboho Clement, and Adriaan Stephanus Luyt (2017)].

6.4.1.3   Nanofiltration

The mean pore diameter for nanofiltration is 0.001 μm–0.01 μm (1–10 nm), and the operating pressure required is between 5 and 50 atm [Liang, Hu (2014)]. A typical nanofiltration process's objective is the removal of dissolved organics, multivalent ions, synthetic dyes, sugars and specific salts [Bowen, W. Richard, and Julian S. Welfoot (2002)]. Similar to ultrafiltration, ENMs are used in the form of TFCs in nanofiltration.

Faccini, Borja et al. prepared a CNT/PAN/tetraethyl orthosilicate (TEOS) membrane, electrospun in DMF solvent for nanofiltration of Ag, Au, and TiO_2 nanoparticles. This membrane was reported to have 99% porosity, as expected of an ENM. An equally efficient system for TiO_2 nanoparticle filtration is silver doped PU ENM. In this membrane, membrane porosity is decreased by a web structure formation. This leads to enhanced membrane selectivity [Pant, Hem Raj (2014)]. Another nanofiltration membrane used for water treatment applications is a polyamide/PAN composite [Yoon, Kyunghwan (2009)]. We have seen that PAN has been extensively used for microfiltration and ultrafiltration. Furthermore, PAN finds application in nanofiltration of salts (desalination) [Kaur, Satinderpal (2011)]. In addition to that, ENMs have been found beneficial for removing organic compounds from wastewater. A nanofiltration membrane made of polystyrene and β-cyclodextrin was used for this purpose by Tamer Uyar, et al. in 2009.

6.4.2 Affinity ENMs for Water Treatment

Contaminants in water can be discriminated by generating an affinity with the membrane surface. Membranes with such preference for certain particles are called "affinity membranes." Here, the affinity typically refers to the chemical adsorption property of the surface. In order to achieve selective absorptivity, the membranes are often functionalized with a new compound/ligand. Hence, these membranes can also be referred to as "functionalized membranes." The idea of adsorptive water filtration has been used to remove a plethora of contaminants, including organics, ions, nanoparticles, and pathogens.

Surface properties of the membrane are crucial for chemical adsorption. The most critical properties considered while selecting an affinity membrane are its adsorption capacity, nature of adsorption kinetics, the ease of contaminant separation, and easy reusability of the membrane. ENMs are incredibly beneficial for adsorptive wastewater filtration, owing to their high surface area. Moreover, ENMs are readily available and economical. Therefore, functionalized ENMs have immense potential as affinity membranes and are worth exploring.

Organic pollutants such as oil, dyes, pesticides, and proteins can make water bodies uninhabitable for aquatic species even at deficient concentrations as they use up the dissolved oxygen in the water. Functionalized polysulfone, PU, PAN, and CA ENMs have the flair to separate proteins such as BSA, lipase and bromelain from water [Homaeigohar, Shahin, and Mady Elbahri (2014)]. In addition to organic pollutants, heavy metals containing industrial wastewater rejected into the river system are highly toxic to living organisms. These carcinogenic substances can severely affect various organ systems, such as the nervous system, and lead to many chronic health issues. Some examples of common heavy metals whose presence in wastewater have to be regulated include arsenic (Ar), copper (Cu), cadmium (Cd), chromium (Cr), mercury (Hg), lead (Pb), and nickel (Ni). Silver doped PU ENMs, previously mentioned for TiO_2 nanofiltration, can also be used for the removal of arsenic and organic dyes [Pant, Hem Raj (2014)]. For removing fluoride ions and nitrates from wastewater, a Al_2O_3/bio-TiO_2 composite with PU ENM can be employed [Suriyaraj, S. P. (2015)]. In a study by Siyuan Xie, et al., uranium-contaminated water was cleaned by a polyamidoxime (PAO)/PVDF composite ENM. Here, PVDF helped to improve the availability of amidoxime groups for the adsorption of uranyl ions.

Currently, PVA ENMs are used more in adsorption than for filtration by size exclusion. Many PVA blends find application in metal removal and organic dyes. A blend of 20% ZnO with PVA isothermally adsorbs U, Cu, and Ni. At the same time, zeolite with PVA filters Cd and Ni from wastewater [Snowdon, Monika R., and Robert L. Liang (2018)]. A combination of hydroxyapatite with PVA is useful for the unrecoverable adsorption of Co [Wang, Hualin (2014)]. Another PVA ENM mainly focused on Ar elimination is PVA/Fe. This membrane is post-treated with ammonia for improved adsorption capacity. Reusable ENMs made of polyacrylic acid (PAA) and PVA blended with specific modifying agents such as cyclodextrin have shown the ability to remove methylene blue, which is one of the most common dyes present in textile wastewaters [Yan, Jiajie (2015)]. In a different research, PAA/PVA polymer blends with MWCNT fillers were surface decorated with zero-valent iron (ZVI) nanoparticles for chemical adsorption of Cu^{2+} ions from wastewater [Xiao, Shili (2011)]. Though PVA is water-soluble, that does not limit a researcher's predilection for PVA, owing to its structure, which can be vastly functionalized to render it water-insoluble. In addition to the above works, a multifaceted PVA-natural gum karaya (GK) ENM shows excellent adsorption efficiency for a series of metal nanoparticles from wastewater, including Au, Ag, Pt, CuO, and Fe_3O_4. The order of adsorption efficiency showed by this membrane is Fe_3O_4<CuO<Au<Ag<Pt. The researchers also treated this PVA-based ENM with methane plasma to further improve the adsorption efficiency and capacity [Padil, V.V.T, and Miroslav Černík (2015)]. Similarly, an ENM using porous CA fibers and dithiothreitol-capped Au nanoclusters (AuNCs) fabricated by Anitha Senthamizhan, et al. can remove divalent ions of Pb, Ni, Mn, Cd, Zn, and Hg with individual ion concentrations of 1 ppm. The highest removal efficiency was found for Pb (99.9%), and it subsequently decreased in the order, as mentioned earlier, with the least removal efficiency of 56.5% for Hg.

Along with PVA ENMs, PAN ENMs have also been extensively studied for adsorptive removal of toxic metals and dyes due to the relative ease in introducing a wide variety of functional groups on PAN nanofibers. A thiol-modified/PAN cellulose composite ENM for decontaminating water from Cr and Pb is a good example [Yang, Rui (2014)]. PAN ENMs are suitable for filtration of both anionic and cationic dyes. PAN grafted with compounds like diethylenetriamine (DETA) have been reported for filtration of cationic dye pollutants [Snowdon, Monika R., and Robert L. Liang (2018)]. With dopants like polyamidoamine (PAMAM) PAN can be used for removing Direct Red-80 and 23, which are anionic dyes [Almasian, A., Olya, M. E., & Mahmoodi, N. M. (2015)]. Alternatively, polystyrene fibers with PDA assisted attachment of β-cyclodextrin displayed enhanced adsorption of anionic pollutants under highly basic conditions [Wu, Huiqing (2015)].

Bio-based materials such as functionalized-CA have been able to eradicate U and Cr from aqueous media. A very efficient system discovered for uranium adsorption is camphor soot cellulose ENMs [Zhou, Weitao (2011)]. Cyclodextrin functionalized CA ENMs finds unique application in the elimination of aromatics from water. A unique "click" mechanism has been reported for the removal of phenanthrene by this membrane [Celebioglu, Asli, Serkan Demirci, and Tamer Uyar (2014)]. Reusable and nonreusable types of chitosan-based ENMs are also available for purifying water from toxic metals. Among the nonreusable category, chitosan ENMs neutralized with K_2CO_3 can carry out monolayer adsorption of Cu and Pb [Haider, Sajjad, and Soo-Young Park (2009)]. Among the reusable category, a chitosan/graphene-oxide composite proved efficient for removing Pb, Cu, and Cr [Najafabadi, Hossein Hadi (2015)]. Polyester ENMs are another class of versatile substances used for adsorption filtration. These can be used for removing organic particles from sewage. Polyester, in combination with iron alkoxides, is capable of surface precipitation of Cr ions [Xu, Guo-Rong, Jiao-Na Wang, and Cong-Ju Li (2012)].

Elbahri, Mady et al. fabricated a smart performing membrane by biofunctionalization using BSA protein (bovine serum albumin) on Poly[acrylonitrile-co-(glycidyl methacrylate)] (PANGMA) nanofibrous membrane. The BSA/PANGMA ENMs have shown higher retention abilities for Au nanoparticles. On wetting, the BSA protein undergoes a conformational change, which engenders protein swelling. Such a change invites higher steric hindrance, making the nanoparticles stick to the membrane surface (Figure 6.4). In addition to being successfully able to segregate Au nanoparticles, BSA/PANGMA ENM can capture biomolecules. Hence, biofunctionalized

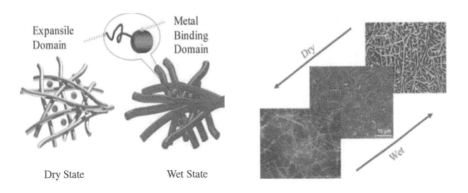

Dry State   Wet State

FIGURE 6.4   Capturing of metal nanoparticles through the wetting-induced conformational change [Homaeigohar, Shahin, and Mady Elbahri (2014)].

nanofibrous membranes could give positive results for removing pathogenic microorganisms from wastewater [Homaeigohar, Shahin, and Mady Elbahri (2014)].

6.4.3   ENMs for Membrane Distillation

Membrane distillation (MD) is a thermally driven water separation process, mainly used for desalination purposes. Generally, salty water will occupy the feed side, and pure water will be on the permeate side. The MD mechanism begins with the diffusion and convection of water vapor across the membrane. This is followed by the condensation of pure water at the cooler side of the membrane. The four main configurations of MD units are as follows: direct contact membrane distillation (DCMD), air gap membrane distillation (AGMD), vacuum membrane distillation (VMD), and sweeping gas membrane distillation (SGMD). These configurations are explained in Figure 6.5. Compared to conventional distillation, MD requires a lower operating pressure and temperature, lower energy requirements, and has around 100% salt rejection. Hydrophobic and superhydrophobic membranes are considered for MD purposes as these can maintain the separation of wastewater and pure water across the membrane. Some other important qualities expected of a membrane used for MD are excellent mechanical, thermal and chemical stability, high porosity and high liquid entry pressure (LEP) [Ray, Saikat Sinha (2016)]. Due to the requirements of efficient membranes, ENMs are highly explored for MD applications. PVDF-based ENMs have been mostly preferred for MD, owing to its spinnability and solubility.

For DCMD, a trilayer PVDF ENM configuration with a diameter gradient was used by Ebrahimi et al. The results of this study show that evaporation and flux are profoundly affected by the varying membrane configuration. In some studies, PVDF ENMs have been modified with TiO_2 and FTCS (fluorododecyl trichlorosilane). An improvement in desalination efficiency has been analyzed for the same [Ren, Long-Fei (2017)]. Another variation is doping silica nanoparticles on PVDF nanofibers. This superhydrophobic ENM shows improved permeate flux, salt rejection and mechanical strength [Su, Chunlei (2017)].

In a study by Leonard D. Tijing et al., a PVDF-co-HFP (hexafluoropropylene) matrix (PcH matrix) was blended with varying CNT concentrations. The inclusion

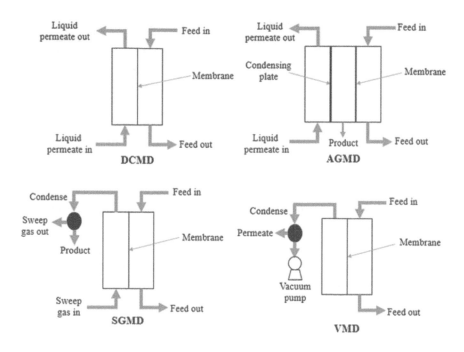

FIGURE 6.5   The four main configurations of membrane distillation units [Yalcinkaya, Fatma (2019)].

of carbonaceous nanofillers improved the mechanical durability and hydrophobicity of the overall ENM. This membrane was fabricated to have a thin top layer of CNT/PcH and a bottom layer made of a thick PcH. In addition to PVDF, few works have also proceeded by using polystyrene, PU, and polytetrafluoroethylene (PTFE) ENMs for DCMD applications [Yalcinkaya, Fatma (2019)].

Recently, researchers have extended their focus to the fabrication of ENMs for AGMD. PVDF ENMs were enhanced with superhydrophobic Al_2O_3 nanoparticles for this purpose. These ENMs were used to remove heavy metals like lead from the water with high flux rates [Attia, Hadi (2017)]. In another approach, Jingcheng Cai used surface fluorinated PAN ENMs for AGMD. The surge in novel research on ENMs for MD applications well assures us that ENMs will be taken to the distillation industry in the near future.

6.5 PRESENT TRENDS AND TECHNOLOGICAL SUSTAINABILITY

For commercial, industrial production of ENMs for water-filtration applications, some technological changes have to be incorporated such that the output rate increases. The successful scale-up of the electrospinning process revolves around increasing the number of Taylor cones. This implies increasing the throughput of the electrospinning solution. Both nozzle-type technologies and free-surface technologies are widely investigated to scale up.

Nozzle-type technologies can be divided into stationary and rotary types. Among the stationary types, multi-needle electrospinning is the most straightforward key to increase productivity. Multi-needle spinnerets can be arranged either in a linear or two-dimensional fashion, according to requirements. The sole challenge while using multi-spinneret systems is to avoid the occurrence of repulsive forces between the liquid jets, which can lower the nanofiber quality. Other stationary nozzle electrospinning methods using tubular and flat spinnerets leaves the user without the worry of jet interaction. Using stationary nozzle-type technology, the highest production rates have been reported for electroblowing processes in a recent work by Péter Lajos Sóti, et al. The probability of spinneret clogging caused by most of these nozzle techniques can be avoided by employing a rotary spinneret, in which the centrifugal force pushes the solution through the nozzle. In addition to that, clogging can be reduced by using spinnerets consisting of orifices in place of needles. Therefore, a combination of high voltage rotating spinnerets with orifices can give high productivities. In a study by Panna Vass, et al., such a configuration reaches productivity of 240 g/hr.

Free surface technologies are also classified based on the stationary and moving type of spinnerets. As the name suggests, we do not use needles in free surface electrospinning. Thus, it is often known as needle-less electrospinning. The difference in electrospinning mechanism between free surface technology and nozzle-type technology is only that, in free-surface technology, the generation and organization of the multiple jets occur from the free surface of the polymer solution. The major benefit of using free surface technology is the elimination of clogging issues, simultaneously increasing productivity. The different types of stationary free-surface spinnerets are wires, conical wires, bowl, plate edge, curved slots and slits. Some varieties of moving free-surface spinnerets are rotary wires, spiral coils, corona, magnetic fluids, bubbles, beaded-chain and balls. The entire idea of various spinneret configurations in nozzle-type and free-surface technologies have been summarized in Figures 6.6, 6.7 and 6.8 for the finer understanding of the reader.

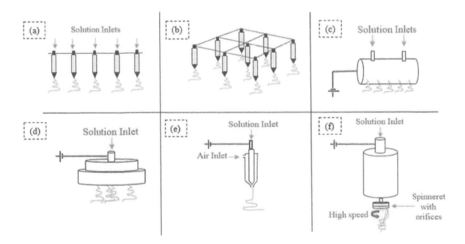

FIGURE 6.6 Nozzle-type electrospinning technologies: (a) linear multi-needle; (b) 2-D multi-needle; (c) porous tube; (d) flat; (e) electro-blowing; and (f) high speed nozzle-based electrospinning [Vass, Panna (2019)].

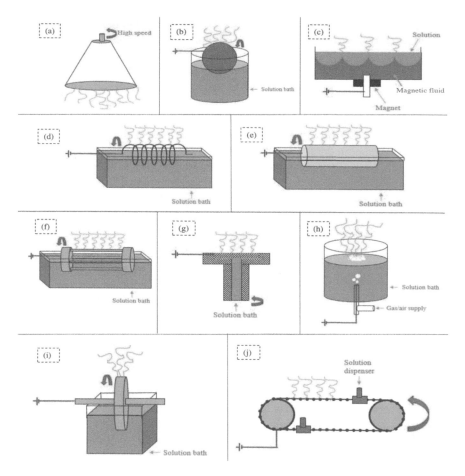

FIGURE 6.7 Free surface technologies with moving-type spinnerets: (a) high-speed free surface; (b) ball; (c) magnetic fluid; (d) spiral coil; (e) cylinder; (f) rotary wire; (g) corona; (h) bubble; (i) rotary disk; and (j) beaded-chain [Vass, Panna (2019)].

6.6 CONCLUSION

Electrospinning is a continuous and low-energy technology, which has been proven the best method for making membranes with a highly porous structure, high surface area and uniform pore distribution. In this chapter, a brief overview of electrospinning, variable parameters of electrospinning, which can decide nanofibers' properties, ENMs for water treatment application, and the future challenges were discussed. Several modifications of the base ENM surface can be carried out to improve the filtration efficiency of the ENM. This could be in doping, coating, thermal and chemical treatment, grafting or interfacial polymerization. With the availability of many polymer combinations and remarkable progress shown in the fabrication of efficient multifunctional ENMs, a promising future can be assured for ENMs to solve water pollution problems. This can lead to the advancement of commercial wastewater

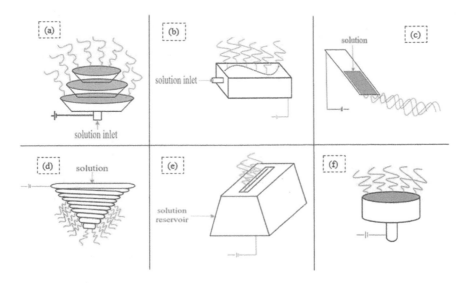

FIGURE 6.8 Free surface technologies with stationary spinnerets: (a) stepped pyramid; (b) curved slot; (c) plate-edge; (d) conical wire; (e) slit; and (f) bowl [Vass, Panna (2019)].

filtration systems in the coming generations. Therefore, the water treatment domain does seem the most predominant research area for polymeric ENMs.

REFERENCES

Akduman, C., E. Perrin Akçakoca Kumabasar, and Ahmet Çay. "Effect of Molecular Weight on the Morphology of Electrospun Poly (vinyl alcohol) Nanofibers." *Proc. XIIIth International Izmir Textile and Apparel Symposium.* 2014.

Al-Hazeem, Nabeel Zabar Abed. "Nanofibers and Electrospinning Method." *Novel Nanomaterials: Synthesis and Applications* (2018): 191.

Almasian, Arash, Mohammad Ebrahim Olya, and Niyaz Mohammad Mahmoodi. "Preparation and adsorption behavior of diethylenetriamine/polyacrylonitrile composite nanofibers for a direct dye removal." *Fibers and Polymers* 16.9 (2015): 1925–1934.

Angammana, Chitral J., and Shesha H. Jayaram. "Fundamentals of electrospinning and processing technologies." *Particulate Science and Technology* 34.1 (2016): 72–82.

Anton, Formhals. "Artificial thread and method of producing same." U.S. Patent No. 2, 187,306. January 16, 1940.

Anton, Formhals. "Method and apparatus for spinning." U.S. Patent No. 2,349,950. May 30, 1944.

Attia, Hadi, et al. "Superhydrophobic electrospun membrane for heavy metals removal by air gap membrane distillation (AGMD)." *Desalination* 420 (2017): 318–329.

Bagbi, Yana, Arvind Pandey, and Pratima R. Solanki. "Electrospun nanofibrous filtration membranes for heavy metals and dye removal." *Nanoscale Materials in Water Purification.* Elsevier, 2019. 275–288.

Beachley, Vince, and Xuejun Wen. "Effect of electrospinning parameters on the nanofiber diameter and length." *Materials Science and Engineering: C* 29.3 (2009): 663–668.

Bowen, W. Richard, and Julian S. Welfoot. "Modelling of membrane nanofiltration—pore size distribution effects." *Chemical Engineering Science* 57.8 (2002): 1393–1407.

Cai, Jingcheng, et al. "Membrane desalination using surface fluorination treated electrospun polyacrylonitrile membranes with nonwoven structure and quasi-parallel fibrous structure." *Desalination* 429 (2018): 70–75.

Casper, Cheryl L., et al. "Controlling surface morphology of electrospun polystyrene fibers: effect of humidity and molecular weight in the electrospinning process." *Macromolecules* 37.2 (2004): 573–578.

Celebioglu, Asli, Serkan Demirci, and Tamer Uyar. "Cyclodextrin-grafted electrospun cellulose acetate nanofibers via "Click" reaction for removal of phenanthrene." *Applied Surface Science* 305 (2014): 581–588.

Choi, Jae Shin, et al. "Effect of organosoluble salts on the nanofibrous structure of electrospun poly (3-hydroxybutyrate-co-3-hydroxyvalerate)." *International Journal of Biological Macromolecules* 34.4 (2004): 249–256.

Cooley, J. F. "Improved methods of and apparatus for electrically separating the relatively volatile liquid component from the component of relatively fixed substances of composite fluids." *United Kingdom Patent* 6385 (1900): 19.

Cooley, John F. "Apparatus for electrically dispersing fluids." U.S. Patent No. 692,631. February 4, 1902.

Daels, Nele, et al. "Structure changes and water filtration properties of electrospun polyamide nanofibre membranes." *Water Science and Technology* 73.8 (2016): 1920–1926.

Deitzel, Joseph M., et al. "The effect of processing variables on the morphology of electrospun nanofibers and textiles." *Polymer* 42.1 (2001): 261–272.

Ditaranto, Nicoletta, et al. "Electrospun nanomaterials implementing antibacterial inorganic nanophases." *Applied Sciences* 8.9 (2018): 1643.

Dobosz, Kerianne M., et al. "Ultrafiltration membranes enhanced with electrospun nanofibers exhibit improved flux and fouling resistance." *Industrial & Engineering Chemistry Research* 56.19 (2017): 5724–5733.

Doshi, Jayesh, and Darrell H. Reneker. "Electrospinning process and applications of electrospun fibers." *Journal of Electrostatics* 35.2–3 (1995): 151–160.

Ebrahimi, Adeleh, Mohammad Karimi, and Farzin Zokaee Ashtiani. "Characterization of triple electrospun layers of PVDF for direct contact membrane distillation process." *Journal of Polymer Research* 25.2 (2018): 50.

Elbahri, Mady, et al. "Smart metal–polymer bionanocomposites as omnidirectional plasmonic black absorber formed by nanofluid filtration." *Advanced Functional Materials* 22.22 (2012): 4771–4777.

Faccini, Mirko, et al. "Electrospun carbon nanofiber membranes for filtration of nanoparticles from water." *Journal of Nanomaterials* 2015 (2015): 2.

Formhals A.: Method and apparatus for the production of fibers. U.S. Patent 2116942, USA (1938).

Formhals, A. Process and apparatus for preparing artificial threads. US Patent, 1975504." Vol 1(1934):7

Ganesh, Venkatesan Anand, et al. "Electrospun differential wetting membranes for efficient oil–water separation." *Macromolecular Materials and Engineering* 301.7 (2016): 812–817.

Gilbert, William, and Edward Wright. *De magnete, magneticisque corporibus, et de magno magnete tellure: physiologia noua, plurimis & argumentis, & experimentis demonstrata.* excudebat Short, 1967.

Gopi, Sreerag, et al. "Chitin nanowhisker–inspired electrospun PVDF membrane for enhanced oil-water separation." *Journal of Environmental Management* 228 (2018): 249–259.

Haider, Sajjad, and Soo-Young Park. "Preparation of the electrospun chitosan nanofibers and their applications to the adsorption of Cu (II) and Pb (II) ions from an aqueous solution." *Journal of Membrane Science* 328.1–2 (2009): 90–96.

Hohman, Moses M., et al. "Electrospinning and electrically forced jets. I. Stability theory." *Physics of Fluids* 13.8 (2001): 2201–2220.

Homaeigohar, S.S., K. Buhr, and K. Ebert. "Polyethersulfone electrospun nanofibrous composite membrane for liquid filtration." *Journal of Membrane Science* 365.1–2 (2010): 68–77.

Homaeigohar, Seyed Shahin, and Mady Elbahri. "Novel compaction resistant and ductile nanocomposite nanofibrous microfiltration membranes." *Journal of Colloid and Interface Science* 372.1 (2012): 6–15.

Homaeigohar, Shahin, and Mady Elbahri. "Nanocomposite electrospun nanofiber membranes for environmental remediation." *Materials* 7.2 (2014): 1017–1045.

Huang, Zheng-Ming, et al. "A review on polymer nanofibers by electrospinning and their applications in nanocomposites." *Composites Science and Technology* 63.15 (2003): 2223–2253.

Islam, Md Shahidul, Jeffrey R. McCutcheon, and Md Saifur Rahaman. "A high flux polyvinyl acctate-coated electrospun nylon 6/SiO2 composite microfiltration membrane for the separation of oil-in-water emulsion with improved antifouling performance." *Journal of Membrane Science* 537 (2017): 297–309.

Kameoka, Jun, et al. "A scanning tip electrospinning source for deposition of oriented nanofibres." *Nanotechnology* 14.10 (2003): 1124.

Kaur, Satinderpal, et al. "Hot pressing of electrospun membrane composite and its influence on separation performance on thin film composite nanofiltration membrane." *Desalination* 279.1–3 (2011): 201–209.

Kaur, Satinderpal, et al. "Influence of electrospun fiber size on the separation efficiency of thin film nanofiltration composite membrane." *Journal of Membrane Science* 392 (2012): 101–111.

Khezli, Sakine, Mojgan Zandi, and Jalal Barzin. "Fabrication of electrospun nanocomposite polyethersulfone membrane for microfiltration." *Polymer Bulletin* 73.8 (2016): 2265–2286.

Koski, Antti, Kate Yim, and S. J. M. L. Shivkumar. "Effect of molecular weight on fibrous PVA produced by electrospinning." *Materials Letters* 58.3–4 (2004): 493–497.

Li, Xiong, et al. "Electrospun superhydrophobic organic/inorganic composite nanofibrous membranes for membrane distillation." *ACS Applied Materials & Interfaces* 7.39 (2015): 21919–21930.

Li, Zongjie, et al. "A novel polyvinylidene fluoride tree-like nanofiber membrane for microfiltration." *Nanomaterials* 6.8 (2016): 152.

Liang, Robert, et al. "Fundamentals on adsorption, membrane filtration, and advanced oxidation processes for water treatment." *Nanotechnology for Water Treatment and Purification.* Springer, Cham, 2014. 1–45.

Zargham, Shamim, et al. "Controlling numbers and sizes of beads in electrospun nanofibers." *Polymer International* 57.4 (2008): 632–636.

Lushnikov, Alex. "Igor' Vasilievich Petryanov-Sokolov (1907–1996)." *Journal of Aerosol Science* 28 (1997): 545–546.

Megelski, Silke, et al. "Micro-and nanostructured surface morphology on electrospun polymer fibers." *Macromolecules* 35.22 (2002): 8456–8466.

Mirtalebi, Elham, et al. "Assessment of atomic force and scanning electron microscopes for characterization of commercial and electrospun nylon membranes for coke removal from wastewater." *Desalination and Water Treatment* 52.34–36 (2014): 6611–6619.

Mokhena, Teboho Clement, and Adriaan Stephanus Luyt. "Development of multifunctional nano/ultrafiltration membrane based on a chitosan thin film on alginate electrospun nanofibres." *Journal of Cleaner Production* 156 (2017): 470–479.

Najafabadi, Hossein Hadi, et al. "Removal of Cu2+, Pb2+ and Cr6+ from aqueous solutions using a chitosan/graphene oxide composite nanofibrous adsorbent." *RSC Advances* 5.21 (2015): 16532–16539.

Norton, Charles L. "Method of and apparatus for producing fibrous or filamentary material." U.S. Patent No. 2,048,651. July 21, 1936.

Padil, Vinod Vellora Thekkae, and Miroslav Černík. "Poly (vinyl alcohol)/gum karaya electrospun plasma treated membrane for the removal of nanoparticles (Au, Ag, Pt, CuO and Fe3O4) from aqueous solutions." *Journal of Hazardous Materials* 287 (2015): 102–110.

Pant, Hem Raj, et al. "One-step fabrication of multifunctional composite polyurethane spider-web-like nanofibrous membrane for water purification." *Journal of Hazardous Materials* 264 (2014): 25–33.

Ramakrishna, Seeram. *An introduction to electrospinning and nanofibers.* World Scientific, 2005.

Ray, Saikat Sinha, et al. "A comprehensive review: electrospinning technique for fabrication and surface modification of membranes for water treatment application." *RSC Advances* 6.88 (2016): 85495–85514.

Ren, Long-Fei, et al. "TiO2-FTCS modified superhydrophobic PVDF electrospun nanofibrous membrane for desalination by direct contact membrane distillation." *Desalination* 423 (2017): 1–11.

Reneker, Darrell H., et al. "Bending instability of electrically charged liquid jets of polymer solutions in electrospinning." *Journal of Applied Physics* 87.9 (2000): 4531–4547.

Sato, Anna, et al. "Novel nanofibrous scaffolds for water filtration with bacteria and virus removal capability." *Journal of Electron Microscopy* 60.3 (2011): 201–209.

Senthamizhan, Anitha, et al. "Nanotraps" in porous electrospun fibers for effective removal of lead (II) in water." *Journal of Materials Chemistry A* 4.7 (2016): 2484–2493.

Shao, Hao, et al. "Effect of electrospinning parameters and polymer concentrations on mechanical-to-electrical energy conversion of randomly-oriented electrospun poly (vinylidene fluoride) nanofiber mats." *RSC Advances* 5.19 (2015): 14345–14350.

Shirazi, A., et al. "Electrospun membranes for desalination and water/wastewater treatment: a comprehensive review." *Journal of Membrane Science and Research* 3.3 (2017): 209–227.

Simon, Eric. (1988). NIH Phase I Final Report: Fibrous Substrates for Cell Culture (R3RR03544A).

Snowdon, Monika R., and Robert L. Liang. "Electrospun polymeric nanofibrous membranes for water treatment." (2018).

Son, Won Keun, et al. "Electrospinning of ultrafine cellulose acetate fibers: studies of a new solvent system and deacetylation of ultrafine cellulose acetate fibers." *Journal of Polymer Science Part B: Polymer Physics* 42.1 (2004): 5–11.

Sóti, Péter Lajos, et al. "Comparison of spray drying, electroblowing and electrospinning for preparation of Eudragit E and itraconazole solid dispersions." *International Journal of Pharmaceutics* 494.1 (2015): 23–30.

Su, Chunlei, et al. "Novel three-dimensional superhydrophobic and strength-enhanced electrospun membranes for long-term membrane distillation." *Separation and Purification Technology* 178 (2017): 279–287.

Suriyaraj, S. P., et al. "Scavenging of nitrate ions from water using hybrid Al 2 O 3/bio-TiO 2 nanocomposite impregnated thermoplastic polyurethane nanofibrous membrane." *RSC Advances* 5.84 (2015): 68420–68429.

Tang, Zhaohui, et al. "UV-cured poly (vinyl alcohol) ultrafiltration nanofibrous membrane based on electrospun nanofiber scaffolds." *Journal of Membrane Science* 328.1–2 (2009): 1–5.

Taylor, Geoffrey Ingram. "Disintegration of water drops in an electric field." *Proceedings of the Royal Society of London. Series A. Mathematical and Physical Sciences* 280.1382 (1964): 383–397.

Taylor, Geoffrey Ingram. "Electrically driven jets." *Proceedings of the Royal Society of London. A. Mathematical and Physical Sciences* 313.1515 (1969): 453–475.

Taylor, Geoffrey Ingram. "The force exerted by an electric field on a long cylindrical conductor." *Proceedings of the Royal Society of London. Series A. Mathematical and Physical Sciences* 291.1425 (1966): 145–158.

Tijing, Leonard D., et al. "Superhydrophobic nanofiber membrane containing carbon nanotubes for high-performance direct contact membrane distillation." *Journal of Membrane Science* 502 (2016): 158–170.

Uyar, Tamer, et al. "Molecular filters based on cyclodextrin functionalized electrospun fibers." *Journal of Membrane Science* 332.1–2 (2009): 129–137.

Vass, Panna, et al. "Continuous alternative to freeze drying: Manufacturing of cyclodextrin-based reconstitution powder from aqueous solution using scaled-up electrospinning." *Journal of Controlled Release* 298 (2019): 120–127.

Vass, Panna, et al. "Scale-up of electrospinning technology: Applications in the pharmaceutical industry." *Wiley Interdisciplinary Reviews: Nanomedicine and Nanobiotechnology* (2019).

Wang, Hualin, et al. "Preparation, characterization of electrospun meso-hydroxylapatite nanofibers and their sorptions on Co (II)." *Journal of Hazardous Materials* 265 (2014): 158–165.

Wang, Ran, et al. "Electrospun nanofibrous membranes for high flux microfiltration." *Journal of Membrane Science* 392 (2012): 167–174.

Wang, Xuefen, et al. "High flux filtration medium based on nanofibrous substrate with hydrophilic nanocomposite coating." *Environmental Science & Technology* 39.19 (2005): 7684–7691.

Wang, Xuefen, et al. "High performance ultrafiltration composite membranes based on poly (vinyl alcohol) hydrogel coating on crosslinked nanofibrous poly (vinyl alcohol) scaffold." *Journal of Membrane Science* 278.1–2 (2006): 261–268.

Wu, Huiqing, et al. "Polydopamine-assisted attachment of β-cyclodextrin on porous electrospun fibers for water purification under highly basic condition." *Chemical Engineering Journal* 270 (2015): 101–109.

Xiao, Shili, et al. "Excellent copper (II) removal using zero-valent iron nanoparticle-immobilized hybrid electrospun polymer nanofibrous mats." *Colloids and Surfaces A: Physicochemical and Engineering Aspects* 381.1–3 (2011): 48–54.

Xie, Siyuan, et al. "Electrospun nanofibrous adsorbents for uranium extraction from seawater." *Journal of Materials Chemistry A* 3.6 (2015): 2552–2558.

Xu, Guo-Rong, Jiao-Na Wang, and Cong-Ju Li. "Preparation of hierarchically nanofibrous membrane and its high adaptability in hexavalent chromium removal from water." *Chemical Engineering Journal* 198 (2012): 310–317.

Yalcinkaya, Fatma. "A review on advanced nanofiber technology for membrane distillation." *Journal of Engineered Fibers and Fabrics* 14 (2019): 1558925018824901.

Yan, Jiajie, et al. "Polydopamine-coated electrospun poly (vinyl alcohol)/poly (acrylic acid) membranes as efficient dye adsorbent with good recyclability." *Journal of Hazardous Materials* 283 (2015): 730–739.

Yang, Qingbiao, et al. "Influence of solvents on the formation of ultrathin uniform poly(vinyl pyrrolidone) nanofibers with electrospinning." *Journal of Polymer Science Part B: Polymer Physics* 42.20 (2004): 3721–3726.

Yang, Rui, et al. "Thiol-modified cellulose nanofibrous composite membranes for chromium (VI) and lead (II) adsorption." *Polymer* 55.5 (2014): 1167–1176.

Yao, Minwei, et al. "Effect of heat-press conditions on electrospun membranes for desalination by direct contact membrane distillation." *Desalination* 378 (2016): 80–91.

Yoon, Kyunghwan, Benjamin S. Hsiao, and Benjamin Chu. "High flux nanofiltration membranes based on interfacially polymerized polyamide barrier layer on polyacrylonitrile nanofibrous scaffolds." *Journal of Membrane Science* 326.2 (2009): 484–492.

Yoon, Kyunghwan, et al. "High flux ultrafiltration membranes based on electrospun nanofibrous PAN scaffolds and chitosan coating." *Polymer* 47.7 (2006): 2434–2441.

You, Hao, et al. "High flux low pressure thin film nanocomposite ultrafiltration membranes based on nanofibrous substrates." *Separation and Purification Technology* 108 (2013): 143–151.

Zargham, Shamim, et al. "The effect of flow rate on morphology and deposition area of electrospun nylon 6 nanofiber." *Journal of Engineered Fibers and Fabrics* 7.4 (2012): 155892501200700414.

Zeleny, John. "The electrical discharge from liquid points, and a hydrostatic method of measuring the electric intensity at their surfaces." *Physical Review* 3.2 (1914): 69.

Zeng, Jun, et al. "Poly-L-lactide nanofibers by electrospinning–influence of solution viscosity and electrical conductivity on fiber diameter and fiber morphology." *e-Polymers* 3.1 (2003).

Zhang, Chunxue, et al. "Study on morphology of electrospun poly (vinyl alcohol) mats." *European Polymer Journal* 41.3 (2005): 423–432.

Zhao, Shengli, et al. "Electrospinning of ethyl–cyanoethyl cellulose/tetrahydrofuran solutions." *Journal of Applied Polymer Science* 91.1 (2004): 242–246.

Zhao, Y. Y., et al. "Study on correlation of morphology of electrospun products of polyacrylamide with ultrahigh molecular weight." *Journal of Polymer Science Part B: Polymer Physics* 43.16 (2005): 2190–2195.

Zhao, Zhiguo, et al. "A novel composite microfiltration membrane: Structure and performance." *Journal of Membrane Science* 439 (2013): 12–19.

Zhou, Weitao, et al. "Studies of electrospun cellulose acetate nanofibrous membranes." *The Open Materials Science Journal* 5.1 (2011).

Zong, Xinhua, et al. "Structure and process relationship of electrospun bioabsorbable nanofiber membranes." *Polymer* 43.16 (2002): 4403–4412.

7 Forward Osmosis as a Sustainable Technology for Water Treatment and Desalination

Noel Jacob Kaleekkal*, Anoopa Ann Thomas and Joel Parayil Jacob
Membrane Separation Group, Department of Chemical
Engineering, National Institute of Technology Calicut, India

CONTENTS

7.1   INTRODUCTION

The last five to six decades have seen rapid population growth and aggressive urbanization that has depleted ~90% of the freshwater resources (de Fraiture and Wichelns 2010). The projections predict an increase in water requirements to 6,900 billion m^3 by 2030 (Misdan, Lau, and Ismail 2012), and as a result, various technologies such as seawater desalination and wastewater reclamation are being explored. According to the International Desalination Association (IDA, 2015), there are 18,426 desalination plants spread out over 150 countries (Figure 7.1) that provide 86.8 million m^3/day

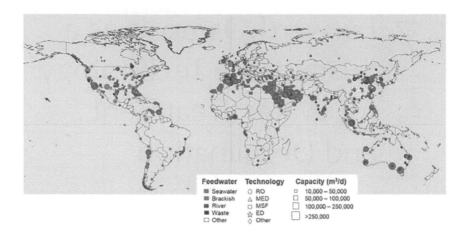

FIGURE 7.1   Global distribution of large desalination plants by feedwater type, desalination technology and capacity (Jones et al. 2019).

potable water to approximately 300 million people (Baawain, Choudri, and Ahmed 2015). Seawater desalination using thermal driven technologies, such as multi-effect distillation (MED) and multi-stage flash (MSF), are predominant in the Middle East, whereas large desalination plants in Australia operate exclusively using reverse osmosis technology (Jones et al. 2019).

Recovery of water from industrial/household wastewater containing trace quantities of emerging pollutants (EPs) is a grave concern (Dharupaneedi et al. 2019). Conventional treatment methods such as adsorption, sedimentation, coagulation, bio-oxidation, and filtration are inadequate for removing these environmentally hazardous and toxic EPs.

Seawater reverse osmosis (SWRO) is a well-established pressure-driven membrane process used for seawater/brackish water desalination and water recovery from industrial effluents. However, the enormous energy requirement is a significant drawback (Alkaisi, Mossad, and Sharifian-Barforoush 2017), and it is estimated that SWRO plants require three to four times more energy than the ΔG_{min} (Sarai Atab, Smallbone, and Roskilly 2016).

Forward osmosis is gaining popularity as an alternative to the RO process (Li, Liu, and Li 2017), and the process utilizes a semipermeable membrane to recover water from the low concentration (low osmotic pressure) feed (seawater/wastewater) using a high concentration (high osmotic pressure) draw solution (Vu et al. 2018). A schematic representation of the forward osmosis principle where the driving force is the difference in osmotic pressure is shown in Figure 7.2.

The low energy requirement, in addition to the low propensity for membrane fouling (no/low applied pressure), high rejection of contaminants, and simple equipment configuration, are the added advantages of the FO process (Siddiqui and Qianhong 2018).

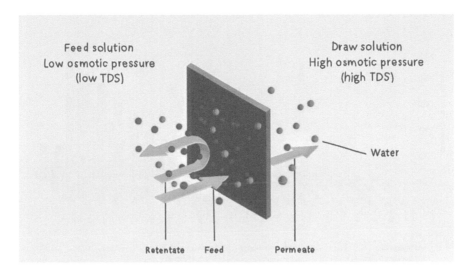

FIGURE 7.2 Forward osmosis process.

Forward osmosis (FO) has been recognized as a feasible process since the 1930s (Subramani, Arun, and Joseph G. Jacangelo. 2015); however, its application, in the fruit juice concentration, was demonstrated after 30 years. The growth in FO research since 1999 is shown in Figure 7.3 with the various aspects covered in the publications (Awad et al. 2019). Despite a large number of research articles (>7000), less than 2.5% of these discuss the large-scale implementation of FO technology.

This technology's commercial development saw a boom in 2002, when Hydration Technology Innovations (HTI) supplied emergency potable water for the U.S. military, using a sugar-based draw solution. In 2008, Oasys Water introduced a thin-film composite polyamide (PA) FO membrane in the market, and the company had developed a fully integrated FO technology based on thermally recoverable ammonium carbonate draw solute in 2013. Two large scale FO-RO installations were set up by Modern Water in Oman and Gibraltar in 2010, which led to them bagging the contract to erect a 200m^3/day desalination plant at Al Naghdah, Oman. The PA membrane developed by Porifera has proved to be successful in wastewater recovery and liquid food processing applications. Since its foundation in 2010, Trevi Systems has demonstrated its cellulose acetate (CA) FO membranes at five pilot studies in the United States (2013) and in the Middle East (2015). In 2017, Aquaporin entered the FO market, launching a new hollow fiber (HF) module.

If the FO can be commercialized to compete with the existing technologies (such as RO or MSF/MEE), it will mark the beginning of a new era in desalination and wastewater treatment. This chapter's prime focus is to explore the importance of forward osmosis and its applications (desalination or wastewater treatment) and to provide a comprehensive review of the development of new membrane materials.

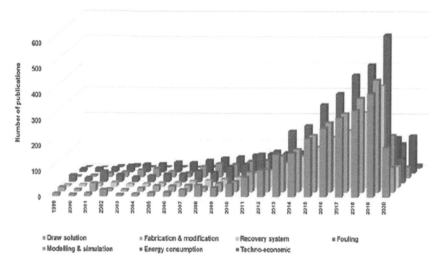

FIGURE 7.3 Total number of publications documented in science direct on FO process from 1999 till date (W. Suwaileh et al. 2020).

7.2 COMPONENTS OF FORWARD OSMOSIS PROCESS

A forward osmosis system mainly consists of three components: a feed solution (lower osmotic pressure), a draw solute, and a semipermeable membrane. A detailed discussion of all three components is given below.

7.2.1 THE FEED SOLUTION

The deionized (DI) water was used as the feed solution to evaluate the membranes' performance for most laboratory-scale studies (Akther et al. 2020; Dutta, Dave, and Nath 2020; Zhao et al. 2018; Kwon et al. 2019). A NaCl solution of 1000–2000 ppm concentration has been explored (Wu et al. 2018; Guo et al. 2018; Ramezani Darabi, Jahanshahi, and Peyravi 2018; Ghaemi and Khodakarami 2019; Ramezani Darabi, Peyravi, and Jahanshahi 2019; Jang et al. 2020; Miao et al. 2020; J. Zhou et al. 2020; Nguyen et al. 2020b). The other feed solutions used include oily wastewater, NH$_4$Cl solution, brackish water, seawater, simulated radioactive wastewater, microalgae broth, biological wastewater treatment effluent, fruit juice, sewage, anaerobic digestion effluent, distillery wastewater, antibiotic solution, fresh synthetic urine (Lee et al. 2019; Sahebi et al. 2020; Bao, Wu, Shi, Wang, Zhu, Zhang, Zhang, Zhang, Zhang, et al. 2019; Jafarinejad et al. 2019; Bao, Wu, Shi, Wang, Zhu, Zhang, Zhang, Zhang, Guo, et al. 2019; Giagnorio, Ricceri, and Tiraferri 2019; Lambrechts and Sheldon 2019; Shen et al. 2019; Chen et al. 2019; Liu et al. 2019; Liu, Wu, and Wang 2019; Nawi et al. 2020; Im et al. 2020; Kim et al. 2019; Singh et al. 2019; Schneider et al. 2019; Singh et al. 2018; Zhang et al. 2020; Volpin et al. 2019), etc. The selection of the feed solution depends solely on the end application of the FO process.

7.2.2   THE DRAW SOLUTE

The choice of the draw solute is critical in determining water recovery and energy costs associated with the process. The draw solute concentration is maintained such that the water permeation across the membrane is a spontaneous process driven by the chemical-potential gradient (Cai and Hu 2016).

7.2.2.1   Parameters for the Selection of Draw Solute

The choice of a suitable draw solute is of paramount importance. An ideal draw solute must fulfill the following conditions (Alejo et al. 2017; Johnson et al. 2018):

i. *High water solubility*
   The strong affinity between the draw solute and water molecules, e.g., via hydration or ionization, favour the solubility of draw solute in water. If the solubility is high, the draw solute can generate a high enough osmotic pressure, which directly increases the FO process's driving force.

ii. *Low/zero reverse solute flux*
   Reverse diffusion of the draw solute refers to its transport back to the feed side, which occurs if the chemical potential difference is too high. As a result, the concentration of draw solute in the feed solution increases with time, which is undesirable (Ge, Ling, and Chung 2013).

iii. *Less viscous*
   A less viscous draw solution is favoured in FO as it lowers the pumping costs.

iv. *High diffusion coefficient*
   A high value of draw solute diffusion coefficient is favoured as it reduces internal concentration polarization ICP, which is desired in FO.

v. *Benign to the FO membrane*
   The draw solute should not damage the selective layer and support layers even after prolonged usage (Cai and Hu 2016).

vi. *Low molecular weight solute*
   Solutes with low molecular weight produce higher osmotic pressure; however, they lead to a higher reverse solute flux.

vii. *Low toxicity*
   Unlike other membrane processes, the FO process's final product is not purified water but rather a dewatered feed solution and a dilute draw solution. As a result, unless the diluted draw solution uses itself or the process is purely being run to dewater the feed rather than produce useful product water, further separation is then necessary. It is not possible to ensure a 100% recovery of draw solute in the subsequent step after FO, so the draw solute has to be selected in such a way that even if it is present in some appreciable amount, it should not cause any adverse effect on human health and environment.

Draw solutes are generally classified based on (i) the properties of draw solute, (ii) the regeneration method used, (iii) compatibility with the feed solution to be treated, and (iv) whether the intended product is a clean permeate or a dewatered feed.

7.2.2.2 Classification of Draw Solutes Based on Physicochemical Property

1. *Gas and volatile compounds*

   The thermolytic NH_3-CO_2 pairs are widely explored as they generate high osmotic pressure due to their high water solubility. In water, these exist as ammonium bicarbonate ions, regenerated by evaporation at a specified temperature and pressure (Johnson et al. 2018). McCutcheon et al. (McCutcheon, McGinnis, and Elimelech 2005) demonstrated using this pair of draw solutes for desalination in a pilot study. The back diffusion of the ammonium salts, the presence of trace quantity of salts in the draw solution (after regeneration), and the difficulty in predicting draw solute (DS) concentration are some of the drawbacks reported. Sato et al. (Sato, Sato, and Yanase 2014) used dimethyl ether (DME), a low boiling, water-inert gas, with a highly volatile liquid as a draw solute, to generate a pressure of 24 bar. In this case, however, energy is consumed not in the regeneration step but in draw solution preparation since external pressure is needed to aid the draw solute's dissolution. Pure water was obtained from the diluted draw solution by allowing DME to volatilize at room temperature. SO_2 is another DS that can be recovered via a thermal process. However, the corrosive nature of SO_2 and the foul odour imparted to the permeate make it less desirable as a DS (Johnson et al. 2018).

2. *Inorganic draw solutes*

   Monovalent inorganic salts such as NaCl, NH_4Cl, and KNO_3 are inexpensive, widely available, and receive attention due to the high water flux they generate. However, these also pose several drawbacks such as high reverse solute flux compared to divalent salts, which leads to a quicker flux decline and increased cost of draw solute replenishment. Multivalent salts such as $MgCl_2$ and $CaCl_2$, each with larger hydration radius, are suitable alternatives as they are capable of achieving higher osmotic pressures at similar concentrations since they produce more ionic species on dissociation (Alejo et al. 2017; Achilli, Cath, and Childress 2010). Besides, the application of ultrasound in a FO process using magnesium and copper sulphate as DS has also been evaluated to alleviate ICP (Qasim et al. 2020).

   The application of ultrasound (40 kHz) was proven successful in mitigating the effects of the ICP for the desalination of synthetic brackish/ seawater. The ultrasound-assisted process enhanced the water flux by 34.6% and 43.9% with magnesium and copper sulphate as DS, respectively(Qasim et al. 2020). A synthetic high nutrient feed was dewatered using sodium phosphate as draw solute coupled with membrane distillation (MD) for regeneration (Nguyen, Chen, et al. 2015; Nguyen et al. 2016). The performance was related to the pH value of the draw solution. The highest water flux was noted at pH 9 (12.5 LMH), while the reverse salt flux stayed low. Interestingly the specific reverse solute flux was lower than that for a NaCl solution of similar concentration (0.1M). This observation was credited to the higher degree of complexation between Na^+ and HPO_4^{2-} at pH 9, leaving a lower number of free Na^+ ions.

3. *Fertilizer as draw solutes*

Fertilizers are favourable draw solutes as the diluted draw solution at the end of the FO run can be directly used for agriculture. This helps lower the energy and costs associated with the regeneration, and recovery of the DS. 1M of fertilizer solution could provide 41% of the water required for fertilizer dilution for irrigation (Zou and He 2016). Sulphur-based seed solution (SBSS) (Tran et al. 2017) is a by-product of the photoelectrochemical (PEC) water-splitting process and has been explored as a draw solute in the fertilizer drawn forward osmosis (FDFO) desalination. It is a mixture of ammonium sulphate and ammonium sulphite that could yield a high flux of 19 LMH from model saline water (5g/L NaCl).

4. *Organic draw solutes*

Several linear polyelectrolytes such as linear poly(sodium acrylate) (PSA), a sodium salt of poly(aspartic acid), are efficient as DSs as they limit the reverse solute flux owing to their large size. The increase in viscosity of the draw solution negatively affects the overall process performance. The introduction of carboxymethyl groups in dextrans' molecular framework contributed to an increase in osmotic pressure while maintaining excellent rejection due to its electrostatic repulsion with the membrane (Lu, Zhong, and Wang 2019). Phosphoric acid-modified organic salts as a DS successfully improved the water flux (47–54 LMH in PRO) and produced negligible reverse solute flux (RSF) due to lower viscosity, higher osmotic pressures, and convenient molecular sizes. (Long, Qingwu, Liang Shen, Rongbiao Chen, Jiaqi Huang, Shu Xiong, and Yan Wang. 2016.)

5. *Switchable polarity solvents*

Switchable polarity solvents (SPS), a new class of draw solutes, show varied miscibility with water depending on the amount of dissolved CO_2. Carbonic acid transformed an insoluble amine N, N-dimethyl cyclohexylamine (DMCHA) to a stable water-soluble complex N, N-dimethyl cyclohexylamine hydrogen carbonate ($DMCHA-HCO_3$). An increase in water solubility from 1.8% by mass to ~77% by mass was reported. On employing $DMCHA-HCO_3$ as DS for desalination, excellent rejection of NaCl was observed (98–99.5%). The permeate was recovered by stripping off CO_2 using heat or gas (Reimund et al. 2016).

6. *Hydrogels*

These are often polymers capable of trapping water within their network; they tend to undergo a volume change or phase change on the application of an external stimulus such as pH, pressure, light, temperature, magnetic field, or applied electric charge (Luo et al. 2020). Since they transition between hydrophilic and hydrophobic when taking up or releasing water, they are effectively exploited in FO processes as draw solutes. Poly (N-isopropyl acrylamide) (PNIPAM) is a cross-linked thermoresponsive hydrogel comprising both hydrophilic and hydrophobic parts. At the lower critical solution temperature (LCST), the PNIPAM undergoes a reversible phase transition from hydrophilic to hydrophobic, thereby releasing the absorbed water.

Electrically-responsive hydrogel (H. Zhang et al. 2015) composed of hyaluronic acid/polyvinyl alcohol (HA/PVA) shrink and release water on the application of electric stress that increases the osmotic pressure of the draw solution. The larger size of the DS ensures zero back diffusion, and the hydrogels are insoluble in water; they can be recovered with low energy expenditure.

7. *Ionic liquids*

Ionic liquids (IL) have attractive characteristics such as high solubility, nonvolatile, high ionic conductivity, and chemical/thermal stability (Dutta and Nath 2018). Due to the increased osmotic potential, ILs can be used to recover water from feed having three times the concentration of seawater. Raising the temperature above LCST causes phase separation, and the ILs can be retrieved by sedimentation or NF. The protonated betaine bis(trifluoromethyl sulfonyl)imide (3.2M), a thermally responsive ionic liquid, was shown to be capable of drawing water from saline water (3.0M) (Zhong et al. 2016). The hindrance to using ILs is their toxicity, environmental impact during production (use of heavy metal precursors), high viscosity, and the massive production cost.

8. *Nanoparticles*

Nanoparticles larger than membrane pore size would theoretically be prevented from back-diffusing. The nanoparticles are generally modified with organic moieties to augment the osmotic potential and to increase their size. However, most nanoparticles are nonresponsive, thus requiring energy-intensive separation techniques. Magnetic nanoparticles such as Fe_3O_4 can be recovered using a magnetic field; however, the separation of these NPs was challenging even when a strong magnetic field was applied. Moreover, the water flux was low even when DI/diluted brackish water was used as the feed stream. The issue of agglomeration was overcome by grafting N-isopropyl acrylamide-co-sodium 2-acrylamido-2-methylpropane sulfonate (A. Zhou et al. 2015).

The various draw solute recovery methods are given in Table 7.1.

7.3 MEMBRANE PARAMETERS

The water flux and reverse solute flux in the FO process, in which a hydraulic pressure difference (ΔP) is absent, can be generally described as follows (Ryu et al., n.d.):

$$J_w = A\left(\pi_i - \pi_{Fm}\right) \tag{7.1}$$

$$J_s = B\left(C_i - C_{fm}\right) \tag{7.2}$$

where A is water permeability; B is solute permeability; π_i and C_i correspond to the osmotic pressure and concentration at the interface between selective layer and support layer/base membrane, respectively. π_{Fm} and C_{fm} represent the osmotic pressure and concentration at the interface between selective layer and feed solution, respectively.

The structural parameter is an intrinsic characteristic of TFC-FO membranes. It quantifies the quality of the underlying support layer. It is determined as the product

TABLE 7.1

Draw solutes classified by the mode of recovery

Mode of Recovery	Classes of Draw Solute	Recovery Technique	Reference
Thermal	SO$_2$ based	Distillation/low grade heat	(Chaoui et al. 2019)
	Dimethyl ether (DME) based	Exposing to air	(Sato, Sato, and Yanase 2014)
Membrane based	Monovalent salts/organic ionic based DS (MW: 100 Da)	Reverse osmosis	(Kim et al. 2017)
	Polyelectrolytes/bivalent and multivalent salts DSs (MW: 100–1000 Da)	Nanofiltration	(Nguyen et al. 2015)
	Polyelectrolyte based (MW: 1000–10^5)	Ultrafiltration	(Nguyen et al. 2015)
	Thermo-responsive copolymer-based DS	Membrane distillation	(Nguyen et al. 2016) (Nguyen et al. 2018)
	Charged DS	Electrodialysis	(Zou and He 2017)
Stimuli responsive	Hydrogels NPs (sodiumN-isopropylacrylamide carbon (SA-NIPAM-C))	Deswelling using solar heating	(Li et al. 2011)
	Organic coated magnetic NPs	Magnetic field	(Mishra et al. 2015)
	Electrically responsive (xerogel particles)	Electric field	(Cui et al. 2018)
No recovery	Fertilizer based DSs	-	(Nasr and Sewilam 2018)
	Emergency drinks (glucose and fructose)	-	(Lee and Hsieh 2019)

of support layer thickness (t) and tortuosity (ζ) and is inversely proportional to porosity (ε) (Suwaileh et al. 2018).

$$S = \frac{t\zeta}{\varepsilon} \qquad (7.3)$$

Membranes with lower S values are preferred to reduce the severity of ICP.

7.4 THE MEMBRANE

Forward osmosis membranes are mostly an asymmetric composite type, meaning that they consist of two layers: a thin rejection layer (typically 100–200 nm thick and a support layer (typically 100–200 μm thick) (Eyvaz, Arslan, and İmer 2018), just as shown in Figure 7.4. In all FO membranes, the nanometer-sized rejection layer is fused with the underlying support layer. The porous support layer is typically formed through a nonsolvent-induced phase separation (NIPS) process. However, those made out of electrospun nanofibers are also in use (Suwaileh et al. 2018). The active rejection layer is typically formed by either interfacial polymerization (IP)

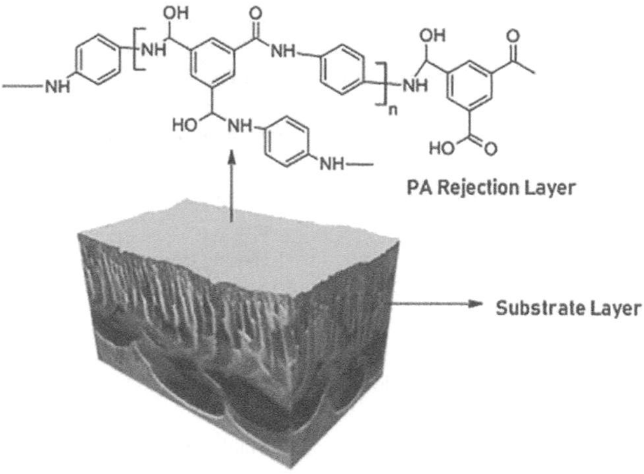

FIGURE 7.4 Schematic representation of a thin-film composite membrane (Ismail et al. 2015).

(Chowdhury, Huang, and McCutcheon 2017) or dip coating, followed by cross-linking. This active skin layer selectively transports water molecules from the feed side to the concentrated draw solution side. It is believed that the support layer is primarily to provide mechanical support for the thin and fragile rejection layer and has little effect on the separation performance of the FO membrane. High porosity, hydrophilicity, chemical resistance, and structural stability are crucial for high-performance composite FO membranes as they significantly influence the membrane permeability (Huang et al. 2019).

In earlier studies, commercial RO membranes were used in the FO process, but the thick sponge-like support layer limited the mass transfer and was responsible for causing severe internal concentration polarization (ICP) within the support layer. This limited the applicability of RO membranes in the FO process (Eyvaz, Arslan, and İmer 2018). It was Hydration Technologies, Inc. (HTI) that introduced the first commercial FO membrane, which has a structure embedding a thin polyester mesh support in cellular triacetate (CTA) (Awad et al. 2019). This HTI's CTA membrane and their TFC FO membranes were later considered a benchmark in developing new FO membranes. However, FO membranes with superior water permeability and salt rejection are still subjects to be focused on the commercialization of FO technology. Some commercial FO membranes available in the market are listed in Table 7.2.

The advantage of using a composite FO membrane is that the outstanding membrane performance can be efficiently realized by tuning the properties of both the active rejection layer and the support layer (Jin et al. 2019; Shakeri, Salehi, et al. 2019).

7.4.1 Modification of the Base Membrane

A significant problem encountered in the FO process is the internal concentration polarization (ICP) (Zuo 2019), which occurs due to the permeation of water from FS to DS, which reduces the net osmotic pressure difference, thereby decreasing the water

TABLE 7.2

Commercially available FO membranes (Awad et al. 2019, Goh et al. 2019)

Supplier	Materials/Commercial Name	Configuration
HTI	CTA-NW	SW
HTI	CTA-ES	SW
HTI	TFC	SW
Aquaporin A/S	AqP	SW, HF
Oasys Water	Thin-film composite (TFC)	SW
Porifera	PFO element	SW
Toray	FO membrane	SW
Fluid technology solution	Cellulose triacetate (CTA)	SW
Modern water	-	SW
Toyobo	-	HF
Trevi System	-	SW

flux. Hydrodynamic parameters cannot modify the ICP occurring within the membrane, and as a result, the modification of the base membrane seems to be the only option. Here the focus is on reducing the solute resistance to diffusivity. The base membrane is generally polymer support (Wang et al. 2018; Ma et al. 2020; Jin et al. 2018; Akther et al. 2020) designed to offer little or no resistance to mass transfer. The support layer of the TFC FO membranes should have high hydrophilicity, sufficient chemical and mechanical stability, and low structural parameter (S) (high porosity, low tortuosity, and low thickness) (Manickam and Mccutcheon 2017). The NIPS method was used for the support-layer fabrication, and pore-forming agents such as PEG, PVP, etc. were incorporated to produce highly porous support membranes (Lee et al. 2019; Bao, Wu, Shi, Wang, Yu, et al. 2019; Shen et al. 2019; Choi et al. 2019). However, highly porous support increases the difficulty in forming a defect-free active layer. On the other hand, a dense, sponge-like morphology of the support can increase the mass transfer resistance.

Electrospinning is a new fabrication technique to develop nanofiber-based TFC with intrinsically low mass transfer resistance (Huang et al. 2019); it has been demonstrated to break the bottleneck of ICP in the FO process (Cheng et al. 2018). This type of membrane is superior in terms of flux and rejection compared to its CTA counterparts. Song and his team were pioneers (Song, Liu, and Sun 2011) in demonstrating TFN membranes prepared on top of the electrospun nanofibrous membranes (ENMs). The calculated S parameter of these membranes was 80 μm, far smaller than the commercial HTI membrane (450 μm). The water flux was ~3 times greater, indicating a highly interconnected porous base layer.

Moreover, this inimitable structure of the base membrane provided an interface for forming a dense polyamide active layer that improved the salt rejection rate of the FO membrane (Liu et al. 2020). Improvement of hydrophilicity of the base membrane improves the water flux during the FO process. Blending with hydrophilic polymers, coaxial spinning, surface grafting, plasma modification, and incorporation of nanomaterials are some of the methods in which the membrane property is

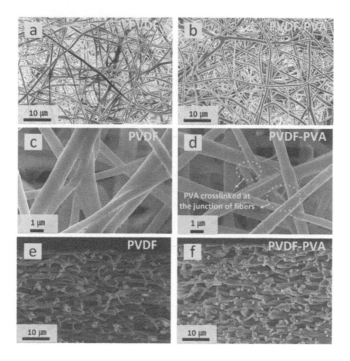

FIGURE 7.5   FESEM images of pure PVDF and PVDF-PVA nanofiber supports for top surface (a&b), top surface at higher magnification (c&d) and cross-section near the top surface (e&f) (Park et al. 2018).

enhanced (Shibuya, Park, et al. 2018; Obaid et al. 2018). The nanomaterials (such as GO) add mechanical strength to the base membrane and provide hydrophilic pathways for the permeation of water (Obaid et al. 2018).

PVDF electrospun membranes modified by dip-coating with PVA followed by cross-linking with GA produced water-stable, highly hydrophilic membranes (Jun et al. 2018). From Figure 7.5, it is observed from the FESEM images that the internal pores remain unhindered even after dip-coating, which is the principal reason for the reduction in ICP and improvement of water flux. Electrospun base membranes of PSF blended with PAN could improve the overall performance of the FO membranes (Shokrollahzadeh and Tajik 2018). Coaxial spinning was used to prepare electrospun fibers with PVDF as the inner core and CA as the sheath. PVDF provided the mechanical strength, and CA improved membrane hydrophilicity, and this electrospun membrane exhibited three times the water recovery compared to a conventional FO membrane (Shibuya, Jun, et al. 2018).

7.4.2 Introduction of a Novel Middle Layer

The balance between membrane permeability and selectivity could be achieved by introducing a sublayer sandwiched between the base membrane (non-woven/nanofibrous) and the active layer (Figure 7.6). High cross-flow velocities are employed to minimize the concentration polarization (Wang et al. 2016), which can deform or break the nanofibrous base membrane. Moreover, the large surface pores and narrow fiber diameter can lead to delamination of the top layer due to the poor interfacial adhesion.

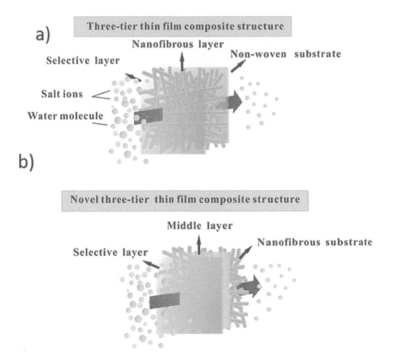

FIGURE 7.6 Different types of FO membrane structures: (a) conventional three-tier, thin-film composite structure; and (b) novel three-tier composite structure (Q. Liu et al. 2020).

A sublayer (middle-layer) introduction provides a suitable platform to anchor the thin polyamide selective layer (Shi et al. 2019). Tian et al. prepared a novel, dense PVDF middle layer on the nanofibrous base membrane using the phase-inversion technique, and the membrane exhibited exceptional mechanical strength with superior water flux (Tian et al. 2017).

7.4.3 MODIFICATION OF THE ACTIVE LAYER

The polyamide is prepared in situ on the base membrane, which acts as the active layer separating the FS and DS compartments. The PA formed by the interfacial polymerization depends on the acyl/amine monomer concentration, reaction time, and curing temperature (Shakeri, Razavi, et al. 2019).

Just as ICP, another major problem encountered in the FO process is the membrane fouling (Chun et al. 2017), although it is reversible and less severe than RO processes. Studies show that the active layer surface with high roughness and large leaf-like structures are more prone to foulant accumulation. The membrane surface charge is another factor that influences membrane-foulant interactions and rejection efficiency. Nowadays, the prime focus is to lower the membrane fouling propensity through the implementation of various modifications like functionalization of the active layer, use of nanomaterials (Kang et al. 2019; Rastgar et al. 2019), and the grafting of zwitterions (Chiao et al. 2019; Zhang et al. 2019; Lee et al. 2019). Table 7.3 shows a list of modifications carried out to improve the property of the membrane active layer.

TABLE 7.3

State-of-the-art literature on forward osmosis membranes

Membrane Configuration- Flat sheet/ hollow fiber/	Membrane preparation Technique- NIPS/ Electrospun-TFC/TFC/	Materials used – Polymer (Wt. %) +	Additives (wt. %)	Feed Properties 1. Concentration 2. Flow Rate	Draw Solution Properties 1. Type of Draw Solute 2. Conc 3. Flow rate	Water Flux – LMH FO/PRO	Reverse Solute Flux (gMH) FO/PRO	*Intrinsic parameters (A/B/S)	References
Flat sheet	NIPS-TFN	PSf(16wt%)/ PEG400(4wt%)- (MPD/GO)/ TMC	GO in aqueous MPD (0.008wt%)	1.DI	1.NaCl 2.2M	34.3/-	1.1/-	3.9/1.1/ 119	(Saeedi-Jurkuyeh et al. 2020)
Flat sheet	NIPS-TFC	(PSf/PSf-g-mPEG)-MPD/ TMC	PSf/PSf-g-mPEG = 9:1 (wt ratio) mPEG – methoxypoly-ethylene glycol (M.W = 500g/mol)	1. 10mM NaCl	1.NaCl 2.2M NaCl	2.5/-	5.3235/-	30.28/ 19.67/ 900	(J. Zhou et al. 2020)
Flat sheet	NIPS-TFC	(CNT/PES/ CNT)-MPD/ TMC Commercial PES MF membrane used	PDA-SWCNT (0.8 g/ m² of membrane area) PDA-SWCNT – Polydopamine- Single walled carbon nanotubes	1.DI	1.NaCl 2.1M	18.1/35.7	1.11/1.42	1.59/0.1/ 324.42	(Deng et al. 2020)
Flat sheet	NIPS TFC Casting solution dipped in coagulation bath containing MPD solution	(PI(14wt%)/NMP/ THF)/(MPD/ HAD/TEA(2wt%)/ SDS(0.1wt%)- (TMC/IL/ Hexylacetate)	NMP/THF = 3:1 IL (ionic liquid)= [C4mim][Tf2N] ([C4mim][Tf2N]/ hexylacetate = 50/50)	1.DI	1.NaCl 2.0.5M	9.1/17	2.3/3.6	0.41/ 0.295/-	(Hartanto et al. 2020)

Flat sheet	Electrospun – TFN	PSf(26wt%)/(TiO2/AgNPs)-MPD/TMC	TiO2/AgNPs – 0.3wt%	1. Secondary-effluent from the Songjiang sewage treatment plan 1.NaCl 2.1M	46.72/56.49	0.75% permeation of TRG	-	(Chen, Zheng, et al. 2020)
Flat sheet	NIPS-TFC	PSf(10wt%)/PVP(15wt%)-(MPD/DMF)-(TMC/TPP)	DMF-10wt/v% TPP – 0.25 wt/v%	1.DI 1.NaCl 2.2M	19/40	8.5/14.4	1.64/0.55/567	(Li et al. 2020)
Flat sheet	NIPS-TFN	PSf(16wt%)/PVP10(17wt%)-(MPD/TEA/ZIF-8)/TMC	ZIF-8(0.2w/v%)/PSS coating TEA (2wt%) in MPD ZIF-8 – zeoliticimidazolate-framework-8 PSS - poly(sodium 4-styrenesulfonate)	1.DI 1.NaCl 2.1M	5.688/-	28.9/-	2.5/-/2565	(Beh et al. 2020)
Flat sheet	NIPS – TFC (PA- mLBL)	PET-(mCNT)-PAN substrate (12wt% PAN)	mixed CNT layer	1.DI 2.25 cm/s 1.NaCl 2.1M 3.25cm/s	29/43.5	9.4/11.1	3.24/0.97/344	(Tsai et al. 2020)
Flat sheet	NIPS-TFC	Mixed cellulose ester (MCE) MF substrate/SWCNTs-MPD/TMC	SWCNTs – 5ml suspension	1.DI 2 6.3cm/s 1.NaCl 2.1M 3.6.3cm/s	62.8/-	19.4-	3.3/0.19/88	(Tang et al. 2020)
Flat sheet	NIPS-TFC (Toluene based IP)	PET fabric/PAN(12wt%)-MPD/TMC in toluene		1.DI 2. 1.NaCl 2.0.5M	22.5/-	0.92(NH4N flux)	3.2/0.22/311	(Kwon et al. 2020)

(Continued)

TABLE 7.3

State-of-the-art literature on forward osmosis membranes (Cont.)

Membrane Configuration- Flat sheet/ hollow fiber/	Membrane preparation Technique- NIPS/ Electrospun- TFC/TFC/	Materials used – Polymer (Wt. %) +	Additives (wt. %)	Feed Properties 1. Concentration 2. Flow Rate	Draw Solution Properties 1. Type of Draw Solute 2. Conc 3. Flow rate	Water Flux –LMH FO/PRO	Reverse Solute Flux (gMH) FO/PRO	*Intrinsic parameters (A/B/S)	References
Flat sheet	TFN	MCE MF commercial membrane-(MPD/COF-COOH)/TMC	COF-COOH (0.5 mg/mL)	1.DI 2.8cm/s	1.NaCl 2.1M 3.8cm/s	64.2/-	6.42/-	2.5/ 0.33/ 58.6	(Xu et al. 2020)
Flat sheet	TFC	Aquaporin membrane		1.DI 2.10cm/s Oily wastewater	1.KCl 2.3M 3.10cm/s 1.KCl 2.3M 3.10cm/s	34/67.22 11.31/-	20/48.3 -	3.1/2.5* 10^{-7}(m/s)/ 735	(Sahebi et al. 2020)
Flat sheet	NIPS TFC	PSf(12wt%)-(MPD/GO)/TMC	GO-0.01wt% (0.01μm²)	1.DI 2.12.6cm/s	1.NaCl 2.0.5M 3.12.6cm/s	24.7/41.9	5.2/7.961	3.7l/ 0.89/-	(Akther et al. 2020)
Flat sheet	NIPS-TFN	PSf(17.5wt%)/ PVP(0.5wt%)/ pCN(0.5wt%)-MPD/(TMC/ CN)	CN(0.05w/v%) in TMC CN-Carbon nitride pCN-protonated carbon nitride	1.DI	1.NaCl 2.2M	0.08/0.031		0.67/7.34* 10^{-8}(m/s)/ 2140	(Aziz et al. 2020)

Flat sheet	Electrospun TFN	PSf(26wt%)/ (TiO2/AgNPs)- MPD/TMC	TiO2/AgNPs – 0.3wt%	1.DI 2.10cm/s	1.NaCl 2.1M 3.10cm/s	1.3/1.75	–	(Chen, Huang, et al. 2020)
Flat sheet	TFC	"HFN 300" (Permionics Membrane Pvt. Ltd.) – PES NF membrane	–	1.DI	1.$MgCl_2$ 2.0.5M	0.087/-	Low pressure NF (<10 bar) OR Chemical route by precipitation method using 1,4-dioxane	4.75/ 0.01/ 2190	(Dutta, Dave, and Nath 2020)
Spiral wound		CTA based SWFO (Spiral wound FO)		Biological wastewater treatment effluents	1.NaCl 2.0.5M	10.076/-	–	(Im et al. 2020)
Flat sheet	NIPS	PPSU(12.5wt%)/ CAP/FMWCNT	0.8wt% FMWCNT(functional multiwalled CNT) vertically aligned at 0.6kV/cm followed by 2T PPSU/CAP=90/10 (wt ratio) PPSU – polyphenylensulfone CAP- cellulose acetate phthalate	DI	–	-/- 70.499	(Samieirad, Mousavi, and Saljoughi 2020)

(Continued)

TABLE 7.3

State-of-the-art literature on forward osmosis membranes (Cont.)

Membrane Configuration- Flat sheet/ hollow fiber/	Membrane preparation Technique- NIPS/ Electrospun- TFC/TFC/	Materials used – Polymer (Wt. %) +	Additives (wt. %)	Feed Properties 1. Concentration 2. Flow Rate	Draw Solution Properties 1. Type of Draw Solute 2. Conc 3. Flow rate	Water Flux – LMH FO/PRO	Reverse Solute Flux (gMH) FO/PRO	*Intrinsic parameters (A/B/S)	References
Flat sheet		CTA membrane		1.DI 2.5000 mg/L NaCl (low salinity brackish water) 3.35000g/L NaCl (synthetic seawater)	0.3 M EDTA-2Na/0.55 M Na$_3$PO$_4$) 0.3 M EDTA-2Na/0.55 M Na$_3$PO$_4$) 0.3 M EDTA-2Na/0.55 M Na$_3$PO$_4$)	0.48601/-	Membrane Distillation		(Nguyen et al. 2020a)
Flat sheet	NIPS-TFN	PES(12.5wt%)/ PEG(35wt%)-(MPD/GQDs@ UiO-66-NH$_2$)/ TMC	GQDs@UiO-66-NH$_2$–250 ppm in MPD solution	1.DI	1.NaCl 2.1M	19.08/36.8		4.88/ 1.356/-	(Bagherzadeh, Bayrami, and Amini 2020)
Flat sheet	NIPS-TFN	Ultrafiltration PAN-(MPD/ PDA)/(TMC/ MOF-801)	MOF-801 (0.01wt%) in MPD		1.NaCl 2.1M	3/-		2.12/ 0.88/-	(He et al. 2020)
Flat sheet	NIPS-TFN	PES/PEG-300-(MPD/ZnO)/ TMC	ZnO-0.5wt% in MPD solution	1.DI	1.NaCl 2.1M	9/9		-	(Amini et al. 2020)

7.5 ECONOMICS AND SUSTAINABILITY OF THE FO PROCESS

The challenges associated with the FO process are mainly the design of a novel DS and the energy costs associated with the draw solute recovery. Typically inorganic salts such as NaCl, KCl, and Na_2SO_4 are available at a low cost, but the energy expended for their recovery prevents FO technology's scale-up and commercialization. Thermally recoverable DS is promising when waste/residual heat can be utilized to recover them. However, the theoretical energy requirements are seldom investigated for the FO process. Zhao et al. reported that the energy consumption involved in a hybrid FO-MD process (29 kWh/m³) is much higher than the FO-RO process or standalone RO process (Zhao et al. 2014).

The theoretical energy consumption using 1.5 M NH_4HCO_3 (McGinnis and Elimelech 2007) as DS was calculated to be 0.84 kWh/m³ (including pumping costs) using the HYSYS software. This projected cost is much lower than the requirement for an RO process (1.6–6 kWh/m³); however, the FO process's energy consumption does not take into account the DS recovery to produce potable water. In most cases, an RO, thermal, or MD process is used for reconcentration of draw solute. However, lower fouling propensity does not rule out the potential of hybrid FO-RO over standalone RO in feeds with high fouling tendencies. Further, FO can be used to dilute the concentrated brine before discharge, thus helping mitigate environmental impact. Water recovered in the FO process can also be used to osmotically backwash the RO membranes, preventing harsh cleaning chemicals.

In applications such as dewatering of feed, fertigation, and the use of commercial hydration packets, the DS requires no regeneration, and these can use the FO process as it is a sustainable technology (Hoover et al. 2011). Thus FO, a low-energy consuming process, has immense potential to be developed as a sustainable technology to recover water from various feed streams.

7.6 OUTLOOKS AND CONCLUSIONS

FO is becoming increasingly attractive due to its low energy consumption, ability to withstand harsh feed/DS conditions, and its suitability to be employed in desalination/wastewater treatment applications. Due to the ICP, reverse solute flux, and low pure water flux, the development of a robust membrane is difficult. The challenge is to develop an ideal FO membrane with minimum ICP, high water permeability, excellent selectivity, and superior mechanical strength. Compared to the traditional phase-inversion membranes, the application of the electrospun base membrane is gaining popularity as it can mitigate or meet some of the desired characteristics. The design of the novel draw solutes and integrating technologies to recover and draw solute or polish water are some of the other aspects of the research that are currently being explored.

This chapter provides an overview of the FO membrane development, the types of draw solutes, and recovery methods. It also discusses the role of electrospinning to prepare novel base membranes and the FO process's sustainability to address the needs of desalination and wastewater treatment.

REFERENCES

Subramani, Arun, and Joseph G. Jacangelo. 2015. "Emerging Desalination Technologies for Water Treatment: A Critical Review." *Water Research*. Elsevier Ltd. https://doi.org/10.1016/j.watres.2015.02.032.

Achilli, Andrea, Tzahi Y Cath, and Amy E Childress. 2010. 'Selection of Inorganic-Based Draw Solutions for Forward Osmosis Applications'. *Journal of Membrane Science* 364 (1–2): 233–41. https://doi.org/10.1016/j.memsci.2010.08.010.

Akther, Nawshad, Ziwen Yuan, Yuan Chen, Sungil Lim, Sherub Phuntsho, Noreddine Ghaffour, Hideto Matsuyama, and Hokyong Shon. 2020. 'Influence of Graphene Oxide Lateral Size on the Properties and Performances of Forward Osmosis Membrane'. *Desalination* 484 (March): 114421. https://doi.org/10.1016/j.desal.2020.114421.

Alejo, Teresa, Manuel Arruebo, Verónica Carcelen, Victor M Monsalvo, and Victor Sebastian. 2017. 'Advances in Draw Solutes for Forward Osmosis: Hybrid Organic-Inorganic Nanoparticles and Conventional Solutes'. *Chemical Engineering Journal* 309: 738–52. https://doi.org/10.1016/j.cej.2016.10.079.

Alkaisi, Ahmed, Ruth Mossad, and Ahmad Sharifian-Barforoush. 2017. 'A Review of the Water Desalination Systems Integrated with Renewable Energy'. *Energy Procedia* 110 (December 2016): 268–74. https://doi.org/10.1016/j.egypro.2017.03.138.

Amini, Mojtaba, Maryam Seifi, Ali Akbari, and Mojtaba Hosseinifard. 2020. 'Polyamide-Zinc Oxide-Based Thin Film Nanocomposite Membranes: Towards Improved Performance for Forward Osmosis'. *Polyhedron* 179: 114362. https://doi.org/10.1016/j.poly.2020.114362.

Awad, Abdelrahman M., Rem Jalab, Joel Minier-Matar, Samer Adham, Mustafa S. Nasser, and S J Judd. 2019. 'The Status of Forward Osmosis Technology Implementation'. *Desalination* 461 (December 2018): 10–21. https://doi.org/10.1016/j.desal.2019.03.013.

Aziz, A Abdul, K C Wong, P S Goh, A F Ismail, and I Wan Azelee. 2020. 'Tailoring the Surface Properties of Carbon Nitride Incorporated Thin Film Nanocomposite Membrane for Forward Osmosis Desalination'. *Journal of Water Process Engineering* 33: 101005. https://doi.org/10.1016/j.jwpe.2019.101005.

Baawain M, BS Choudri, M Ahmed, and A Purnama. 2015. 'An Overview: Desalination, Environmental and Marine Outfall Systems'. *In Recent Progress in Desalination, Environmental and Marine Outfall Systems*, 3–10.

Bagherzadeh, Mojtaba, Arshad Bayrami, and Mojtaba Amini. 2020. 'Enhancing Forward Osmosis (FO) Performance of Polyethersulfone/Polyamide (PES/PA) Thin-Film Composite Membrane via the Incorporation of GQDs@UiO-66-NH2 Particles'. *Journal of Water Process Engineering* 33: 101107. https://doi.org/10.1016/j.jwpe.2019.101107.

Bao, Xian, Qinglian Wu, Wenxin Shi, Wei Wang, Huarong Yu, Zhigao Zhu, Xinyu Zhang, Zhiqiang Zhang, Ruijun Zhang, and Fuyi Cui. 2019. 'Polyamidoamine Dendrimer Grafted Forward Osmosis Membrane with Superior Ammonia Selectivity and Robust Antifouling Capacity for Domestic Wastewater Concentration'. *Water Research* 153: 1–10. https://doi.org/10.1016/j.watres.2018.12.067.

Bao, Xian, Qinglian Wu, Wenxin Shi, Wei Wang, Zhigao Zhu, Zhiqiang Zhang, Ruijun Zhang, Xinyu Zhang, Bing Zhang, et al. 2019. 'Insights into Simultaneous Ammonia-Selective and Anti-Fouling Mechanism over Forward Osmosis Membrane for Resource Recovery from Domestic Wastewater'. *Journal of Membrane Science*. https://doi.org/10.1016/j.memsci.2018.11.072.

Bao, Xian, Qinglian Wu, Wenxin Shi, Wei Wang, Zhigao Zhu, Zhiqiang Zhang, Ruijun Zhang, Bing Zhang, Yuan Guo, and Fuyi Cui. 2019. 'Dendritic Amine Sheltered Membrane for Simultaneous Ammonia Selection and Fouling Mitigation in Forward Osmosis'. *Journal of Membrane Science* 584 (March): 9–19. https://doi.org/10.1016/j.memsci.2019.04.063.

Beh, J J, B S Ooi, J K Lim, E P Ng, and H Mustapa. 2020. 'Development of High Water Permeability and Chemically Stable Thin Film Nanocomposite (TFN) Forward Osmosis (FO) Membrane with Poly(Sodium 4-Styrenesulfonate) (PSS)-Coated Zeolitic Imidazolate Framework-8 (ZIF-8) for Produced Water Treatment'. *Journal of Water Process Engineering* 33: 101031. https://doi.org/10.1016/j.jwpe.2019.101031.

Eyvaz, Murat, Serkan Arslan, Derya İmer, Ebubekir Yüksel, and İsmail Koyuncu. 2018. 'Forward Osmosis Membranes – A Review: Part I'. In *Osmotically Driven Membrane Processes - Approach, Development and Current Status.*

Cai, Yufeng, and Xiao Matthew Hu. 2016. 'A Critical Review on Draw Solutes Development for Forward Osmosis'. *Desalination* 391: 16–29. https://doi.org/10.1016/j.desal.2016.03.021.

Chaoui, Imane, Souad Abderafi, Sébastien Vaudreuil, and Tijani Bounahmidi. 2019. 'Water Desalination by Forward Osmosis: Draw Solutes and Recovery Methods – Review'. *Environmental Technology Reviews* 8 (1): 25–46. https://doi.org/10.1080/21622515.2019.1623324.

Chen, Haisheng, Manhong Huang, Zhiwei Wang, Pin Gao, Teng Cai, Jialing Song, Yueyang Zhang, and Lijun Meng. 2020. 'Enhancing Rejection Performance of Tetracycline Resistance Genes by a TiO2/AgNPs-Modified Nanofiber Forward Osmosis Membrane'. *Chemical Engineering Journal* 382: 123052. https://doi.org/10.1016/j.cej.2019.123052.

Chen, Haisheng, Shengyang Zheng, Lijun Meng, Gang Chen, Xubiao Luo, and Manhong Huang. 2020. 'Comparison of Novel Functionalized Nanofiber Forward Osmosis Membranes for Application in Antibacterial Activity and TRGs Rejection'. *Journal of Hazardous Materials* 392: 122250. https://doi.org/10.1016/j.jhazmat.2020.122250.

Chen, Qiaozhen, Qingchun Ge, Wenxuan Xu, and Wenbin Pan. 2019. 'Functionalized Imidazolium Ionic Liquids Promote Seawater Desalination through Forward Osmosis'. *Journal of Membrane Science* 574 (November 2018): 10–16. https://doi.org/10.1016/j.memsci.2018.11.078.

Cheng, Cheng, Xiong Li, Xufeng Yu, Min Wang, and Xuefen Wang. 2018. *Electrospun Nanofibers for Water Treatment. Electrospinning: Nanofabrication and Applications.* Elsevier Inc. https://doi.org/10.1016/B978-0-323-51270-1.00014-5.

Chiao, Yu Hsuan, Arijit Sengupta, Shu Ting Chen, Shu Hsien Huang, Chien Chieh Hu, Wei Song Hung, Yung Chang, et al. 2019. 'Zwitterion Augmented Polyamide Membrane for Improved Forward Osmosis Performance with Significant Antifouling Characteristics'. *Separation and Purification Technology* 212 (September 2018): 316–25. https://doi.org/10.1016/j.seppur.2018.09.079.

Choi, Hyeon Gyu, Aatif Ali Shah, Seung Eun Nam, You In Park, and Hosik Park. 2019. 'Thin-Film Composite Membranes Comprising Ultrathin Hydrophilic Polydopamine Interlayer with Graphene Oxide for Forward Osmosis'. *Desalination* 449 (March 2018): 41–49. https://doi.org/10.1016/j.desal.2018.10.012.

Chowdhury, Maqsud R, Liwei Huang, and Jeffrey R McCutcheon. 2017. 'Thin Film Composite Membranes for Forward Osmosis Supported by Commercial Nanofiber Nonwovens'. *Industrial and Engineering Chemistry Research* 56 (4): 1057–63. https://doi.org/10.1021/acs.iecr.6b04256.

Chun, Youngpil, Dennis Mulcahy, Linda Zou, and In S Kim. 2017. 'A Short Review of Membrane Fouling in Forward Osmosis Processes'. *Membranes* 7 (2): 1–23. https://doi.org/10.3390/membranes7020030.

Cui, Hongtao, Hanmin Zhang, Mingchuan Yu, and Fenglin Yang. 2018. 'Performance Evaluation of Electric-Responsive Hydrogels as Draw Agent in Forward Osmosis Desalination'. *Desalination* 426: 118–26. https://doi.org/10.1016/j.desal.2017.10.045.

Deng, Luyao, Qun Wang, Xiaochan An, Zhuangzhi Li, and Yunxia Hu. 2020. 'Towards Enhanced Antifouling and Flux Performances of Thin-Film Composite Forward Osmosis Membrane via Constructing a Sandwich-like Carbon Nanotubes-Coated Support'. *Desalination* 479: 114311. https://doi.org/10.1016/j.desal.2020.114311.

Dharupaneedi, Suhas P, Sanna Kotrappanavar Nataraj, Mallikarjuna Nadagouda, Kakarla Raghava Reddy, Shyam S Shukla, and Tejraj M Aminabhavi. 2019. 'Membrane-Based Separation of Potential Emerging Pollutants'. *Separation and Purification Technology* 210 (September 2018): 850–66. https://doi.org/10.1016/j.seppur.2018.09.003.

Dutta, Supritam, Pragnesh Dave, and Kaushik Nath. 2020. 'Journal of Water Process Engineering Performance of Low Pressure Nanofiltration Membrane in Forward Osmosis Using Magnesium Chloride as Draw Solute'. *Journal of Water Process Engineering* 33 (November 2019): 101092. https://doi.org/10.1016/j.jwpe.2019.101092.

Dutta, Supritam, and Kaushik Nath. 2018. 'Prospect of Ionic Liquids and Deep Eutectic Solvents as New Generation Draw Solution in Forward Osmosis Process'. *Journal of Water Process Engineering* 21: 163–76. https://doi.org/10.1016/j.jwpe.2017.12.012.

Siddiqui, Farrukh Arsalan, Qianhong She, Anthony G. Fane, and Robert W Field. 2018. 'Exploring the Differences between Forward Osmosis and Reverse Osmosis Fouling'. *Journal of Membrane Science* 565: 241–53.

Fraiture, Charlotte de, and Dennis Wichelns. 2010. 'Satisfying Future Water Demands for Agriculture'. *Agricultural Water Management*. 2010. https://doi.org/10.1016/j.agwat.2009.08.008.

Ge, Qingchun, Mingming Ling, and Tai Shung Chung. 2013. 'Draw Solutions for Forward Osmosis Processes: Developments, Challenges, and Prospects for the Future'. *Journal of Membrane Science* 442: 225–37. https://doi.org/10.1016/j.memsci.2013.03.046.

Ghaemi, Negin, and Zahra Khodakarami. 2019. 'Nano-Biopolymer Effect on Forward Osmosis Performance of Cellulosic Membrane: High Water Flux and Low Reverse Salt'. *Carbohydrate Polymers* 204 (October 2018): 78–88. https://doi.org/10.1016/j.carbpol.2018.10.005.

Giagnorio, Mattia, Francesco Ricceri, and Alberto Tiraferri. 2019. 'Desalination of Brackish Groundwater and Reuse of Wastewater by Forward Osmosis Coupled with Nanofiltration for Draw Solution Recovery'. *Water Research* 153: 134–43. https://doi.org/10.1016/j.watres.2019.01.014.

Goh, Pei Sean, Ahmad Fauzi Ismail, Be Cheer Ng, and Mohd Sohaimi Abdullah. 2019. 'Recent Progresses of Forward Osmosis Membranes Formulation and Design for Wastewater Treatment'. *Water (Switzerland)* 11 (10): 2043.

Guo, Hao, Zhikan Yao, Jianqiang Wang, Zhe Yang, Xiaohua Ma, and Chuyang Y Tang. 2018. 'Polydopamine Coating on a Thin Film Composite Forward Osmosis Membrane for Enhanced Mass Transport and Antifouling Performance'. *Journal of Membrane Science* 551: 234–42. https://doi.org/10.1016/j.memsci.2018.01.043.

Hartanto, Yusak, Maxime Corvilain, Hanne Mariën, Julie Janssen, and Ivo F J Vankelecom. 2020. 'Interfacial Polymerization of Thin-Film Composite Forward Osmosis Membranes Using Ionic Liquids as Organic Reagent Phase'. *Journal of Membrane Science* 601: 117869. https://doi.org/10.1016/j.memsci.2020.117869.

He, Miaolu, Lei Wang, Yongtao Lv, Xudong Wang, Jiani Zhu, Yan Zhang, and Tingting Liu. 2020. 'Novel Polydopamine/Metal Organic Framework Thin Film Nanocomposite Forward Osmosis Membrane for Salt Rejection and Heavy Metal Removal'. *Chemical Engineering Journal* 389: 124452. https://doi.org/10.1016/j.cej.2020.124452.

Hoover, Laura A, William A Phillip, Alberto Tiraferri, Ngai Yin Yip, and Menachem Elimelech. 2011. 'Forward with Osmosis: Emerging Applications for Greater', 9824–30. https://doi.org/10.1021/es202576h.

Huang, Manhong, Lijun Meng, Beibei Li, Feihu Niu, Yan Lv, Qian Deng, and Jin Li. 2019. 'Fabrication of Innovative Forward Osmosis Membranes via Multilayered Interfacial Polymerization on Electrospun Nanofibers'. *Journal of Applied Polymer Science* 136 (12): 1–9. https://doi.org/10.1002/app.47247.

Im, Sung-Ju, Ganghyeon Jeong, Sanghyun Jeong, Jaeweon Cho, and Am Jang. 2020. 'Fouling and Transport of Organic Matter in Cellulose Triacetate Forward-Osmosis Membrane for Wastewater Reuse and Seawater Desalination'. *Chemical Engineering Journal* 384: 123341. https://doi.org/10.1016/j.cej.2019.123341.

Ismail, A F, M Padaki, N Hilal, T Matsuura, and W J Lau. 2015. 'Thin Film Composite Membrane — Recent Development and Future Potential'. *Desalination* 356 (1): 140–48. https://doi.org/10.1016/j.desal.2014.10.042.

Jafarinejad, Shahryar, Hosung Park, Holly Mayton, Sharon L Walker, and Sunny C Jiang. 2019. 'Concentrating Ammonium in Wastewater by Forward Osmosis Using a Surface Modified Nanofiltration Membrane'. *Environmental Science: Water Research and Technology* 5 (2): 246–55. https://doi.org/10.1039/c8ew00690c.

Jang, Jaewon, Insu Park, Sang-Soo Chee, Jun-Ho Song, Yesol Kang, Chulmin Lee, Woong Lee, Moon-Ho Ham, and In S Kim. 2020. 'Graphene Oxide Nanocomposite Membrane Cooperatively Cross-Linked by Monomer and Polymer Overcoming the Trade-off between Flux and Rejection in Forward Osmosis'. *Journal of Membrane Science* 598: 117684. https://doi.org/10.1016/j.memsci.2019.117684.

Jin, Limei, Zhongying Wang, Sunxiang Zheng, and Baoxia Mi. 2018. 'Polyamide-Crosslinked Graphene Oxide Membrane for Forward Osmosis'. *Journal of Membrane Science* 545 (February 2017): 11–18. https://doi.org/10.1016/j.memsci.2017.09.023.

Jin, Soon, Sang-hee Park, Min Gyu, Min Sang, Kiho Park, Seungkwan Hong, Hosik Park, You-in Park, and Jung-hyun Lee. 2019. 'Fabrication of High Performance and Durable Forward Osmosis Membranes Using Mussel-Inspired Polydopamine-Modi Fi Ed Polyethylene Supports'. *Journal of Membrane Science* 584 (March): 89–99. https://doi.org/10.1016/j.memsci.2019.04.074.

Johnson, Daniel James, Wafa Ali Suwaileh, Abdul Wahab Mohammed, and Nidal Hilal. 2018. 'Osmotic's Potential: An Overview of Draw Solutes for Forward Osmosis'. *Desalination* 434 (August 2017): 100–120. https://doi.org/10.1016/j.desal.2017.09.017.

Jones, Edward, Manzoor Qadir, Michelle T H van Vliet, Vladimir Smakhtin, and Seong mu Kang. 2019. 'The State of Desalination and Brine Production: A Global Outlook'. *Science of the Total Environment.* Elsevier B.V. https://doi.org/10.1016/j.scitotenv.2018.12.076.

Jun, Myoung, Ralph Rolly, Ahmed Abdel-wahab, Sherub Phuntsho, and Ho Kyong. 2018. 'Hydrophilic Polyvinyl Alcohol Coating on Hydrophobic Electrospun Nano Fiber Membrane for High Performance Thin Film Composite Forward Osmosis Membrane'. *Desalination* 426 (October 2017): 50–59. https://doi.org/10.1016/j.desal.2017.10.042.

Kang, Hui, Wei Wang, Jie Shi, Zhiwei Xu, Hanming Lv, Xiaoming Qian, Liyan Liu, Miaolei Jing, Fengyan Li, and Jiarong Niu. 2019. 'Interlamination Restrictive Effect of Carbon Nanotubes for Graphene Oxide Forward Osmosis Membrane via Layer by Layer Assembly'. *Applied Surface Science* 465 (September 2018): 1103–6. https://doi.org/10.1016/j.apsusc.2018.09.255.

Kim, David Inhyuk, Gimun Gwak, Min Zhan, and Seungkwan Hong. 2019. 'Sustainable Dewatering of Grapefruit Juice through Forward Osmosis: Improving Membrane Performance, Fouling Control, and Product Quality'. *Journal of Membrane Science.* https://doi.org/10.1016/j.memsci.2019.02.031.

Kim, Geon-Youb, Jun-Won Jeong, Min-Gue Kim, Hyung-Soo Kim, Ji-Hoon Kim, and Hyung-Sook Kim. 2017. 'A Study on Scale Determination of FO-RO Hybrid Process Based on FO Recovery Rate'. *Journal of Coastal Research*, no. 79 (10079) (January): 75–79. https://doi.org/10.2112/SI79-016.1.

Kwon, Daeeun, Soon Jin Kwon, Jeonghwan Kim, and Jung-Hyun Lee. 2020. 'Feasibility of the Highly-Permselective Forward Osmosis Membrane Process for the Post-Treatment of the Anaerobic Fluidized Bed Bioreactor Effluent'. *Desalination* 485: 114451. https://doi.org/10.1016/j.desal.2020.114451.

Kwon, Hyo Eun, Soon Jin Kwon, Sung Joon Park, Min Gyu Shin, Sang Hee Park, Min Sang Park, Hosik Park, and Jung Hyun Lee. 2019. 'High Performance Polyacrylonitrile-Supported Forward Osmosis Membranes Prepared via Aromatic Solvent-Based Interfacial Polymerization'. *Separation and Purification Technology* 212 (November 2018): 449–57. https://doi.org/10.1016/j.seppur.2018.11.053.

Lambrechts, R, and M S Sheldon. 2019. 'Performance and Energy Consumption Evaluation of a Fertiliser Drawn Forward Osmosis (FDFO) System for Water Recovery from Brackish Water'. *Desalination* 456 (November 2018): 64–73. https://doi.org/10.1016/j.desal.2019.01.016.

Lee, Duu-Jong, and Meng-Huan Hsieh. 2019. 'Forward Osmosis Membrane Processes for Wastewater Bioremediation: Research Needs'. *Bioresource Technology* 290: 121795. https://doi.org/10.1016/j.biortech.2019.121795.

Lee, W J, P S Goh, W J Lau, C S Ong, and A F Ismail. 2019. 'Antifouling Zwitterion Embedded Forward Osmosis Thin Film Composite Membrane for Highly Concentrated Oily Wastewater Treatment'. *Separation and Purification Technology* 214 (June 2018): 40–50. https://doi.org/10.1016/j.seppur.2018.07.009.

Li, Dan, Xinyi Zhang, Jianfeng Yao, Yao Zeng, George P Simon, and Huanting Wang. 2011. 'Composite Polymer Hydrogels as Draw Agents in Forward Osmosis and Solar Dewatering'. *Soft Matter* 7 (21): 10048–56. https://doi.org/10.1039/C1SM06043K.

Li, Lan, Xing-peng Liu, and Hui-qiang Li. 2017. 'A Review of Forward Osmosis Membrane Fouling: Types, Research Methods and Future Prospects'. (January). https://doi.org/10.1080/21622515.2016.1278277.

Li, Yu, Guoyuan Pan, Jing Wang, Yang Zhang, Hongwei Shi, Hao Yu, and Yiqun Liu. 2020. 'Tailoring the Polyamide Active Layer of Thin-Film Composite Forward Osmosis Membranes with Combined Cosolvents during Interfacial Polymerization'. *Industrial & Engineering Chemistry Research* 59 (17): 8230–42. https://doi.org/10.1021/acs.iecr.0c00682.

Liu, Qin, Ziwei Chen, Xiaoyuan Pei, Changsheng Guo, Kunyue Teng, Yanli Hu, Zhiwei Xu, and Xiaoming Qian. 2020. 'Review: Applications, Effects and the Prospects for Electrospun Nanofibrous Mats in Membrane Separation'. *Journal of Materials Science* 55 (3): 893–924. https://doi.org/10.1007/s10853-019-04012-7.

Liu, Xiaojing, Jinling Wu, Li an Hou, and Jianlong Wang. 2019. 'Removal of Co, Sr and Cs Ions from Simulated Radioactive Wastewater by Forward Osmosis'. *Chemosphere* 232: 87–95. https://doi.org/10.1016/j.chemosphere.2019.05.210.

Liu, Xiaojing, Jinling Wu, and Jianlong Wang. 2019. 'Removal of Nuclides and Boric Acid from Simulated Radioactive Wastewater by Forward Osmosis'. *Progress in Nuclear Energy* 114 (March): 155–63. https://doi.org/10.1016/j.pnucene.2019.03.014.

Lu, Ang, Fangrui Zhong, and Yan Wang. 2019. 'Chemical Engineering Research and Design Application of Polysaccharide Derivatives as Novel Draw Solutes in Forward Osmosis for Desalination'. *Chemical Engineering Research and Design* 146: 211–20. https://doi.org/10.1016/j.cherd.2019.04.005.

Luo, Huayong, Kelin Wu, Qin Wang, Tian C Zhang, Hanxing Lu, Hongwei Rong, and Qian Fang. 2020. 'Forward Osmosis with Electro-Responsive P(AMPS-Co-AM) Hydrogels as Draw Agents for Desalination'. *Journal of Membrane Science* 593: 117406. https://doi.org/10.1016/j.memsci.2019.117406.

Ma, Jinjin, Tonghu Xiao, Nengbing Long, and Xing Yang. 2020. 'The Role of Polyvinyl Butyral Additive in Forming Desirable Pore Structure for Thin Film Composite Forward Osmosis Membrane'. *Separation and Purification Technology* 242 (March): 116798. https://doi.org/10.1016/j.seppur.2020.116798.

Manickam, Seetha S, and Jeffrey R McCutcheon. 2017. 'Understanding Mass Transfer through Asymmetric Membranes during Forward Osmosis: A Historical Perspective and Critical Review on Measuring Structural Parameter with Semi-Empirical Models and Characterization Approaches'. *Desalination* 421: 110–26. https://doi.org/10.1016/j.desal.2016.12.016.

McCutcheon, Jeffrey R, Robert L McGinnis, and Menachem Elimelech. 2005. 'A Novel Ammonia–Carbon Dioxide Forward (Direct) Osmosis Desalination Process'. *Desalination* 174: 1–11.

McGinnis, Robert L, and Menachem Elimelech. 2007. 'Energy Requirements of Ammonia–Carbon Dioxide Forward Osmosis Desalination'. *Desalination* 207 (1): 370–82. https://doi.org/10.1016/j.desal.2006.08.012.

Miao, Lei, Tingting Jiang, Shudong Lin, Tao Jin, Jiwen Hu, Min Zhang, Yuanyuan Tu, and Guojun Liu. 2020. 'Asymmetric Forward Osmosis Membranes from P-Aramid Nanofibers'. *Materials & Design* 191: 108591. https://doi.org/10.1016/j.matdes.2020.108591.

Misdan, N, W J Lau, and A F Ismail. 2012. 'Seawater Reverse Osmosis (SWRO) Desalination by Thin-Film Composite Membrane-Current Development, Challenges and Future Prospects'. *Desalination* 287: 228–37. https://doi.org/10.1016/j.desal.2011.11.001.

Mishra, Tripti, Sudipta Ramola, Anil Kumar Shankhwar, and R K Srivastava. 2015. 'Use of Synthesized Hydrophilic Magnetic Nanoparticles (HMNPs) in Forward Osmosis for Water Reuse'. *Water Supply* 16 (1): 229–36. https://doi.org/10.2166/ws.2015.131.

Nasr, Peter, and Hani Sewilam. 2018. '13 - Fertilizer Drawn Forward Osmosis for Irrigation'. In, edited by Gnaneswar B T Veera, Emerging Technologies for Sustainable Desalination Handbook Guide, 433–60. Butterworth-Heinemann. https://doi.org/10.1016/B978-0-12-815818-0.00013-8.

Nawi, Normi Izati Mat, Siti Nor Hidayah M Arifin, Shafiq M Hizam, Erdina Lulu Atika Rampun, Muhammad Roil Bilad, Muthia Elma, Asim Laeeq Khan, Yusuf Wibisono, and Juhana Jaafar. 2020. 'Chlorella Vulgaris Broth Harvesting via Standalone Forward Osmosis Using Seawater Draw Solution'. *Bioresource Technology Reports* 9: 100394. https://doi.org/10.1016/j.biteb.2020.100394.

Nguyen, Hau Thi, Shiao-Shing Chen, Nguyen Cong Nguyen, Huu Hao Ngo, Wenshan Guo, and Chi-Wang Li. 2015. 'Exploring an Innovative Surfactant and Phosphate-Based Draw Solution for Forward Osmosis Desalination'. *Journal of Membrane Science* 489: 212–19. https://doi.org/10.1016/j.memsci.2015.03.085.

Nguyen, Hau Thi, Nguyen Cong Nguyen, Shiao-Shing Chen, and Shu-Ying Wu. 2015. 'Concentrate of Surfactant-Based Draw Solutions in Forward Osmosis by Ultrafiltration and Nanofiltration'. *Water Supply* 15 (5): 1133–39. https://doi.org/10.2166/ws.2015.060.

Nguyen, Nguyen Cong, Shiao-Shing Chen, Su-Thing Ho, Hau Thi Nguyen, Saikat Sinha Ray, Nhat Thien Nguyen, Hung-Te Hsu, Ngoc Chung Le, and Thi Tinh Tran. 2018. 'Optimising the Recovery of EDTA-2Na Draw Solution in Forward Osmosis through Direct Contact Membrane Distillation'. *Separation and Purification Technology* 198: 108–12. https://doi.org/10.1016/j.seppur.2017.02.001.

Nguyen, Nguyen Cong, Hung Cong Duong, Hau Thi Nguyen, Shiao-Shing Chen, Huy Quang Le, Huu Hao Ngo, Wenshan Guo, Chinh Cong Duong, Ngoc Chung Le, and Xuan Thanh Bui. 2020a. 'Forward Osmosis–Membrane Distillation Hybrid System for Desalination Using Mixed Trivalent Draw Solution'. *Journal of Membrane Science* 603: 118029. https://doi.org/10.1016/j.memsci.2020.118029.

Nguyen, Nguyen Cong, Hau Thi Nguyen, Su-Thing Ho, Shiao-Shing Chen, Huu Hao Ngo, Wenshan Guo, Saikat Sinha Ray, and Hung-Te Hsu. 2016. 'Exploring High Charge of Phosphate as New Draw Solute in a Forward Osmosis–Membrane Distillation Hybrid System for Concentrating High-Nutrient Sludge'. *Science of the Total Environment* 557–558: 44–50. https://doi.org/10.1016/j.scitotenv.2016.03.025.

Obaid, M., Yesol Kang, Sungrok Wang, Myung Han Yoon, Chang Min Kim, Jun Ho Song, and In S. Kim. 2018. 'Fabrication of Highly Permeable Thin-Film Nanocomposite Forward Osmosis Membranes: Via the Design of Novel Freestanding Robust Nanofiber Substrates'. *Journal of Materials Chemistry A* 6 (25): 11700–713. https://doi.org/10.1039/c7ta11320j.

Park, Myoung Jun, Ralph Rolly Gonzales, Ahmed Abdel-Wahab, Sherub Phuntsho, and Ho Kyong Shon. 2018. 'Hydrophilic Polyvinyl Alcohol Coating on Hydrophobic Electrospun Nanofiber Membrane for High Performance Thin Film Composite Forward Osmosis Membrane'. *Desalination*. https://doi.org/10.1016/j.desal.2017.10.042.

Qasim, Muhammad, Faisal W Khudhur, Ahmed Aidan, and Naif A Darwish. 2020. 'Ultrasonics - Sonochemistry Ultrasound-Assisted Forward Osmosis Desalination Using Inorganic Draw Solutes'. *Ultrasonics - Sonochemistry* 61 (September 2019): 104810. https://doi.org/10.1016/j.ultsonch.2019.104810.

Ramezani Darabi, Rezvaneh, Mohsen Jahanshahi, and Majid Peyravi. 2018. 'A Support Assisted by Photocatalytic Fe3O4/ZnO Nanocomposite for Thin-Film Forward Osmosis Membrane'. *Chemical Engineering Research and Design* 133: 11–25. https://doi.org/10.1016/j.cherd.2018.02.029.

Ramezani Darabi, Rezvaneh, Majid Peyravi, and Mohsen Jahanshahi. 2019. 'Modified Forward Osmosis Membranes by Two Amino-Functionalized ZnO Nanoparticles: A Comparative Study'. *Chemical Engineering Research and Design* 145: 85–98. https://doi.org/10.1016/j.cherd.2019.02.019.

Rastgar, Masoud, Ali Bozorg, Alireza Shakeri, and Mohtada Sadrzadeh. 2019. 'Substantially Improved Antifouling Properties in Electro-Oxidative Graphene Laminate Forward Osmosis Membrane'. *Chemical Engineering Research and Design* 141: 413–24. https://doi.org/10.1016/j.cherd.2018.11.010.

Reimund, Kevin K, Benjamin J Coscia, Jason T Arena, Aaron D Wilson, and Jeffrey R McCutcheon. 2016. 'Characterization and Membrane Stability Study for the Switchable Polarity Solvent N, N-Dimethylcyclohexylamine as a Draw Solute in Forward Osmosis'. *Journal of Membrane Science* 501: 93–99. https://doi.org/10.1016/j.memsci.2015.10.039.

Ryu, Hoyoung, Azeem Mushtaq, Eunhye Park, Kyochan Kim, Yong Keun Chang, and Jong-in Han. n.d. 'Dynamical Modeling of Water Flux in Forward Osmosis with Multistage Operation and Sensitivity Analysis of Model Parameters'. *Water*, 1–20. https://doi.org/10.3390/w12010031

Saeedi-Jurkuyeh, Alireza, Ahmad Jonidi Jafari, Roshanak Rezaei Kalantary, and Ali Esrafili. 2020. 'A Novel Synthetic Thin-Film Nanocomposite Forward Osmosis Membrane Modified by Graphene Oxide and Polyethylene Glycol for Heavy Metals Removal from Aqueous Solutions'. *Reactive and Functional Polymers* 146: 104397. https://doi.org/10.1016/j.reactfunctpolym.2019.104397.

Sahebi, Soleyman, Mohammad Sheikhi, Bahman Ramavandi, Mehdi Ahmadi, Shuaifei Zhao, Adeyemi S Adeleye, Zhara Shabani, and Toraj Mohammadi. 2020. 'Sustainable Management of Saline Oily Wastewater via Forward Osmosis Using Aquaporin Membrane'. *Process Safety and Environmental Protection* 138: 199–207. https://doi.org/10.1016/j.psep.2020.03.013.

Samieirad, Saeed, Seyed Mahmoud Mousavi, and Ehsan Saljoughi. 2020. 'Alignment of Functionalized Multiwalled Carbon Nanotubes in Forward Osmosis Membrane Support Layer Induced by Electric and Magnetic Fields'. *Powder Technology* 364: 538–52. https://doi.org/10.1016/j.powtec.2020.02.017.

Sarai Atab, M, A. J. Smallbone, and A P Roskilly. 2016. 'An Operational and Economic Study of a Reverse Osmosis Desalination System for Potable Water and Land Irrigation'. *Desalination* 397: 174–84. https://doi.org/10.1016/j.desal.2016.06.020.

Sato, Nagahisa, Yuya Sato, and Satoshi Yanase. 2014. 'Forward Osmosis Using Dimethyl Ether as a Draw Solute'. *Desalination* 349: 102–5. https://doi.org/10.1016/j.desal.2014.06.028.

Schneider, Carina, Rajath Sathyadev Rajmohan, Agata Zarebska, Panagiotis Tsapekos, and Claus Hélix-Nielsen. 2019. 'Treating Anaerobic Effluents Using Forward Osmosis for Combined Water Purification and Biogas Production'. *Science of the Total Environment* 647: 1021–30. https://doi.org/10.1016/j.scitotenv.2018.08.036.

Long, Qingwu, Liang Shen, Rongbiao Chen, Jiaqi Huang, Shu Xiong, and Yan Wang. 2016. "Synthesis and Application of Organic Phosphonate Salts as Draw Solutes in Forward Osmosis for Oil–Water Separation." *Environmental Science & Technology* 50 (21): 12022–29. https://doi.org/10.1021/acs.est.6b02953.

Shakeri, Alireza, Reza Razavi, Hasan Salehi, Maryam Fallahi, and Tala Eghbalazar. 2019. 'Thin Film Nanocomposite Forward Osmosis Membrane Embedded with Amine-Functionalized Ordered Mesoporous Silica'. *Applied Surface Science* 481 (February): 811–18. https://doi.org/10.1016/j.apsusc.2019.03.040.

Shakeri, Alireza, Hasan Salehi, Farnaz Ghorbani, Mojtaba Amini, and Hadi Naslhajian. 2019. 'Journal of Colloid and Interface Science Polyoxometalate Based Thin Film Nanocomposite Forward Osmosis Membrane: Superhydrophilic, Anti-Fouling, and High Water Permeable'. *Journal of Colloid and Interface Science* 536: 328–38. https://doi.org/10.1016/j.jcis.2018.10.069.

Shen, Liang, Lian Tian, Jian Zuo, Xuan Zhang, Shipeng Sun, and Yan Wang. 2019. 'Developing High-Performance Thin-Film Composite Forward Osmosis Membranes by Various Tertiary Amine Catalysts for Desalination'. *Advanced Composites and Hybrid Materials* 2 (1): 51–69. https://doi.org/10.1007/s42114-018-0070-1.

Shi, Jie, Hui Kang, Nan Li, Kunyue Teng, Wanying Sun, Zhiwei Xu, Xiaoming Qian, and Qin Liu. 2019. 'Chitosan Sub-Layer Binding and Bridging for Nanofiber-Based Composite Forward Osmosis Membrane'. *Applied Surface Science* 478 (November 2018): 38–48. https://doi.org/10.1016/j.apsusc.2019.01.148.

Shibuya, Masafumi, Myoung Jun Park, Sungil Lim, Sherub Phuntsho, Hideto Matsuyama, and Ho Kyong Shon. 2018. 'Novel CA/PVDF Nanofiber Supports Strategically Designed via Coaxial Electrospinning for High Performance Thin-Film Composite Forward Osmosis Membranes for Desalination'. *Desalination* 445: 63–74. https://doi.org/10.1016/j.desal.2018.07.025.

Shokrollahzadeh, Soheila, and Saharnaz Tajik. 2018. 'Fabrication of Thin Film Composite Forward Osmosis Membrane Using Electrospun Polysulfone/Polyacrylonitrile Blend Nano Fibers as Porous Substrate'. *Desalination* 425 (May 2017): 68–76. https://doi.org/10.1016/j.desal.2017.10.017.

Singh, N, S Dhiman, S Basu, M Balakrishnan, I Petrinic, and C Helix-Nielsen. 2019. 'Dewatering of Sewage for Nutrients and Water Recovery by Forward Osmosis (FO) Using Divalent Draw Solution'. *Journal of Water Process Engineering* 31 (April): 100853. https://doi.org/10.1016/j.jwpe.2019.100853.

Singh, N, I Petrinic, C Hélix-Nielsen, S Basu, and M Balakrishnan. 2018. 'Concentrating Molasses Distillery Wastewater Using Biomimetic Forward Osmosis (FO) Membranes'. *Water Research* 130: 271–80. https://doi.org/10.1016/j.watres.2017.12.006.

Song, Xiaoxiao, Zhaoyang Liu, and Darren Delai Sun. 2011. 'Nano Gives the Answer: Breaking the Bottleneck of Internal Concentration Polarization with a Nanofiber Composite Forward Osmosis Membrane for a High Water Production Rate'. *Advanced Materials* 23 (29): 3256–60. https://doi.org/10.1002/adma.201100510.

Suwaileh, Wafa Ali, Daniel James Johnson, Sarper Sarp, and Nidal Hilal. 2018. 'Advances in Forward Osmosis Membranes: Altering the Sub-Layer Structure via Recent Fabrication and Chemical Modification Approaches'. *Desalination* 436 (November 2017): 176–201. https://doi.org/10.1016/j.desal.2018.01.035.

Suwaileh, Wafa, Nirenkumar Pathak, Hokyong Shon, and Nidal Hilal. 2020. 'Forward Osmosis Membranes and Processes: A Comprehensive Review of Research Trends and Future Outlook'. *Desalination* 485 (April): 114455. https://doi.org/10.1016/j.desal.2020.114455.

Tang, Yuanyuan, Shan Li, Jia Xu, and Congjie Gao. 2020. 'Thin Film Composite Forward Osmosis Membrane with Single-Walled Carbon Nanotubes Interlayer for Alleviating Internal Concentration Polarization'. *Polymers*. https://doi.org/10.3390/polym12020260.

Tian, Enling, Xingzu Wang, Yuntao Zhao, and Yiwei Ren. 2017. 'Middle Support Layer Formation and Structure in Relation to Performance of Three-Tier Thin Film Composite Forward Osmosis Membrane'. *Desalination*. https://doi.org/10.1016/j.desal.2017.02.014.

Tran, Van Huy, Sherub Phuntsho, Hyunwoong Park, Dong Suk Han, and Ho Kyong Shon. 2017. 'Sulfur-Containing Air Pollutants as Draw Solution for Fertilizer Drawn Forward Osmosis Desalination Process for Irrigation Use'. *Desalination* 424 (September): 1–9. https://doi.org/10.1016/j.desal.2017.09.014.

Tsai, Meng-ting, Li-han Chung, Guan-you Lin, Min-chao Chang, Chi-young Lee, and Nyan-hwa Tai. 2020. 'Layered Carbon Nanotube/Polyacrylonitrile Thin-Film Composite Membrane for Forward Osmosis Application'. *Separation and Purification Technology* 241: 116683. https://doi.org/10.1016/j.seppur.2020.116683.

Volpin, F., L. Chekli, S. Phuntsho, N. Ghaffour, J. S. Vrouwenvelder, and Ho Kyong Shon. 2019. 'Optimisation of a Forward Osmosis and Membrane Distillation Hybrid System for the Treatment of Source-Separated Urine'. *Separation and Purification Technology* 212 (November 2018): 368–75. https://doi.org/10.1016/j.seppur.2018.11.003.

Vu, Minh T, Ashley J Ansari, Faisal I Hai, and Long D Nghiem. 2018. 'Performance of a Seawater-Driven Forward Osmosis Process for Pre-Concentrating Digested Sludge Centrate: Organic Enrichment and Membrane Fouling'. *Environmental Science: Water Research and Technology* 4 (7): 1047–56. https://doi.org/10.1039/c8ew00132d.

Wang, Jin, Tingting Xiao, Ruyi Bao, Tao Li, Yanqiang Wang, Dengxin Li, Xuemei Li, and Tao He. 2018. 'Zwitterionic Surface Modification of Forward Osmosis Membranes Using N-Aminoethyl Piperazine Propane Sulfonate for Grey Water Treatment'. *Process Safety and Environmental Protection* 116: 632–39. https://doi.org/10.1016/j.psep.2018.03.029.

Wang, Yue, Mengke Zhang, Yanqiu Liu, Qinqin Xiao, and Shichang Xu. 2016. 'Quantitative Evaluation of Concentration Polarization under Different Operating Conditions for Forward Osmosis Process'. *Desalination* 398: 106–13. https://doi.org/10.1016/j.desal.2016.07.015.

Wu, Xing, Mahdokht Shaibani, Stefan J D Smith, Kristina Konstas, Matthew R Hill, Huanting Wang, Kaisong Zhang, and Zongli Xie. 2018. 'Microporous Carbon from Fullerene Impregnated Porous Aromatic Frameworks for Improving the Desalination Performance of Thin Film Composite Forward Osmosis Membranes'. *Journal of Materials Chemistry A* 6 (24): 11327–36. https://doi.org/10.1039/C8TA01200H.

Xu, Lina, Tingting Yang, Mingdi Li, Jiaqi Chang, and Jia Xu. 2020. 'Thin-Film Nanocomposite Membrane Doped with Carboxylated Covalent Organic Frameworks for Efficient Forward Osmosis Desalination'. *Journal of Membrane Science* 610: 118111. https://doi.org/10.1016/j.memsci.2020.118111.

Zhang, Hanmin, Jianjun Li, Hongtao Cui, Haijun Li, and Fenglin Yang. 2015. 'Forward Osmosis Using Electric-Responsive Polymer Hydrogels as Draw Agents: Influence of Freezing–Thawing Cycles, Voltage, Feed Solutions on Process Performance'. *Chemical Engineering Journal* 259: 814–19. https://doi.org/10.1016/j.cej.2014.08.065.

Zhang, Xinyu, Ming Xie, Zhe Yang, Hao Chen Wu, Chuanjie Fang, Langming Bai, Li Feng Fang, Tomohisa Yoshioka, and Hideto Matsuyama. 2019. 'Antifouling Double-Skinned Forward Osmosis Membranes by Constructing Zwitterionic Brush-Decorated MWCNT Ultrathin Films'. Research-article. *ACS Applied Materials and Interfaces* 11 (21): 19462–71. https://doi.org/10.1021/acsami.9b03259.

Zhang, Yueyang, Tianwei Mu, Manhong Huang, Gang Chen, Teng Cai, Haisheng Chen, Lijun Meng, and Xubiao Luo. 2020. 'Nanofiber Composite Forward Osmosis (NCFO) Membranes for Enhanced Antibiotics Rejection: Fabrication, Performance, Mechanism, and Simulation'. *Journal of Membrane Science* 595: 117425. https://doi.org/10.1016/j.memsci.2019.117425.

Zhao, Dieling, Peng Wang, Qipeng Zhao, Ningping Chen, and Xianmao Lu. 2014. 'Thermoresponsive Copolymer-Based Draw Solution for Seawater Desalination in a Combined Process of Forward Osmosis and Membrane Distillation'. *Desalination* 348: 26–32. https://doi.org/10.1016/j.desal.2014.06.009.

Zhao, Wang, Huiyuan Liu, Yue Liu, Meipeng Jian, Li Gao, Huanting Wang, and Xiwang Zhang. 2018. 'Thin-Film Nanocomposite Forward-Osmosis Membranes on Hydrophilic Microfiltration Support with an Intermediate Layer of Graphene Oxide and Multiwall Carbon Nanotube'. *ACS Applied Materials & Interfaces* 10 (40): 34464–74. https://doi. org/10.1021/acsami.8b10550.

Zhong, Yujiang, Xiaoshuang Feng, Wei Chen, Xinbo Wang, Kuo-Wei Huang, Yves Gnanou, and Zhiping Lai. 2016. 'Using UCST Ionic Liquid as a Draw Solute in Forward Osmosis to Treat High-Salinity Water'. *Environmental Science & Technology* 50 (2): 1039–45. https://doi.org/10.1021/acs.est.5b03747.

Zhou, Aijiao, Huayong Luo, Qin Wang, Lin Chen, Tian C Zhang, and Tao Tao. 2015. 'Magnetic Thermoresponsive Ionic Nanogels as Novel Draw Agents in Forward Osmosis'. *RSC Advances* 5 (20): 15359–65. https://doi.org/10.1039/C4RA12102C.

Zhou, Jin, Heng-Li He, Fei Sun, Yu Su, Hai-Yin Yu, and Jia-Shan Gu. 2020. 'Structural Parameters Reduction in Polyamide Forward Osmosis Membranes via Click Modification of the Polysulfone Support'. *Colloids and Surfaces A: Physicochemical and Engineering Aspects* 585: 124082. https://doi.org/10.1016/j.colsurfa.2019.124082.

Zou, Shiqiang, and Zhen He. 2016. 'Enhancing Wastewater Reuse by Forward Osmosis with Self-Diluted Commercial Fertilizers as Draw Solutes'. *Water Research* 99: 235–43. https://doi.org/10.1016/j.watres.2016.04.067.

———. 2017. 'Electrodialysis Recovery of Reverse-Fluxed Fertilizer Draw Solute during Forward Osmosis Water Treatment'. *Chemical Engineering Journal* 330 (12): 550–58. https://doi.org/10.1016/j.cej.2017.07.181.

Zuo, Min. 2019. 'Preparation of Super-Hydrophilic Polyphenylsulfone Nanofiber Membranes for Water Treatment', *RSC Advances,* 278–86. https://doi.org/10.1039/C8RA06493H.

Part III

Energy-Water Nexus

8 Microbial Desalination Cells: Opportunities and Challenges

J. Jayapriya*

Department of Applied Science and Technology,
Anna University, Chennai, India

CONTENTS

8.1 INTRODUCTION

The quest for clean and sustainable energy generated from renewable sources through new technological processes has become necessary in recent decades. The substantial rise in energy demand is caused by the expeditious dwindling of fossil fuels, and the resulting adverse effects on climate, particularly global warming, have urged us to focus on renewable energy sources. Moreover, the rapid decline in access to clean water due to global warming necessitates sustainable options, such as wastewater recycling. The increasing volume of wastewater from sources, including sewage and industrial effluents, may overwhelm the conventional cost- and energy-intensive wastewater treatment facilities and contaminate our valuable water resources unless alternate energy resources with potential sustainability are widely employed. It is only recently that the idea of utilizing microbial fuel cells (MFCs) for wastewater treatment is being entertained due to their increased power output (Liu and Logan, 2004).

Another issue is the globally increasing water crisis in which desalination plants have gained popularity lately since they serve as alternative freshwater sources. The average salinity of seawater is in the range of 35,000–45,000 mg L^{-1} TDS (Ali et al., 2016) before being desalinated using different techniques, such as reverse osmosis, multi-stage flash distillation, and multi-effect distillation, electrodialysis, electrode ionization, and hybrid technologies. One critical issue associated with desalination is high energy consumption, which accounts for over one-third of the total cost (Semiat, 2008). A microbial desalination cell (MDC) is a modified version of MFC that has emerged as a low-cost solution for wastewater treatment (Liang et al., 2009).

8.2 PRINCIPLE OF MICROBIAL DESALINATION CELL

Microbial fuel cells (MFC) are special fuel cells that use microorganisms as catalysts to oxidize substrates and produce electrical energy. The typical scheme of an MFC is shown in Figure 8.1a. The anodic and cathodic chambers of the MFC are divided by a proton exchange membrane (PEM). Under anaerobic conditions, bacteria in the anodic chamber oxidize organic matter and transfer electrons to the cathode through an external circuit, generating an electric current. Meanwhile, water is produced at the cathode as protons diffuse through the PEM and react with oxygen and electrons. The MFC's power production depends on various factors, such as microbes, proton selective membrane, electrode materials, cathode/anode catalysts, electrode spacing, and system design (He and Minteer, 2005; Jayapriya and Ramamurthy, 2013).

Microbial desalination cells are modified MDCs with a three-chambered reactor. Typically, an anion exchange membrane (AEM) and a cation exchange membrane (CEM) are inserted next to the anode and the cathode of the MFC, respectively, thus creating a water desalination chamber in the middle, as shown in Figure 8.1b. When the bacteria oxidize the substrates, they release electrons and protons into the anolyte, respectively. Protons cannot pass through the AEM, whereas anions (such as Cl$^-$) from the desalination chamber can move to the anode. At the cathode, protons are released from water for the oxygen reduction reaction, while cations (Na$^+$) move from the desalination chamber across the CEM to balance the charge in the cathode compartment. As a result, the salinity of the saltwater in the middle chamber is reduced without any external power source. The advantages of MDC are

- Desalination is achieved without any external power supply.
- Wastewater is turned into a source of renewable energy in the desalination process in MDCs.
- MDCs form an integrated system that simultaneously accomplishes wastewater treatment, desalination, and power generation.

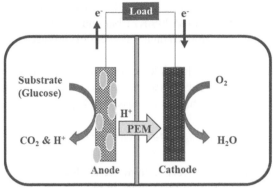

PEM - Proton exchange membrane

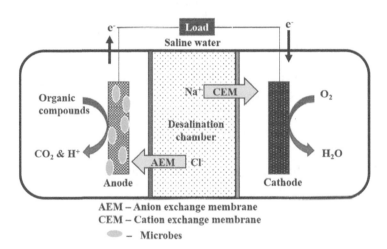

AEM – Anion exchange membrane
CEM – Cation exchange membrane

FIGURE 8.1   Schematic diagram of a) microbial fuel cell b) microbial desalination cell.

8.3   PERFORMANCE ANALYSIS OF MDC

For MFC calculations, the reaction is generally evaluated in terms of the overall cell electromotive force (EMF) (Bard et al., 1985; Reitz, 2007; Robert A. Alberty, 2005), which is calculated as Equation 8.1:

$$E_{emf} = E°_{emf} - RT/nF \ \ln(\gamma) \tag{8.1}$$

where E_{emf} (V) is the potential difference between the cathode and anode and n is the number of electrons transferred per mole. F is the Faraday constant (96481 C mol^{-1}), γ (unitless) is the reaction quotient, which is calculated as the ratio of the product activity to that of the reactants, and $E°_{emf}$ (V) is the electromotive force in the cell under standard conditions (298 K, 1 bar and 1 M). In principle, under open-circuit conditions, the measured voltage (open circuit voltage) must be the same as the equilibrium voltage (E_{emf}).

The actual cell potential is always lower than the ideal potential because of irreversible losses, and the difference between the two values is referred to as the over-voltage (Logan et al., 2006), which is expressed as Equation 8.2:

$$E_{cell} = E_{emf} - \eta_{act} - \eta_{iR} - \eta_{diff} \qquad (8.2)$$

where η_{act} (V) is the activation overpotential due to slow electrode reactions, η_{iR} (V) is the overpotential due to ohmic resistances in the cell, and η_{diff} (V) is the overpotential due to mass diffusion limitations. Cell voltage has a linear relationship with the current and is described simply as $E_{cell} = OCV - R_{Int}$, where R_{Int} is the sum of all the system's internal losses. Under this electric potential gradient, anions move to the anode and cations are driven to the cathode across the ion-exchange membranes, thereby desalinating the water in the middle chamber of the MDC.

The performance of an MDC is evaluated in terms of the following parameters (Ramírez-Moreno et al., 2019)

a. Current density (I_j) is defined as the amount of charge that flows per unit area of a chosen cross-section per unit time

$$I_j \left(mA/cm^2 \right) = I/A_e \qquad (8.3)$$

where I is the electric current (mA), and A_e is the electrode's surface area.

b. Salt removal is defined as the amount of salt removed per desalination cycle.

$$SR \ (\%) = (C_i - C_f)/C_i \qquad (8.4)$$

C_i and C_f are the initial and final molar concentrations of salt (mol/m³) in the MDC. The conductivity of desalinated water should be below 1 mS cm⁻¹.

c. Nominal desalination rate (NDR) is expressed as the volume of desalted water per unit area of the membrane and unit time (Lm⁻² time⁻¹).

$$NDR = V_D / (t_d x \ A_m) \qquad (8.5)$$

V_D is the volume of desalinated water; t_d - time for the desalination process; A_m -membrane area

d. The total circulated charge, Q (Coulomb), is calculated as

$$Q = \int I \ (t)dt \ x \ t_d \qquad (8.6)$$

e. Specific energy (SE) production is defined as the energy produced by the MDC per cubic meter of desalinated water produced.

$$SE = 1/N_D \int E \; cell \; I(t)d \tag{8.7}$$

f. Coulombic efficiency (η_{Cb}) is the ratio of the total electric charge transferred to the anode from the consumed organic substrate and can be calculated as:

$$\eta C_b = \frac{M \int I(t)\,dt}{F.b.V. \; \Delta \; COD} \tag{8.8}$$

where M is the molecular weight of oxygen ($32 \; g \cdot mol^{-1}$), b is the number of electrons exchanged per mole of oxygen ($b = 4$), V is the volume of the anolyte, ΔCOD is the change in chemical oxygen demand (COD) during the experiment ($mg \; O_2 \; L^{-1}$), and F is the Faraday constant ($96,485 \; C \cdot mol^{-1}$).

g. Current efficiency, η_c (%), is defined as the number of ions separated per electron transferred at the electrodes. It is the ratio of the total charge associated with salt removal from the saline compartment to the amount of electric charge transferred (ECT) across the membranes (coulombs $\cdot m^{-3}$) over a complete cycle of the desalination process.

$$\eta_c = \frac{v \; z \; F \; (Ci - Cf)}{ECT} \tag{8.9}$$

$$ECT = \frac{1}{V} \int I(t)\,dt \tag{8.10}$$

where v and z represent the stoichiometric coefficient and the valency of the salt ions, respectively, and F is the Faraday constant ($96,485 \; C \cdot mol^{-1}$).

8.4   FACTORS THAT AFFECT MDC OPTIMIZATION

The performance of MDCs is mainly influenced by pH variations, reactor configuration, electrode material, bacterial species used, catholyte and membranes.

8.4.1   pH

During the desalination process, the protons moving from the anode chamber to the cathode are inhibited by the presence of an AEM in the MDC. Due to these H$^+$ ions in the anode chamber, the pH value may reach as low as 4, resulting in an acidic environment (Jayapriya and Ramamurthy, 2012). This inhibits the growth of microbes, as well as affects the microbial oxidation rate. Similarly, the catholyte pH can become high (more alkaline), leading to a considerable reduction in the cell voltage. Kim and Logan, 2013 reported that for a unit increase in the catholyte pH above neutrality (pH 7.0), a 59 mV reduction in the cell voltage is produced in MFCs. Similarly, Luo and Jenkins, 2011 reported that the pH of the wastewater anolyte dropped to 4.2 and that of the catholyte increased to 11–13 in a single cycle when no additional buffer was added to the anolyte. This pH fluctuation significantly inhibits the bacterial oxidation rate, which in turn affects the desalination efficiency.

8.4.2 REACTOR CONFIGURATION

The basic MDC design was based on a three-chambered device consisting of a middle desalination chamber separated from the anodic and cathodic compartments by selective partitions, namely the AEM and CEM, respectively (Figure 8.1b). To increase the desalination efficiency and reduce the internal resistance in the MDC, different MDC configurations were designed. Some of the factors that were considered in designing the MDCs were (i) the distance between the CEM and AEM, (ii) high ohmic losses and solution resistance due to large electrode spacing, (iii) exposed area of the electrodes and membranes, and (iv) the distance between the membranes. Vertical reactors are preferred over horizontal reactors to reduce the intermembrane distance. Various studies have reported different types of MDCs, such as stacked microbial desalination cells (SMDC), MDCs with recirculation, ion-exchange resin-packed desalination celld, upflow microbial desalination cells (UMDC), capacitive MDC, bipolar membrane MDC, forward osmosis MDC, and separator-coupled stacked MDC with circulation. The construction and working principle of different MDCs are reviewed below, and their characteristics are listed in Table 8.1.

TABLE 8.1
Different types of MDCs and their characteristics

S. No.	Type of MDC	Characteristics
1	Stacked MDC	Increase in the number of ion-exchange membrane (IEM) pairs in the stack leads to increased charge transfer efficiency and desalination rate.
2	MDC with recirculation	Maintaining stable pH in the cell increases power density.
3	Ion exchange resin MDC	Low ohmic resistance and high conductivity of saline water increases the desalination rate.
4	Upflow MDC	• Since agitation is not required, the energy input is low • Availability of more surface area for biofilm adhesion
5	Capacitive MDC	Stabilization of pH prevents salt ion migration across the chambers
6	Bipolar membrane MDC	• BPM leads to notable reduction in pH fluctuation • Produce value-added products such as HCl and NaOH
7	Forward osmosis MDC	• An FO membrane is cheaper than IEM • Improved degradation of organic matter in the anolyte
8	Separator-coupled stacked MDC with recirculation	Inhibition of membrane fouling

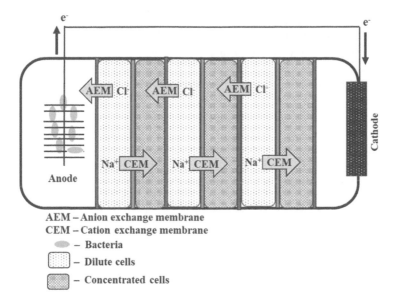

FIGURE 8.2   Schematic diagram of a stacked MDC.

8.4.2.1   Stacked Microbial Desalination Cell (SMDC)

Stacked MDCs consist of an anode chamber, a cathode chamber, and stacked desalination cells. The stacked cells comprise a series of dilute and concentrated chambers compartmentalized by AEMs and CEMs (Figure 8.2). An AEM separates each dilute chamber on the side towards the anode and a CEM on the side towards the cathode. Between two adjacent dilute chambers lies a concentrated chamber, which collects the ions moving out of the desalination chambers. The different chambers and ion exchange membranes are clamped together with gaskets to provide water-sealing between the chambers. Thus, the saline water flows through a series of MDCs to increase desalination efficiency. Chen et al., 2011 developed the first prototype of stacked MDCs with two desalination and one concentrated chamber using two pairs of CEMs and AEMs. A notable 1.4 times increase in the total desalination rate (TDR) was achieved compared with that of a single-desalination-chambered MDC. Kim and Logan, 2011 operated four stacked MDCs in series, each containing five desalination chambers, and reduced the salinity by 44%, producing a total desalination rate (TDR) of 77 mg/h. The power density generated in this MDC was high and ranged from 800 to 1140 mW/m^2, which was much greater than that obtained using a three-chambered MDC (424 mW/m^2).

Choi and Ahn, 2013 studied the effect of the variations in electrode connection and hydraulic flow mode on stacked MDCs' desalination performance. Two MDCs were connected in different modes, such as series in parallel flow, parallel in parallel flow, series in series flow, parallel in series flow, and individual in series flow. The combination of hydraulic flow in series and parallel electrode connection exhibited the highest power density of 420 mWm^{-2}, 44% COD removal, and an hydraulic retention time (HRT) of 0.33 hours.

When more chambers are utilized between the electrodes in a stacked MFC, a pH reduction at the anode can inhibit the microbial growth. In stacked MFCs, the AEM does not allow the diffusion of protons generated in the anolyte, causing an increase in the cathode's pH value that leads to an increase in the overpotential losses (Qu et al., 2012). Zuo et al., 2014 replaced the anolyte and catholyte every 24 hours to maintain a stable pH during the stacked MDC operation, achieving a desalination efficiency of 93.4% in treating low-saline (<10 g/L) water and simultaneously producing a maximum power of 11.8 W/m^3 and the maximum current of 202.1 mA.

The catholyte usually has higher conductivity because of the salinity of the treated water (~70 mS/cm with 35 g/L NaCl). Therefore, the addition of the catholyte to the anolyte helps improve the anolyte solution's conductivity and lowers the internal resistance. Davis et al., 2013 solved the anolyte pH imbalance issue using the non-buffered saline catholyte effluent from the previous cycle in a large-scale stacked MDC with three desalination chambers and four anode brushes. A TDR of 0.1074 g/h and a subsequent decrease in anodic substrate loss from 11% to 2.6% was achieved.

Chen et al., 2011 examined MDCs containing one, two, and three cell pairs (CPs) with an intermembrane distance of 10 mm and found that the salt removal rate increased when more CPs were added in stacked MFCs. However, the voltage generated between the anode and cathode was low and might not drive salt migration through multiple CPs, resulting in a larger salt gradient between the dilute and concentrated chambers, i.e., back diffusion would occur. A more considerable intermembrane distance in the stacked MFC causes a greater internal resistance and a drop in the current generation and desalination efficiency. Furthermore, water loss is an important issue in MDCs containing multiple CPs. Ge et al., 2014 designed a stacked MDC containing 10 pairs of desalination chambers with an intermembrane distance of about 0.5 mm and showed an increase in the charge transfer efficiency by 450% and a TDR 90.8 mg/h. Reduction in the intermembrane distance and the optimization of the number of CPs in stacked MFCs could reduce the hydraulic retention time and thereby water losses, i.e., water transport from the dilute to the concentrated cells due to osmosis (Kim and Logan, 2011; Qu et al., 2013).

8.4.2.2   Microbial Desalination Cell with Recirculation (rMDC)

Recirculating the anolyte and catholyte improves the desalination performance and power density by reducing pH fluctuations besides reducing the cost of buffers that are usually added to stabilize the pH (Figure 8.3). In an rMDC, the anode and cathode chambers' solutions are recirculated between each other through a thin tube and an external pump to avoid drastic pH changes. Anolyte recirculation has been shown to alleviate the inhibition of bacterial activity due to pH and high salinity and further increase the current density of the system from 87.2 to 140 A/m^3, enhancing the desalination rate by 80% and H_2 production by 30% (Luo and Jenkins, 2011). Qu et al., 2012 tested an MDC using 20 g/L NaCl and generated the power densities of 931 ± 29 mW/m^2 and 698 ± 10 mW/m^2 with and without recirculation, respectively. It reduced the salinity by $34 \pm 1\%$ with recirculation (rMDC) and 39 ± 1 without recirculation (MDC). Qu et al., 2013 connected the MDCs in series with a cyclic flow of the anodic solution between the anode and cathode chambers, which increased NaCl removal to 97% at an HRT of two days while 76% removal was achieved in one day.

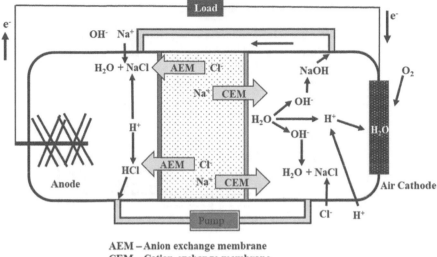

AEM – Anion exchange membrane
CEM – Cation exchange membrane

FIGURE 8.3   Schematic diagram of an MDC with recirculation.

8.4.2.3   Ion-Exchange Resin-Packed MDC

In stacked MDCs, widely spaced electrodes (1 cm) produce high ohmic resistance (18 Ω per membrane pair). Sometimes, the ohmic resistance in the cell decreases when the conductivity of the saline water and salt concentration decrease. This leads to a decline in the generation of electricity and desalination efficiency. MDCs packed with ion-exchange resins (usually a mixture of anion- and cation-exchange resins with high conductivity) are used in desalination chambers to circumvent this issue. Morel et al., 2012 studied the effect of using an ion-exchange resin (IER) in MDCs (Figure 8.4) and observed that the ohmic resistances of rMDCs stabilized at lower values (3.0–4.7 Ω) compared with those of conventional MDCs (5.5–12.7 Ω), leading to increased desalination rate and charge efficiency. However, the calcium and magnesium salts deposited on the resin surfaces may cause scaling, which reduces the coulombic efficiency. Shehab et al., 2014 constructed a microbial electro deionization cell (MEDIC) stack using MDCs packed with IERs to eliminate the spacers between the membranes and enhance water desalination.

8.4.2.4   Upflow Microbial Desalination Cell

An upflow microbial desalination cell (UMDC) is a tubular reactor with two compartments divided by ion-exchange membranes. Jacobson et al., 2011 designed the UMDC shown in Figure 8.5. The surface area for bacterial oxidation in the anode chamber was enhanced by filling with graphite granules, and graphite rods were immersed as electrodes (the anode) to mediate the electron transfer. An AEM separated the anode chamber and the desalination chamber. The desalination chamber was further protected by a CEM tube. A catalyst mixture (Pt/C powder with water) was applied on the outer surface of the CEM (Pt loading rate = 0.2 mg Pt/cm^2) and covered by two layers of carbon cloth, which acted as the cathode. This cathode was connected to the

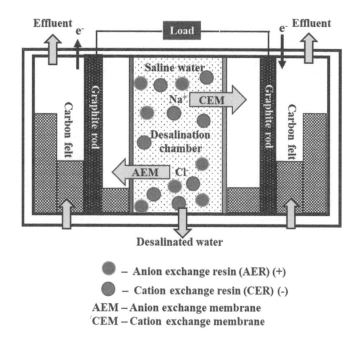

FIGURE 8.4   Schematic diagram of an ion-exchange resin packed MDC.

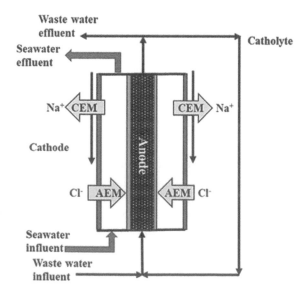

FIGURE 8.5   Schematic diagram of an upflow microbial desalination cell (UMDC).

external electric circuit using a Pt wire. Seawater was introduced from the bottom, and the desalinated water was collected at the top of the unit. The microbes remained suspended and efficiently carried out the oxidation of organic matter. Further, the system did not require mechanical agitation because of the upflow mechanism. This UMDC could remove more than 99% NaCl from a salt solution with an initial concentration of 30 g total dissolved solids (TDS)/L.

8.4.2.5 Capacitive MDC

In conventional MDCs, the Na^+ and Cl^- ions that move from the desalination chamber into the anode and cathode chambers result in increased pH imbalance and TDS levels in the anolyte and catholyte. A capacitive MDC (cMDC) that incorporates the capacitive deionization (CDI) module in the MDC scheme can circumvent this problem (Figure 8.6). A CDI module is composed of double-layer capacitors assembled on electrodes with high surface area, and they adsorb the ions when the saline solution flows between the two charged electrodes. Forrestal et al., 2012 designed a CDI assembly in an MDC using a CEM, a Ni/Cu mesh current collector, and three layers of activated carbon cloth (ACC) together. The salt ions from the desalinated water were adsorbed on the electric double-layer capacitors at the solution-ACC interface, which prevented the transfer of ions to the catholyte. This capacitive assembly stabilized the pH in the cell and improved the desalination process by 7 to 25 times compared with the traditional capacitive deionization processes. When the ACC was saturated with the adsorbed ions, the electrical potential was removed to regenerate the assembly, and the retained salts could be fully recovered in situ for disposal or further recovery.

8.4.2.6 Bipolar Membrane MDC

A bipolar membrane (BPM) is a layered membrane with one surface acting as a cation exchange layer, while the opposite surface acts as an anion exchange layer.

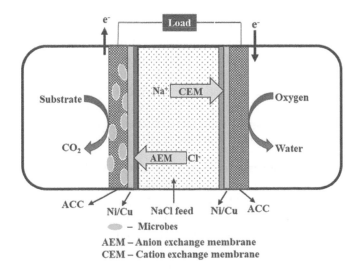

FIGURE 8.6 Schematic diagram of a capacitive MDC.

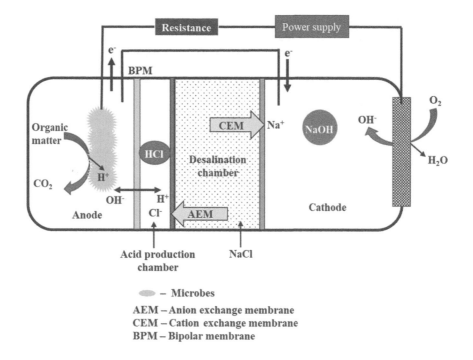

FIGURE 8.7   Schematic diagram of a bipolar membrane MDC.

Water passing through the membrane breaks up into hydroxyl ions and protons, thus developing a high potential gradient. This membrane is usually placed next to the anodic chamber forming four compartments in the cell, as depicted in Figure 8.7. The hydroxyl ions produced at the BPM move towards the anode, where the oxidation of organic matter occurs. The hydrogen ions flow into the acid production chamber, forming HCl and NaOH and thereby stabilizing the pH in the anode chamber. There are two limitations of a bipolar membrane MDC: (I) external power supply is required for water splitting and (ii) the BPM is exposed to chemical fouling because of the organic matter in the anolyte. Chen et al., 2012a reported that a salinity removal efficiency of 86% was achieved at an applied voltage of 1 V, while only ~50% removal was achieved at 0.3 V. On removing the external voltage, salinity removal dropped to ~5%, indicating that the water-splitting reaction at the bipolar membrane played a critical role in the desalination of water (Kim and Logan, 2013).

8.4.2.7   Forward Osmosis Microbial Desalination Cell

In osmosis, water moves from the high-concentration region to the low-concentration region through a semipermeable membrane. Zhang and He, 2013 hydraulically coupled an osmotic microbial fuel cell (OsMFC) with a forward-osmosis (FO) membrane to an MDC with ion-exchange membranes (Figure 8.8). In this system, water moved from the anode to the cathode chamber of the OsMFC, which diluted the salinity in the catholyte. Then, the diluted saline water was passed into

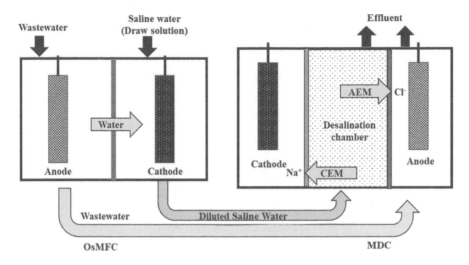

FIGURE 8.8   Schematic diagram of an OsMFC coupled with MDC.

the desalination chamber in the conventional MDC and the wastewater from the anode chamber (OsMFC) to the anode chamber of the MDC. The coupled system significantly improved the desalination efficiency to 95% and produced energy of 0.160 kWh/m^3. Zhang et al., 2011 designed an MFC with an FO membrane that split the MFC into anode and cathode chambers. The two chambers were filled with wastewater and saline water, respectively. The water flux through the FO membrane also transported the ions (e.g., protons) from the anode to the cathode and diluted the saline water present as the catholyte (Figure 8.9).

8.4.2.8   Separator-Coupled Stacked Microbial Desalination Cell with Circulation

This configuration integrates a separator, recirculation, and stacking in an MDC to prevent pH imbalances in the anolyte and catholyte without buffer solutions. Chen et al., 2012b designed an SMDC (Figure 8.10) by attaching a piece of glass fiber

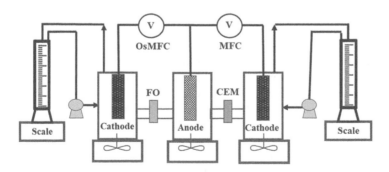

FIGURE 8.9   Schematic diagram of a three-compartment OsMFC coupled with MDC.

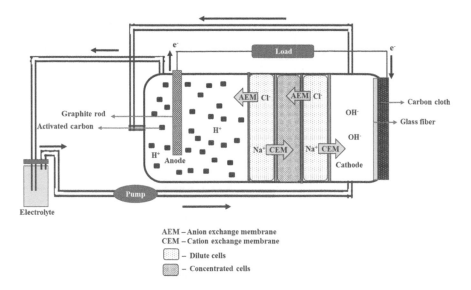

FIGURE 8.10   Schematic diagram of a separator-coupled stacked MDC with circulation.

to the water-facing side of the air cathode to act as a separator. The main role of the separator was to avoid fouling on the air cathodes. Thus, the separator-coupled SMDC with circulation achieved desalination ratios of 37-65% better COD removal than regular SMDCs and higher CE values than SMDCs with circulation but without a separator.

8.4.3   CATHOLYTE

A cathodic electron acceptor with a higher redox potential can reduce the cathode activation overpotential. In general, oxygen acts as the terminal electron acceptor in the reduction reaction at the cathode in MFCs since it is virtually inexhaustible and has a high standard potential value ($O_2 + 2H_2O + 4e^- \rightarrow 4 OH^-$, $E'^\circ = 0.815$ V, pH = 7). In acidic solutions, the oxygen reduction reaction (ORR) at the cathode occurs via two different processes (Kong et al., 2010): 1) a four-electron pathway: $O_2 + 4H+ + 4e- \rightarrow 2H_2O$ ($E'^\circ = 0.816$V) 2) and a two-electron pathway: $O_2 + 2H^+ + 2e- \rightarrow H_2O_2$ ($E'^\circ = 0.295$ V). Unfortunately, though the two-electron pathway is thermodynamically favorable, hydrogen peroxide produced during this reaction is poisonous for the microorganisms and electrodes and causes a decline in the MFC performance. Therefore, promoting the four-electron reaction by introducing a catalyst at the cathode is crucial. However, the development of air cathodes with long-term stability at different pH values and low-cost metal catalysts for ORR is challenging.

One of the major limitations of MDCs is the low available potential for desalination when an air cathode is used for O_2 reduction. When acetate was used as the anolyte substrate, ORR gave a potential difference of up to 1.1 V in the open-circuit condition ($E_{MDC} = E_c -E_a = 0.81$ V-(-0.3 V) = 1.11 V). However, the available voltage for desalination was only 0.5–0.6 V because of the electrode overpotential losses (Wei et al., 2012). When ferri-ferrocyanide reduction was used in the cathode compartment, the $E_{MDC} = E_c - E_a = 0.36$ V- (−0.3 V) = 0.66 V. Liang et al., 2009 proved that ferricyanide was a good catholyte for MDCs as it resulted in salt removal up to 94% and generated a higher power density of 2 W/m² compared with those achieved using ORR as the cathodic reaction.

Ramírez-Moreno et al., 2019 tested potassium ferricyanide as the catholyte in MDCs that was used to desalinate brackish water and seawater and was compared to air cathodes. A desalination efficiency above 93% was achieved for brackish and seawater when air cathode/ferro-ferricyanide redox was used in the MDC. However, only partial desalination of seawater was achieved when oxygen acted as the electron acceptor, and 98% removal was achieved with the Fe^{3+}/Fe^{2+} redox couple. These results summarize that the partial desalination is due to (i) the low potential available for the air cathodes to drive the migration of ions and (ii) the migration of ions from the middle desalination chamber to the anode/cathode chamber during the desalination process. This increases the electrical conductivity in the anode and cathode chambers and the concentration gradient drops due to back diffusion (Ping et al., 2016), which hinders the desalination process further. However, due to the irreversibility of Fe^{3+}/Fe^{2+}, ferricyanide should be replenished, and hydrogen cyanide is formed when the pH drops to below 4, which is a very toxic class 4 health hazard.

Sodium percarbonate has also been tested as a catholyte in MDCs (Forrestal et al., 2014; Puspitarini et al., 2019). The molecular structure of sodium percarbonate showed hydrogen peroxide bound to sodium carbonate. The addition of sodium percarbonate led to the quick release of hydrogen peroxide into the solution, according to Eq. (8.11):

$$2Na_2CO_3 . 3H_2O_2 \rightarrow 2Na_2CO_3 + 3H_2O_2 \qquad (8.11)$$

The hydrogen peroxide then reacted with the cathode and received the electrons transferred from the anode, producing water and sodium carbonate species, which also acted as a pH buffer in the cathode compartments. Puspitarini et al., 2019 assessed different concentrations of sodium percarbonate (NP) as catholytes (0.05M, 0.1M, 0.15M, and 0.2M) in the desalination process. An increase in the catholyte concentration increased the salt removal rate from the desalination chamber. The salt removal was high at 0.15 M (NP), but it dropped low at 0.2 M, and it was inferred that back diffusion occurred during the desalination cycle at such a high concentration of the electrolyte. The percarbonate cathode resulted in higher pH values due to conversion to carbonate, which has a pKa value of 10.3. Therefore, the pH stabilized around this value when it was formed, and this catholyte served as a natural antifoulant for the cathode (Zhang et al., 2011). Other catholytes, such as acidified water and phosphate buffer solution, have also been examined in MDCs (Chen et al., 2011; Jacobson et al., 2011).

Hypochlorite serves as an excellent cathodic electron acceptor because of its standard redox potential ($E^0 = 2.01$ V vs. SHE). Pradhan and Ghangrekar, 2019 evaluated hypochlorite as a catholyte for MDCs using saline water with a TDS concentration of 30 g/L and found that the TDS removal rate and power production were 89±2%

and 347 mW/m², respectively. The high redox potential of hypochlorite favored a superior desalination performance to other electron acceptors, such as oxygen (72%), permanganate (78%), and dichromate (82%).

8.4.4 BIOCATALYSTS

Usually, electron transfer between electron carriers in the bacterial transport chain and the anode/cathode is low, which leads to high internal resistance, thus reducing the power generation in MFCs. The current research focus is primarily on the dynamics of the electrochemically active bacterial communities (EAB) in biofilms and the mechanism underlying the role of bacterial species in extracellular electron transfer. EABs are also known as exoelectrogens, electricigens, and anode-respiring or anodophilic species (Marsili et al., 2010). Exoelectrogens are microorganisms that oxidize organic matter and produce electrical energy by transferring the electrons to an external electron acceptor (an anode that lies outside the cells). The exoelectrogenic bacteria transfer electrons to the anode in the anodic chamber via three main pathways, as described below.

i. Using electron shuttles

Electron shuttles are low-molecular-weight organic molecules that participate in redox reactions. Bacterial cells utilize both self-produced (endogenous) shuttle compounds and those from external sources (exogenous) for extracellular electron transport. Some bacterial species are known to produce their own electron shuttles. For example, *P. aeruginosa* produces pyocyanin and several other electron shuttling compounds (Jayapriya and Ramamurthy, 2012; Rabaey et al., 2005). Von Canstein et al., 2008 observed that *Shewanella oneidensis* MR-1 released riboflavin and flavin mononucleotide (FMN) and used them as electron shuttles during the reduction of Fe (III) oxides.

ii. Direct contact of outer surface c-type cytochromes

Microbial outer-membrane-bound proteins, such as multi-heme proteins (c type cytochromes), shuttle the electrons directly from the inside of the cell to a solid electron acceptor. Several isolates of *Shewanella putrefaciens* (Kim et al., 2002), *Geobacter metallireducens* (Bond and Lovley, 2003), *Rhodoferax ferrireducens* (Chaudhuri and Lovley, 2003) have been shown to generate electricity in mediator-less MFC systems.

iii. Electrically conductive pili or nanowires

Several bacteria produce electrically conductive appendages, which are referred to as bacterial nanowires or microbial nanowires. These nanowires measure 3–5 nm in width and up to tens of micrometers in length and facilitate long-range electron transfer across thick biofilms. Most notably, those found in the phototrophic cyanobacterium *Synechocystis* PCC6803, the thermophilic fermentative bacterium *Pelotomaculum thermopropionicum,* and the *Shewanella* genera have been shown to enhance the power output of MFCs (Gorby et al., 2006). These electrochemically active mixed cultures are sourced either from aquatic sediments (marine or freshwater) or activated sludge in wastewater treatment plants. MDCs that involve mixed bacterial cultures have shown remarkable advantages over MFCs driven by axenic cultures. These include better resistance against process disturbances, higher substrate consumption rates, wider substrate specificity, and improved power output (Rabaey et al., 2004). The microorganisms used in MDCs are reviewed below in Table 8.2.

8.4.5 ELECTRODES

Among the factors known to influence power generation in MDCs, the electrodes' composition is likely the most important. The enhancement of electrode performance in microbial fuel cells can be achieved mainly by reducing the internal resistance, while the material cost should also be considered before large-scale implementation of the MDC technology. The anodic materials must have a non-corrosive scalable and highly porous conductive surface with enhanced active surface area and redox behavior conducive to microbial metabolism and biofilm formation. A wide range of materials from non-corrosive stainless steel to versatile carbons in different forms and shapes have been explored as anodes (Logan et al., 2006). Carbon electrodes are widely used in MFC/MDCs since they have unique advantages, such as a wide potential window, low residual current, long-term stability, excellent biocompatibility, high electrical conductivity, relatively high chemical stability, high corrosion resistance, low density, low thermal expansion and low cost (Kalcher et al., 2006; Uslu and Ozkan, 2007). Both homogeneous forms, such as glassy carbon, vitreous carbon, screen-printed carbon, graphite, fullerenes, carbon brushes, carbon fiber, carbon nanotubes, and heterogeneous forms such as carbon-paste electrodes and carbon-fiber-reinforced sheets, have been investigated for use as electrodes (Whittaker and Kintner, 1969).

Cathode materials also influence the power capacity of MFCs greatly. They should have a high redox potential to accept the electrons. Some commonly used cathode materials include graphite, carbon paper and carbon cloth. However, they generally require metal catalysts to achieve high cathodic potentials. Since precious metal catalysts are expensive, non-precious metals linked with organics, such as Co-tetramethylphenylporphyrin (CoTMPP) and iron (II) phthalocyanine (FePc) are employed (Cheng et al., 2006; Zhao et al., 2005). These non-precious metal catalysts have exhibited excellent performance; however, performance similar to Pt could be achieved only with higher loading rates (Freguia et al., 2007).

Electrode modification is an effective approach to enhance the surface area for better biofilm adhesion (anode) and catalytic activity for reduction (cathode), which, in turn, improves MFC performance. Carbon, being one of the highly allotropic materials, presents various physicochemical properties, electrochemical behavior, and structural characteristics. Hence, properties that are suitable for MDCs can be engineered by modifying their surface. Chemical modification methods can effectively immobilize metals, metal oxides, and other active compounds on carriers, such as carbon materials or conductive polymers, and enhance the output power of MFCs (Jayapriya et al., 2012; Narayanasamy and Jayaprakash, 2018). These modification methods include soaking, chemical vapor deposition, sintering, and carburization. Different combinations of electrode materials that have been used in MDCs so far are given in Table 8.2.

TABLE 8.2

Summary of different exoelectrogens and commercial electrodes used in MDCs

Exoelectrogens	Type of MDC	Desalination efficiency	Power output	Substrate	Anode	Cathode	Ref.
Aerobic sludge from wastewater treatment plant	Conventional MDC	60.3	7.5 mW/m²	Sodium acetate	Graphite rod	Graphite rod	Jegathambal et al., 2019
Bacillus subtilis moh3	Conventional MDC	62.2 ± 0.4%	0.15 ± 0.05 W/m³	0.1% yeast extract with malachite green dye	Carbon cloth	Carbon cloth	Kalleary et al., 2014
Bacillus subtilis moh3	Conventional MDC	57.6 ± 0.2%	0.14 ± 0.03 W/m³	0.1% yeast extract with sunset yellow	Carbon cloth	Carbon cloth	Kalleary et al., 2014
Pseudomonas putida and activated sludge	Multistage MDC	68 ± 1.5%	10.2 mW/m²	Steel plant wastewater	Carbon felt	Carbon felt	Shinde et al., 2018
Biofilm with predominantly *Actinobacteria*	Conventional MDC	66%	8.01 W/m³	Municipal wastewater	Heat treated graphite brushes	Carbon cloth	Luo et al., 2012
Biofilm with predominantly *Proteobacteria*	Conventional MDC	66–69%	3.6 W/m³	Domestic wastewater	Heat treated graphite brushes	Carbon cloth	Luo and Jenkins, 2011
Acetate-fed MFC	Stacked MDC	97.6%	800–1140 mW/m²	Sodium acetate	Graphite fiber brush	Air cathode with Pt catalysts	Kim and Logan, 2011
Aerobic and anaerobic sludge from the South Shore Water Reclamation Facility	Osmosis MDC	95.9	0.160 kWh/m³	Sodium acetate	Carbon brush	Carbon cloth	Zhang and He, 2013

Primary clarifier effluent	SMEDIC	61–72%		Sodium acetate	Graphite fiber brush	Air cathode with Pt catalysts	Shehab et al., 2014
Klebsiella ornithinolytica, Propionibacterium sp. (9.7%) and *Bacteroides rodentium* (6.5%) and *Geobacter* species	Multiple MDCs in series with recirculation	76–97%	712 ± 32–860 ± 11 mW m^{-2}	Xylose	Graphite fiber brush	Air cathode with Pt catalysts	Qu et al., 2013
Aerobic and anaerobic sludge from ETP	Upflow MDC	99%	30.8 w/m^3	Sodium acetate	Graphite granules	Carbon cloth	Jacobson et al., 2011
Anaerobic sludge from ETP	Capacitive MDC	$25.2 + 3.6$	-	Sodium acetate	Graphite fiber brush	Air cathode with Pt catalysts	Forrestal et al., 2012
Anaerobic consortia collected from septic tank bottom	Five-chambered MDC	$72.6 \pm 3.0\%$	62.3 mW/m^2	Synthetic wastewater	Carbon felt	Carbon felt/ Silver-based SnO$_2$ composite	Anusha et al., 2018

Notably, carbon cloth, graphite fiber brush, and carbon felt have been tested as anodes in MDCs. Carbon cloth is a fabric with significant porosity woven out of extremely thin carbon fibers. A graphite fiber brush consists of graphite fiber coiled around one or more conductive corrosion-resistant metal, such as titanium wires. This material is desirable for its high surface area and low electrode resistance. However, a limitation of the brush architecture is that the brush size constrains the minimum electrode spacing, leading to high ohmic resistance. Therefore, air cathodes are mainly used in MDCs. Conducting materials, such as carbon cloth, carbon paper, and stainless-steel mesh, are used as the base for cathode preparation, which usually comprises a gas diffusion layer on the air-facing side and a catalyst layer on the water-facing side with PTFE as the binder. The major limitations of air cathodes are (i) slow redox kinetics at ambient conditions and (ii) the high cost of metal catalysts used for oxygen reduction.

Biocathodes (biological cathodes) use microorganisms as electron acceptors (biocatalysts), thereby avoiding the use of noble or non-noble metals in oxygen reduction. Various microbial consortia, such as nitrifying and denitrifying bacteria and algae, can be applied as biocatalysts (He and Angenent, 2009). Zamanpour et al., 2017 developed an MDC with the microalgae *Chlorella vulgaris*, desalinated saline water at a 35 g/L concentration and achieved a 20.25 mW/m² power density and 0.341 g/L/d salinity removal, besides remarkably high algal growth (38%).

8.4.6 MEMBRANES

Ion-exchange membranes (IEMs) are of two types based on their charged functional groups.

i. Cation exchange membranes (CEMs) have negative-charged functional groups and allow cations to pass through them but exclude anions. E.g., sulfonic acid ($-SO_3^-$) and carboxylic acid ($-COO^-$).
ii. Anion exchange membranes (AEMs) have positive-charged functional groups and therefore allow the passage of anions but prohibit cations. E.g., ammonium ($-NH_3^+$), secondary amine ($-NRH_2^+$), tertiary amine ($-NR_2H^+$), and quaternary amine ($-NR_3^+$) (Mei and Tang, 2018).

The number of fixed charged groups in the matrix per unit weight of dry membrane decides the ion-exchange capacity (IEC) of the membrane. Membrane ionic resistance is the capacity of the membrane to prevent the passage of the ionic current. Therefore, when the ionic resistance increases, the ability to conduct ions would decrease in the MDC. The water content or the extent of membrane absorption under a given solution condition is termed as swelling degree (SD). The SD of a membrane varies with its structure and hydrophilicity and the anolyte and catholyte properties. If the membrane has high SD, its ionic resistance is usually low, especially in AEMs (Długoł et al., 2008) and affects the membrane's permselectivity. The most critical membrane characteristics required for use in MDCs are (i) durability in harsh pH conditions, (ii) sufficient mechanical strength to endure the osmotic pressure, as well as the hydraulic pressure caused by the water flow, (iii) resistance to the adsorption of organic materials from wastewater, as well as biofouling of the membrane, and (iv) high membrane

TABLE 8.3
Commercially available membranes tested in MDCs

AEM	CEM
• DF120/Tianwei Membrane	• CEM, Ultrex CMI-7000
• AMI-7001S/Membranes International	• Selemion CMV
• AMV/Asahi glass, Japan	• Assembly: Ni or Cu/ACC/CEM (Ni or Cu mesh current/ activated carbon cloth/CMXSB, Astom Corporation, Japan)

permselectivity and low ionic resistance that drive power generation in MDCs. Some of the commercial AEMs and SEMs used in MDCs are listed in Table 8.3.

8.5 SUSTAINABILITY OF MDC AT LARGE SCALE

The maximum capacity of MDC tested so far is 105 L by Zhang and He, 2015, and the system generated 2000 mA of electricity with a salt removal rate of 9.2 kg m^{-3} d^{-1}. These studies suggest that MDC technology is effective in the dual duty of desalination of industrial effluent/seawater and simultaneous power production. However, the volume of MDCs is too small to meet the industrial requirements and contribute to the rising population's energy needs. Hence, significant technical enhancements are needed to scale up the process to industry standards in a cost-effective way. The most important subset of the parameters that influence MDC performance on a large scale is system design, characteristics of the electrodes and membrane, and microorganisms' roles. Tackling the pH imbalance, membrane fouling, electrode stability, and cost of ORR catalysts used in industrial applications is highly challenging. Only 20–30% of the energy is converted into power, and the rest diverted to (i) microbial synthesis, (ii) microbial metabolism, (iii) activation losses at the anode/cathode, (iv) ohmic losses/ mass transfer resistance and (iv) catholyte regeneration. Hence, the MDC is not economically viable when used as a stand-alone process. Nevertheless, when integrated as a pre- or post-treatment step with other processes employed for the desalination of wastewater/seawater, it can substantially reduce the capital cost and energy consumption of the overall wastewater treatment process.

REFERENCES

Ali, H.M., Gadallah, H., Ali, S.S., Sabry, R., Gadallah, A.G., 2016. Pilot-scale investigation of forward/reverse osmosis hybrid system for seawater desalination using impaired water from steel industry. Int. J. Chem. Eng. 2016, 8745943, 1–9.

Anusha, G., Noori, T., Ghangrekar, M.M., 2018. Application of silver-tin dioxide composite cathode catalyst for enhancing performance of microbial desalination cell. Mater. Sci. Energy Technol. 1, 188–195.

Bard, A.J., Parsons, R., Jordan, J., 1985. Standard Potentials in Aqueous Solution, Monographs in Electroanalytical Chemistry and Electrochemistry. Taylor & Francis, New York.

Bond, D.R., Lovley, D.R., 2003. Electricity production by Geobacter sulfurreducens attached to electrodes. Appl. Environ. Microbiol. 69, 1548–1555.

Chaudhuri, S.K., Lovley, D.R., 2003. Electricity generation by direct oxidation of glucose in mediatorless microbial fuel cells. Nat. Biotechnol. 21, 1229–1232.

Chen, S., Liu, G., Zhang, R., Qin, B., Luo, Y., 2012. Development of the microbial electrolysis desalination and chemical-production cell for desalination as well as acid and alkali productions.

Chen, X., Liang, P., Wei, Z., Zhang, X., Huang, X., 2012. Sustainable water desalination and electricity generation in a separator coupled stacked microbial desalination cell with buffer free electrolyte circulation. Bioresour. Technol. 119, 88–93.

Chen, X., Xia, X., Liang, P., Cao, X., Sun, H., Huang, X., 2011. stacked microbial desalination cells to enhance water desalination efficiency 2465–2470.

Cheng, S., Liu, H., Logan, B.E., 2006. Power densities using different cathode catalysts (Pt and CoTMPP) and polymer binders (Nafion and PTFE) in single chamber microbial fuel cells. Environ. Sci. Technol. 40, 364–369.

Choi, J., Ahn, Y., 2013. Continuous electricity generation in stacked air cathode microbial fuel cell treating domestic wastewater. J. Environ. Manage. 130, 146–152.

Davis, R.J., Kim, Y., Logan, B.E., 2013. Increasing desalination by mitigating anolyte pH imbalance using catholyte effluent addition in a multi-anode, bench scale microbial desalination cell Increasing desalination by mitigating anolyte pH imbalance using catholyte effluent addition in a multi-anode, bench scale microbial desalination cell present address. Department of Civil Engineering, McMaster University, Hamilton, Canada.

Długoł, P., Nymeijer, K., Metz, S., Wessling, M., 2008. Current status of ion exchange membranes for power generation from salinity gradients J. Memb. Sci. 319, 214–222.

Forrestal, C., Huang, Z., Ren, Z.J., 2014. Percarbonate as a naturally buffering catholyte for microbial fuel cells bioresource technology percarbonate as a naturally buffering catholyte for microbial fuel cells. Bioresour. Technol. 172, 429–432.

Forrestal, C., Pei xu, Ren, Z., 2012. Sustainable desalination using a microbial capacitive desalination cell. Energy Environ. Sci. 5, 7161–7167.

Freguia, S., Rabaey, K., Yuan, Z., Keller, J., 2007. Non-catalyzed cathodic oxygen reduction at graphite granules in microbial fuel cells. Electrochim. Acta 53, 598–603.

Ge, Z., Dosoretz, C.G., He, Z., 2014. Effects of number of cell pairs on the performance of microbial desalination cells. Desalination 341, 101–106.

Gorby, Y.A., Yanina, S., McLean, J.S., Rosso, K.M., Moyles, D., Dohnalkova, A., Beveridge, T.J., Chang, I.S., Kim, B.H., Kim, K.S., Culley, D.E., Reed, S.B., Romine, M.F., Saffarini, D.A., Hill, E.A., Shi, L., Elias, D.A., Kennedy, D.W., Pinchuk, G., Watanabe, K., Ishii, S., Logan, B., Nealson, K.H., Fredrickson, J.K., 2006. Electrically conductive bacterial nanowires produced by Shewanella oneidensis strain MR-1 and other microorganisms. Proc. Natl. Acad. Sci. 103, 11358–11363.

He, Z., Angenent, L.T., 2006. Application of bacterial biocathodes in microbial fuel cells Electroanalysis. 18, 2009–2015.

He, Z., Minteer, S.D., 2005. Electricity generation from artificial wastewater using an upflow microbial fuel cell Environ. Sci. Technol. 39, 5262–5267.

Jacobson, K.S., Drew, D.M., He, Z., 2011. Efficient salt removal in a continuously operated upflow microbial desalination cell with an air cathode Bioresour. Technol. 102, 376–380.

Jayapriya, J., Gopal, J., Ramamurthy, V., Kamachi Mudali, U., Raj, B., 2012. Preparation and characterization of biocompatible carbon electrodes. Compos. Part B Eng. 43, 1329–1335.

Jayapriya, J., Ramamurthy, V., 2013. The role of electrode material in capturing power generated in pseudomonas catalysed fuel cells Can. J. Chem. Eng. 92, 610–614.

Jayapriya, J., Ramamurthy, V., 2012. Use of non-native phenazines to improve the performance of Pseudomonas aeruginosa MTCC 2474 catalysed fuel cells. Bioresour. Technol. 124, 23–28.

Jegathambal, P., Nisha, R., Parameswari, K., Subathra, M., 2019. Desalination and removal of organic pollutants using electrobiochemical reactor. Appl. Water Sci. 9, 1–10.

Kalcher, K., Svancara, I., Metelka, R., Vytras, K., Walcarius, A., 2006. Heterogeneous carbon electrochemical sensors. Encycl. Sensors 4, 283–430.

Kalleary, S., Abbas, F. Mohammed, Ganesan, A., Meenatchisundaram, S., Srinivasan, B., Packirisamy, A.S.B., Krishnan Kesavan, R., Muthusamy, S., 2014. Biodegradation and bioelectricity generation by microbial desalination cell. Int. Biodeterior. Biodegradation 92, 20–25.

Kim, H.J., Park, H.S., Hyun, M.S., Chang, I.S., Kim, M., Kim, B.H., 2002. A mediator-less microbial fuel cell using a metal reducing bacterium, Shewanella putrefaciens. Enzyme Microb. Technol. 30, 145–152.

Kim, Y., Logan, B.E., 2013. Microbial desalination cells for energy production and desalination. Desalination 308, 122–130.

Kim, Y., Logan, B.E., 2011. Series assembly of microbial desalination cells containing stacked electrodialysis cells for partial or complete seawater desalination Sci. Technol. 45, 5840–5845.

Kong, X., Sun, Y., Yuan, Z., Li, D., Li, L., Li, Y., 2010. Effect of cathode electron-receiver on the performance of microbial fuel cells. Int. J. Hydrogen Energy 35, 7224–7227.

Liang, P., Xiao, K., Zhou, Y., Zhang, X., Logan, B.E., 2009. A new method for water desalination using microbial desalination cells Environ. Sci. Technol. 43, 7148–7152.

Liu, H., Logan, B.E., 2004. Electricity generation using an air cathode single chamber microbial fuel cell in the presence and absence of a proton exchange membrane Environ. Sci. Technol. 38, 4040–4046.

Logan, B.E., Hamelers, B., Rozendal, R., Schröder, U., Keller, J., Freguia, S., Aelterman, P., Verstraete, W., Rabaey, K., 2006. Microbial fuel cells: Methodology and technology. Environ. Sci. Technol. 40, 5181–5192.

Luo, H., Jenkins, P.E., 2011. Concurrent desalination and hydrogen generation using microbial electrolysis and desalination cells Environ. Sci. Technol. 45, 340–344.

Luo, H., Xu, P., Ren, Z., 2012. Long-term performance and characterization of microbial desalination cells in treating domestic wastewater. Bioresour. Technol. 120, 187–193.

Marsili, E., Sun, J., Bond, D.R., 2010. Voltammetry and growth physiology of geobacter sulfurreducens biofilms as a function of growth stage and imposed electrode potential. Electroanalysis 22, 865–874.

Mei, Y., Tang, C.Y., 2018. Recent developments and future perspectives of reverse electrodialysis technology: A review. Desalination 425, 156–174.

Morel, A., Zuo, K., Xia, X., Wei, J., Luo, X., Liang, P., Huang, X., 2012. Microbial desalination cells packed with ion-exchange resin to enhance water desalination rate. Bioresour. Technol. 118, 43–48.

Narayanasamy, S., Jayaprakash, J., 2018. Improved performance of Pseudomonas aeruginosa catalyzed MFCs with graphite/polyester composite electrodes doped with metal ions for azo dye degradation. Chem. Eng. J. 343, 258–269.

Ping, Q., Edwards, M.A., Achenie, L.E., Keen, O.S., 2016. Advancing Microbial Desalination Cell towards Practical Applications Advancing Microbial Desalination Cell towards Practical Applications. PhD Thesis, Virginia Polytechnic Institute and State University, Blacksburg, Virginia, 2016.

Pradhan, H., Ghangrekar, M.M., 2019. Effect of cathodic electron acceptors on the performance of microbial desalination cell. Springer, Singapore.

Puspitarini, P.A., Damayanti, N.W., Hfidzah, M.A., 2019. The usage of UI lake and coconut charcoal as a bio electrode with microbial desalination cell for optimizing a concentration of sodium percarbonate, 2092. AIP Conference Proceedings.

Qu, Y., Feng, Y., Liu, J., He, W., Shi, X., Yang, Q., Lv, J., Logan, B.E., 2013. Salt removal using multiple microbial desalination cells under continuous flow conditions Desalination. 317, 17–22.

Qu, Y., Feng, Y., Wang, X., Liu, J., Lv, J., He, W., Logan, B.E., 2012. Simultaneous water desalination and electricity generation in a microbial desalination cell with electrolyte recirculation for pH control. Bioresour. Technol. 106, 89–94.

Rabaey, K., Boon, N., Höfte, M., Verstraete, W., 2005. Microbial phenazine production enhances electron transfer in biofuel cells. Environ. Sci. Technol. 39, 3401–3408.

Rabaey, K., Boon, N., Siciliano, S.D., Verhaege, M., Verstraete, W., 2004. Biofuel cells select for microbial consortia that self-mediate electron transfer. Appl. Environ. Microbiol. 70, 5373–5382.

Ramírez-moreno, M., Rodenas, P., Aliaguilla, M., Bosch-jimenez, P., 2019. Comparative performance of microbial desalination cells using air diffusion and liquid cathode reactions: Study of the salt removal and desalination efficiency Front. Energy Res. 7, 1–12.

Reitz, D.W., 2007. Handbook of fuel Cells: Fundamentals, technology, and applications (Volume 2), W. Vielstich, A. Lamm, and H. A. Gasteiger (eds.). Mater. Manuf. Process. 22, 789.

Robert A. Alberty, 2005. Frontmatter, in Thermodynamics of Biochemical Reactions. John Wiley & Sons, Ltd., New Jersy pp. i–ix.

Semiat, R., 2008. Energy issues in desalination processes Environ. Sci. Technol. 42, 8193–8201.

Shehab, N.A., Amy, G.L., Logan, B.E., Saikaly, P.E., 2014. Enhanced water desalination ef fi ciency in an air-cathode stacked microbial electrodeionization cell (SMEDIC). J. Memb. Sci. 469, 364–370.

Shinde, O.A., Bansal, A., Banerjee, A., Sarkar, S., 2018. Bioremediation of steel plant wastewater and enhanced electricity generation in microbial desalination cell. Water Sci. Technol. 77, 2101–2112.

Uslu, B., Ozkan, S.A., 2007. Electroanalytical application of carbon based electrodes to the pharmaceuticals. Anal. Lett. 40, 817–853.

Von Canstein, H., Ogawa, J., Shimizu, S., Lloyd, J.R., 2008. Secretion of flavins by Shewanella species and their role in extracellular electron transfer. Appl. Environ. Microbiol. 74, 615 LP – 623.

Wei, L., Han, H., Shen, J., 2012. Effects of cathodic electron acceptors and potassium ferricyanide concentrations on the performance of microbial fuel cell Int. J. Hydrogen Energy, 37, 12980–12986.

Whittaker, A.G., Kintner, P.L., 1969. Carbon: Observations on the new allotropic form. Science 165, 589–591.

Zamanpour, M.K., Kariminia, H., Vosoughi, M., 2017. Electricity generation, desalination and microalgae cultivation in a biocathode-microbial desalination cell. J. Environ. Chem. Eng. 5, 843–848.

Zhang, B., He, Z., 2013. Improving water desalination by hydraulically coupling an osmotic microbial fuel cell with a microbial desalination cell. J. Memb. Sci. 441, 18–24.

Zhang, F., Brastad, K.S., He, Z., 2011. Integrating forward osmosis into microbial fuel cells for wastewater treatment, water extraction and bioelectricity generation. Environ. Sci. Technol. 45, 6690–6696.

Zhang, F., He, Z., 2015. Scaling up microbial desalination cell system with a post-aerobic process for simultaneous wastewater treatment and seawater desalination. Desalination 360, 28–34.

Zhao, F., Harnisch, F., Schröder, U., Scholz, F., Bogdanoff, P., Herrmann, I., 2005. Application of pyrolysed iron(II) phthalocyanine and CoTMPP based oxygen reduction catalysts as cathode materials in microbial fuel cells. Electrochem. Commun. 7, 1405–1410.

Zuo, K., Cai, J., Liang, S., Wu, S., Zhang, C., Liang, P., Huang, X., 2014. A Ten Liter Stacked Microbial Desalination Cell Packed with Mixed Ion-Exchange Resins for Secondary Effluent Desalination. Environ. Sci. Technol. 165, 589–591.

9 Microbial Fuel Cell: A Potential Solution for Desalination

Harsha Nagar*
Department of Chemical Engineering, Chaitanya
Bharathi Institute of Technology, Hyderabad, India

Vineet Aniya
Process Engineering and Technology Transfer Division,
CSIR-Indian Institute of Chemical Technology,
Hyderabad, India

CONTENTS

9.1 INTRODUCTION

Clean water and energy are essential for the survival of humankind, flora, and fauna, but rapid industrialization and urbanization with increased population causes their depletion worldwide (Baek, J. *et al.* 2009). Water covers 75% of the earth's surface in which a significant portion (97%) is seawater that is unfit for drinking. The remaining water is present as snow or glaciers, and only the remaining 1% is accessible for human consumption (Al-Mamun, A. *et al.* 2016). Additionally, the energy generation from non-renewable resources is the primary source of environmental pollution, resource depletion, and climate change. Therefore, there is a need to mitigate the water and energy crisis by developing a sustainable process that fulfills the present demand without compromising the future. In this prospect, microbial desalination cell (MDC) is an emerging technology that simultaneously generates clean water and energy through wastewater treatment and seawater desalination (Oh, S.E. *et al.* 2006). MDC is an integrated unit of bio-electrochemical systems (BESs) with membrane technology that provides high energy recovery with wastewater treatment efficiencies compared to the conventional process such as distillation, reverse osmosis, etc. (Brown, R.K. *et al.* 2015). MDC is an extended form of the microbial fuel cell (MFC) and was first introduced by Cao *et al.* (2009). The fundamental difference between MFCs and MDC configuration is the presence of an additional compartment in MDC. This additional compartment in MDC contains saltwater that is placed between the anode and cathode chambers. In MDC, bacterial activity is used to generate electricity, treat wastewater, and simultaneously desalinate saltwater. Presently MDC is used in water softening, heavy metal and multivalent ion removal, bioremediation, nutrient recovery, and value-added chemical productions (Euntae, Y., *et al.* 2019, Kim, K., *et al.* 2013). Considering MDC's wide application, the chapter provides an insight into the basic principle with their different design configurations. A recent trend of materials that have shown promising performance for MDC fabrication is also discussed. A mathematical model is also described for the optimization of process parameters. Finally, the chapter provides insight into the sustainability of the MDC process with life cycle assessment with different scenarios with its challenges and future opportunities.

9.2 CONCEPT AND BASIC TERMINOLOGY IN MDC

9.2.1 BASIC PRINCIPLE

The MDC reactor's basic design is shown in Figure 9.1. It consists of three chambers: anode, desalination, and cathode. The anode chamber is filled with organic wastewater along with the microbial culture. This microbial culture digests the organic matter and releases the electrons via an oxidation reaction. The released electrons are further transferred to the cathode's external circuit and combined with oxygen and protons to form water and electric potential via a reduction reaction. The middle (desalination) chamber contains the saltwater, which contains various cations and anions. The anion exchange (AEM) is sandwiched between the anode and desalination chambers while the cation exchange membrane (CEM) is kept between the desalination and cathode chambers. The generated potential between the two

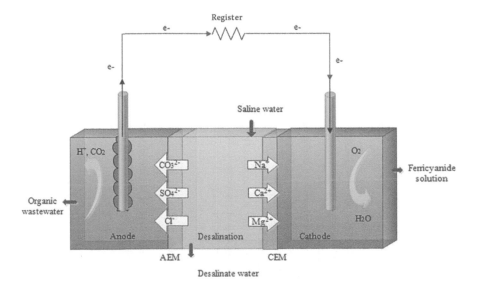

FIGURE 9.1   Basic design of MDC reactor.

electrodes attracts the cations and anions present in the saltwater towards the anode and the cathode side. This phenomenon leads to the desalination in seawater and follows the reaction mechanism, as shown below:

$$Anode: (CH_2O)_n + nH_2O \rightarrow nCO_2 + 4ne^- + 4nH^+ \tag{9.1}$$

$$Cathode: O_2 + 4ne^- + 4nH^+ \rightarrow 2H_2O \tag{9.2}$$

To further enhance the MDC performance, stack arrangement can be used wherein many CEM and AEM are alternatively arranged to form a repeating pair of desalting and concentrate cells. In such stacks, the separation efficiency is based on the number of monovalent ion pairs separated from the saltwater per electron transferred and is equal to the number of repeated cell pairs (or membrane pairs) in the stack. The only drawback is the high internal resistance in MDC that is generated due to the increase in the number of membranes that ultimately impact the voltage generation (Jacobson, K. S., *et al.* 2011b). This drawback can be minimized by keeping less intermembrane distance, which improves the ion separation in stacks by generating sufficient potential. Over the years, different MDC design configurations have been developed for specific effluent treatment. Table 9.1 details the different MDC configurations along with their key features, the associated problems, and solutions to improving their performance.

9.2.2 Driving Force

The electric potential gradient is the main driving force for separating ions from the saltwater present in the desalination chamber towards the anode and cathode

TABLE 9.1

Classification of MDC based on different configurations

No.	Configuration	Key features	Associated Problems with Solutions	Reference
1.	Air cathode MDC	• In this configuration oxygen acts as a terminal electron acceptor. • Achieved salinity removal of ~63% with increase in reduction potential. • Cost-effective and environmentally friendly in nature due to absence of oxidizing agent.	• The slow redox kinetic reaction at ambient conditions due to the presence of air catholyte. • Need of expensive catalytic material for oxygen reduction. • High energy requirement for maintaining the dissolved oxygen concentration. • The performance can be improved by using the carbon substrate, which has a high surface area that provides acceptable levels of oxygen reduction.	(Gude, V.G., et al, 2013)
2.	Biocathode MDC	• In this configuration microbial populations present on the surface are used as biocathode. • Microorganism catalyzes the reduction reaction at the cathode chamber. • Reduced around 92% salt content of seawater.	• Low performance of MDC. • The performance of the cell can be improved by providing oxygen biofilm at the cathode compartment that enhances the oxygen concentration; this causes the easy flow of electrons, improves cathodic efficiency, and yields high current flow.	(Choi, J., et al 2013)
3.	Stack structure MDC	• In this configuration multiple pairs of ion-exchange membranes (IEMs) are inserted between the anode and cathode chambers. • A series of MDCs allows high salt removal with more charge transfer. • In a series flow model with a parallel electrode, connection resulted in 15 to 20% higher electricity generation and 80 to 85% more coulombic efficiency as compared to series connection.	• High internal resistance of MDC due to an increase in IEMs. • The performance can be improved by modifying the cell design configuration and selection of desired properties of the IEM.	(Choi, J., et al. 2013,)

#		Description	Limitations	Reference
4.	Recirculation MDC	• In this configuration anolyte and catholyte solutions are re-circulating sequentially to maintain the pH in both anode and cathode compartment • The recirculation improves the electricity generation and desalination rate.	• Low coulombic efficiency is obtained and is mainly due to the recirculation of the anolyte and catholyte solutions. • This problem can be overcome by replacing the reduction of oxygen at the cathode by hydrogen evolution.	(Kim, Y., et al. 2013.)
5.	Microbial electrolysis desalination and chemical-production cell (MEDCC)	• This configuration combines electrodialysis and microbial electrolysis cell (MEC) with an acid-production chamber and a bipolar membrane (BPM). • Simultaneously removes salt with a production of HCl and NaOH. • Production of OH^- at the anode chamber resolves the problems of pH fluctuations and Cl^- accumulation in a conventional MDC.	• The impact of different parameters like membrane spacing, type of stack structure and the amount of voltage supplied on the rate of desalination and chemical production. • Therefore, a detailed study is required to evaluate the acid and alkali production rates by different factors.	(Euntae, Y., et al. 2019)
6.	Capacitive MDC	• In this configuration a double-layer capacitor is formed on the high surface electrodes and is used for salt removal by capacitive deionization. • When the saline solution flows between two electrodes. ions are absorbed by the capacitive electrode whereas removal of potential allows ions to flow back into the solution. • The deionization of salt is carried out by electrochemical salt adsorption without its contamination in the cathode and anode chambers.	• High electrode material cost and their regeneration. • The problem can be addressed by investigating different cost-effective electrode material with a high surface area.	(Zhang, B., et al. 2012.)
7.	Upflow MDC	• It consists of a tubular unit with two chambers separated by IEMs; with the inner compartment is the anode chamber filled with graphite granules. • Graphite rods are used as a current collector by immersing them into a graphite granule. • In this configuration, saline water enters at the bottom while the desalinated water leaves at the top of the chamber. • Easy mixing of fluids within chambers and a high percentage of water recovery. • The integration of upflow MDC with RO reduces the energy cost by 22%.	• The limitation of ion transfers due to the resistance between anode and cathode. • Difficult to maintain the pH between the chambers. • The performance can be enhanced by optimizing the reaction rate, selecting proper anodic bacteria and employing efficient catalysts to reduce the resistance. • Proton losses can also be compensated by circulating the anode effluent.	(Jacobson, K.S., et al. 2011)

(Continued)

TABLE 9.1

Classification of MDC based on different configurations (Cont.)

No.	Configuration	Key features	Associated Problems with Solutions	Reference
8.	Osmotic MDC	• This configuration integrates forward osmosis (FO) with MDC by replacing the IEM with FO membranes. • FO membrane allows easy flow of water but restricts the transfer of ions. • The coupled system of MDC and osmotic MFC enhances the salt removal and wastewater treatment efficiency.	• The FO membrane has high fouling that increases the internal resistance in MDC. • The replacement of AEM with FO membrane reduces the current efficiency. • To address the above problems, in-depth studies are required to understand the FO membrane performance at different conditions.	(Zhang, B., et al. 2013)
9.	Bipolar membrane MDC	• In this configuration bipolar membrane are used by laminating the AEM and CEM selectively in a single layer. • Integrating the bipolar membrane with MDC improves the desalination rate with balanced pH.	• The bipolar membrane has a high fouling tendency and also requires an external power source for ion dissociation. • Selection of a suitable polymer with low fouling tendency will overcome the above challenges.	(Alvarez, F., et al. 1997)
10.	Decoupled MDC	• In this configuration anode and cathode chambers are placed in the salt solution. • A plate configuration consists of AEM and CEM on both sides for anode and cathode, respectively, with a stainless-steel mesh wrapped with carbon cloth used as an electrode.	• Fouling issue and low performance.	(Ping, Q., et al. 2013)
11.	Separator coupled stacked circulation MDC	• In this configuration all anode, cathode and desalination chambers have an axial cylindrical cavity. • AEM and CEM membranes are used for separation of anode and cathode chamber. • Activated carbon particles are used in the anode chamber for growth of bacterial activity.	• High cost and circulation.	(Bouhidel, K.E., et al. 2006)
12.	Ion-exchange resin coupled MDC	• In this configuration ion exchange resins are integrated with conventional MDC. • It provides a high desalination rate as compared to conventional MDC.	• Regeneration of ion exchange resins.	(Morel, A., et al. 2012)

chambers via IEMs. Apart from IEMs' junction potential, water transport also plays an important role and affects salt removal efficiency. The difference in the dissolved solid concentration between the desalination and anodic chamber generates a steep potential gradient (junction potential) that further helps in the movement of salt ions towards anodic and cathodic chambers. This junction potential can be defined as:

$$\left|\varphi_j\right| = \left(\frac{RT}{F}\right)\left|\sum_i \frac{t_i}{z_i} \ln\left(\frac{a_{i,seawater}}{a_{i,wastewater}}\right)\right| \tag{9.3}$$

R, T, and F are gas constant, absolute temperature, and Faraday's constant, respectively. z, a, and t are ionic charge, the activity of the ionic species I, and transport number, respectively. Moreover, the difference in concentration between the seawater and wastewater induces the osmotic pressure that allows water to move into the middle chamber and dilute salt concentration. The osmotic pressure (π) is defined as follows:

$$\pi = \Delta CRT \tag{9.4}$$

Here ΔC is the concentration gradient.

Additionally, electro-osmosis is another driving force that helps in the transport of water across the IEM. Here an ionic movement that occurs through nano-scale pores present in the IEM is responsible for water transportation through their hydrated ions towards the middle chamber and diluting salt concentration. In MDC, the impact of osmotic pressure is high compared to electro-osmosis, which ultimately increases the net volume of water in the middle chamber. The electro-osmotic water transport directly relates to the current density through the IEMs, while osmotic water transport is directly proportional to the hydraulic residence time. Thereby both osmotic pressure for water transport and IEM junction potentials help in the desalination for three-chambered MDC. However, in stacked MDC, the impact of IEM junction potential and osmotic pressure is negative due to potential loss and concentration difference.

9.2.3 MDC PERFORMANCE PARAMETERS

9.2.3.1 Salinity Removal

In MDC, the desalination rate can be defined in terms of salinity removal, and it is found that 90% of salinity removal can be achieved in three-chambered MDC. This high performance can be achieved by using a large volume of non-salty anolyte and catholyte. To reduce large volumes of anolyte and catholyte, stacked MDC is a viable option that further enhances the salt removal efficiency. Furthermore, salinity removal is also dependent on the initial concentration of salt present in the middle chamber that ultimately decides the volume of wastewater that needs to be used for a particular operation (Euntae, Y., et al. 2019).

9.2.3.2 Polarization Curve

The polarization curve defines the relationship between the current and power density. In MDC, a trade-off is required between the current and power density for its higher performance. The increase in current density improves MDC's desalination rate, thereby selecting a proper operating condition that reduces the reactor's hydraulic retention time and increases the desalination rate. The low internal resistance in MDC improves the current density, which ultimately enhances salt removal. Nevertheless, in some cases, low resistance can deprive the current density due to the unfavorable anode potential (Mostafa, R., *et al.* 2019). Therefore, the polarization curve measurement is essential and is obtained by measuring the voltage (V) across the anode and cathode using an external resistor (Rex). While the current (I) measurement is carried out through the electrical circuit from the measured voltage using equation 9.5 and power density (p) through equation 9.6:

$$I = V\!\!\left/\!\!R_{ex}\right.$$

(9.5)

$$p = \frac{V^2}{(R_{ex} \times v)}$$

(9.6)

where v is the volume of the anode chamber.

9.2.3.3   Current and Coulombic Efficiency

Current efficiency (IE) is defined as the number of ions separated per electron transferred at the electrodes and is given by the following equation:

$$IE = \frac{Fzv\Delta c}{N_{cp} \int idt}$$

(9.7)

where Δc, v, i *and* N_{cp} represent the reduction in saltwater concentration, desalinated volume, current, and number of cell pairs, respectively. In MDC, osmotic pressure (driving force) improves the current efficiency by transporting the water in the middle chamber that dilutes the seawater and influences the ionic separation. The IE also provides the information about the IEMs performance wherein IE below 90% indicates a failure or a substantial current loss along the feed channels, and those above 100% indicate a high amount of water transport through the IEMs (Mostafa, R., *et al.* 2019).

Coulombic efficiency (CE) is defined as the fraction of the total number of electrons transferred from the anode to the maximum number of electrons generated by the substrate's bacterial oxidation reactions and calculated as follows.

$$CE = \frac{MO_2 \int idt}{n_e FV_{an}\Delta COD}$$

(9.8)

where V_{an}, ΔCOD, n_e *and* MO_2 are anode chamber volume, change in chemical oxygen demand of the wastewater, number of electrons to reduce oxygen to water, and oxygen molecular mass, respectively. Here, the CE provides information about the utilization of substrate fraction by the bacteria for the current production. Maintained anaerobic conditions in an anode chamber using multiple IEMs enhances the CE (Chih, Y. M., *et al.* 2019).

9.2.3.4 Chemical Oxygen Demand (COD) Removal

The MDC performance is also characterized by determining their ability to reduce the COD of the wastewater. The percentage reduction in COD is estimated by the change in initial and final COD of wastewater to that of initial. The wastewater with low COD can be discharged safely, and in the case of MDC, the high COD reduction is obtained for substrate like sucrose and xylose as compared to domestic wastewater (Oh, S.E., *et al.* 2006).

9.2.3.5 Salinity Effects on Exoelectrogenic Bacteria

The activity of bacterial is dependent on the presence of anion concentration in the anodic chamber of MDC. It has been reported that up to a certain level of anion concentration, the power generation has no significant effect, while higher concentrations of anions have depleted the power generation due to the lowered bacterial performance. It has also been observed that different salt concentrations in the anolyte impact the activity of different bacteria, thereby suggesting that the selection of microbes is based on the salinity level (Oh, S.E., *et al.* 2006).

9.3 APPLICATIONS OF MDC

9.3.1 Pre-Treatment and Post-Treatment of Desalination Processes

MDC can be integrated with reverse osmosis (RO) and electrodialysis systems as a pre-treatment and post-treatment process. This integration helps to reduce their energy requirements along with an increase in salt removal efficiency. An addition of upflow MDCs coupled with RO systems reduces their energy requirement to 20% and a high reduction in TDS in addition to the power generation (Jacobson, K. S., *et al.* 2011b). MDC also treats the RO concentrate in an energy-effective way through its MEDCC configuration compared to other traditional processes. Additionally, MDC also desalinate the FO treated leachate by reducing their salinity and chemical oxygen demand (COD) (Euntae, Y., *et al.* 2019).

9.3.2 Valuable Chemical Production

The hydrogen gas or hydrogen peroxide can be produced using MDC, which includes the modification in cathode reduction reaction and by applying the external power sources. The integration of the bipolar membrane in MDC shows their potential applicability in acid and alkali production. Utilization of electrically selective ion separation in MEDCCs can be used to produce hydrochloric acid, caustic soda, malic acid, and formic acid (Euntae, Y., *et al.* 2019).

9.3.3 Harmful Substance Removal and Water Softening

MDC can remove the harmful heavy metals, such as hexavalent chromium and copper, along with hardness by their cathodic reduction reaction and electrical ion migration in desalination chambers. An *et al.* 2014 investigated the heavy metal copper-containing water treatment in MDCs via the cathodic reduction reaction that converts copper (II) to copper (I) oxide. Dong *et al.* 2017 used stacked MDCs for hydroxide precipitation using the alkali effluent in a cathode chamber for copper removal. MDC can also treat the wastewater containing hardness and heavy metals, such as arsenic, nickel, and mercury, via electric field-driven ion migration. Removal of boron can also be possible through the integration of MDC with the Donnan dialysis cell. Phenolic-containing wastewater also treated effectively by the MDC using P. aeruginosa and mixed anaerobic consortia as inoculum. MDC is also useful in removing nickel and lead from industrial wastewater generated by the electroplating industry and steel company (Euntae, Y., *et al.* 2019).

9.3.4 NUTRIENT REMOVAL AND RECOVERY

Modified MDC configuration has shown its potential applicability for denitrification and nutrient recovery. The electric field collects the nitrates from the groundwater at the anode chamber, which is further reduced at the cathode chamber electrochemically to provide the nitrate-free groundwater. Researchers Chen *et al.* 2017 designed the modified stacked MDC that sequentially treats the nitrogen, phosphorous-containing wastewater through anaerobic biodegradation at an anode chamber while aerobic biodegradation occurs in the cathode chamber and electrical nutrient ion migration occurred at desalination chambers. A hybrid submersible microbial desalination cell (SMDC) and a continuous stirred tank reactor (CSTR) system were investigated for the simultaneous in situ ammonia recovery and electricity production in biogas plants. The implication of SMDC resulted in 112% extra biogas production due to the ammonia recovery and wastewater treatment.

9.3.5 BIOSENSORS

MDCs are also useful for monitoring the volatile fatty acids concentration in real-time applications produced in anaerobic digestion processes. In this process, anaerobic digestion effluent was kept at the desalination chamber wherein the movement of ionized volatile fatty acids into the anode chamber occurred using AEM. These ions were further utilized by the electrochemically active bacteria and generate current. The generated current is directly proportional to the concentration of volatile fatty acids in the effluent fed into the middle chamber (Jin, X., *et al.* 2016).

9.4 RECENT TRENDS OF MATERIAL FOR MDC COMPONENTS

The most important components in the MDC are electrodes, membrane, and microbial culture. The recently investigated MDC electrodes include a three-dimensional macroporous carbon nanotube-chitosan sponge as the anode. Bare graphite electrodes were employed as both anode and cathode material due to their high surface area and low biofouling properties. Polypyrrole coated and silver tin dioxide composite coated carbon felt has also been explored as a catalyzed cathode. A novel air diffusion cathode using carbon nanofibers with iron nanoparticles as a catalyst is also reported (Chih, Y. M., *et al.* 2019, Havan, H., *et al.* 2020, Gui, H., *et al.* 2018, Anusha, G., *et al.* 2018, Marina, R.M., *et al.* 2019). Sulfonated polyetheretherketone CEM and non-patterned poly (2,6-dimethyl 1,4-phenylene) oxide-based AEM for MDC has been reported and shown a promising result (Francisco, L. M., *et al.* 2018). Microorganism consortium plays a significant role in MDC performance; various electrochemically active bacteria such as Shewanella putrefaciens, Geobacter sulfurreducens, Bacillus subtilis, Aeromonas hydrophila, etc. have been successfully tested in MDC. Along with that, mixed bacterial culture has also shown its potential capability for desalination and wastewater treatment (Sabina, K., *et al.* 2014).

9.5 MATHEMATICAL MODEL

Qingyun P. *et al.* 2014 has proposed a mathematical model to understand MDC's performance at different operating conditions such as different substrate concentrations, salt concentrations, and external electrical resistance. The mass balance was established for the substrate, microorganisms, and electron mediators in the anodic chamber. Assumptions were made for developing the mathematical model that includes ideal mixing in all three chambers of MDCs; microbial growth follows the multiplicative Monod kinetics, and an intracellular redox mediator was assumed. The concentration was estimated by ordinary differential equations in the anodic chamber. The mass balance equation for the substrate can be given as follows:

$$\frac{ds}{dt} = D_{anode}\left(S_{in} - S\right) - k_{s,a,max}\frac{S}{K_a + S}\frac{M_{OX}}{K_M + K_{OX}}C_a - k_{s,m,max}\frac{S}{K_m + S} \quad (9.9)$$

Where S *and* S_{in} are the concentration and influent substrate concentration, respectively (mgSL^{-1}); C_a and C_m are the anodophilic and methanogenic microorganisms concentrations (mg CL^{-1}), respectively; $k_{s,a,max}$ and $k_{s,m,max}$ are the anodophilic and methanogenic microorganisms maximum substrate consumption rates (mgSmg-a^{-1}day^{-1}), respectively; M_{OX} is the oxidized mediator fraction per anodophillic microorganism (mgMmga^{-1}); K_a, Km, and K_M are the half saturation concentrations for the anodophillic microorganisms, methanogenic microorganisms, and the redox mediator (mgSL^{-1}, mgSL^{-1}, and mgMmga^{-1}), respectively. The concentrations of anodophilic and methanogenic microorganisms in the anodic chamber are given by the following differential equations:

$$\frac{dC_a}{dt} = k_a C_a - k_{d,a}C_a - D_{anode} + \frac{1 + tahn\left(k_{a,x}\left(C_a + C_m - C_{a,max}\right)\right)}{2}C_a \quad (9.10)$$

$$\frac{dC_m}{dt} = k_m C_m - k_{d,m}C_m - D_{anode} + \frac{1 + tahn\left(k_{m,x}\left(C_a + C_m - C_{m,max}\right)\right)}{2}C_a \quad (9.11)$$

D_{anode} is the dilution rate, which is the ratio of initial flow rate of the substrate to anode volume (L). $k_{d,a}$ and $k_{d,m}$ are the decay rates of the microorganisms (day^{-1});

$k_{a,x}$ and $k_{m,x}$ are the steepness factors for anodophillic microorganisms (Lmg-a^{-1}) and methanogenic microorganisms (Lmgm^{-1}) for the biofilm retention; $C_{a,max}$ and $C_{m,max}$ are the maximum attainable concentrations for anodophillic microorganisms and methanogenic microorganisms (mgmL^{-1});

Here k_a and k_m are the growth rates of the anodophillic microorganisms and methanogenic microorganisms (day^{-1}) are estimated by the following equations.

$$k_a = k_{a,max} \frac{S}{K_a + S} \frac{M_{OX}}{K_M + M_{OX}} \tag{9.12}$$

$$k_m = k_{m,max} \frac{S}{K_m + S} \tag{9.13}$$

where $k_{a,max}$ and $k_{m,max}$ are the maximum microorganism growth rates (day^{-1}). The intracellular material balance for the oxidized mediator can be described as follows:

$$\frac{dM_{OX}}{dt} = -Y_M k_{s,a} + \frac{\gamma}{V_{anode} C_a} \frac{I_{MDC}}{n_e F} \tag{9.14}$$

$$M_{Total} = M_{OX} + M_{red} \tag{9.15}$$

where M_{Total} and M_{red} is the total mediator and reduced mediation fraction per microorganisms (mgMmga^{-1}), respectively; Y_M, γ and I_{MDC} are the mediator yield (mgMmgS^{-1}); mediator molar mass (mgMmoleM^{-1}); and the current through the circuit of MDCs (A), respectively; F and n_e is the Faraday constant (Adaymol^{-1}); and the number of electrons transferred per mole of a mediator, respectively.

The ionic current generated between the electrodes is responsible for removing salts in the desalination chamber and the concentration gradient developed between the chamber also helps in salt diffusion. The salt concentration can be quantified by using the following ordinary differential equations:

$$\frac{dC_{salt,m}}{dt} = D_{salt} \left(C_{salt,in} - C_{salt,m} \right) - d \left(C_{salt,m} - C_{salt,\,a} \right) - d \left(C_{salt,m} - C_{salt,c} \right) - \frac{I_{MDC}}{FV_{salt}} \tag{9.16}$$

$$\frac{dC_{salt,a}}{dt} = d \left(C_{salt,m} - C_{salt,a} \right) - D_{anode,} C_{salt,a} \tag{9.17}$$

$$\frac{dC_{salt,c}}{dt} = d \left(C_{salt,m} - C_{salt,c} \right) \tag{9.18}$$

where $C_{salt,a}$, $C_{salt,m}$, and $C_{salt,c}$, are the salt concentrations in the anodic, desalination and cathodic chambers (mol-salt·L^{-1}), respectively; D_{salt} is the salt dilution rate (day^{-1}), and d is the membrane salt transfer coefficient. Thereby the produced overall cell voltage in MDC process can be defined by following equation:

$$I_{MDC} R_{ext} = V_{OC} - OP_{anode} - OP_{cathode} - OP_{conc} - I_{MDC} R_{int} \tag{9.19}$$

where R_{ext} is the external resistance (Ω); V_{OC}, OP_{anode}, $OP_{cathode}$ and OP_{conc} are the open-circuit voltage; anode overpotential; cathode overpotential and overpotential concentration (V), respectively; R_{int} is the internal resistance of the cell (Ω), which is the summation of mass transfer resistance (membrane resistance), ohmic resistance (anolyte solution and salt solution resistance), and activation resistance (internal resistance due to the bacterial growth). Moreover, the dependence of cell voltage on the catalyst loading shows the exponential relationship between anodophilic micro-organism concentration and the open circuit voltage (V_{OC}):

$$V_{OC} = V_{min} + \left(V_{max} - V_{min}\right)e^{-\frac{1}{k_f C_s}} \tag{9.20}$$

where V_{min} and V_{max} are the lowest and highest observed V_{OC} values (V).

When sufficient buffer solution is used in anode and cathode, the overpotential loss can be assumed to be negligible. Based on that assumption MDC current is calculated as follows:

$$I_{MDC} = \frac{V_{OC} - OP_{conc}}{\left(R_{ext} + R_{int} + R_{salt}\right)} \tag{9.21}$$

The modelling of concentration overpotential can be determined by using equation 9.22, which is associated with electron mediators:

$$OP_{conc} = \frac{RT}{F}\ln\left(\frac{M_{total}}{M_{red}}\right) \tag{9.22}$$

The above-mentioned mathematical model can be used to control and optimize MDC performance at different operating conditions.

9.6   SUSTAINABILITY OF THE PROCESS

The advantageous dual features of MDC technology of treating wastewater and salt-water shows the potential applicability. To understand its sustainability for future commercialization, a life cycle assessment (LCA) is an important tool to determine the negative environmental impacts during an MDC's whole life cycle by material consumption and energy use. In LCA, the manufacturing data can be taken directly from the lab-scale experiments or the literature. The MDC system's environmental impacts can be carried out from cradle-to-grave or cradle-to-gate and character-ized by the International Life Cycle Data System method using GaBi software. The LCA for MDC consists of seven phases: raw material extraction, material process-ing, component manufacturing, pre-treatment, operation, post-treatment, and end of life (EoL). System manufacturing includes raw material extraction, material process-ing, and component manufacturing, whereas the operational phase involves MDC operation, solid waste treatment, membrane maintenance, and electricity generation. In the case of EoL for MDC, system collection and disassembly, recycling/reuse, and disposal need to be considered. The component manufacturing can be modeled with experimental data, while the other phases can be modeled using the database provided by the GaBi 6.1 professional database (ecoinvent V3.3) and literature. In the case of the unavailability of disassembly data at EoL for MDC, a regular battery disassembly process can be considered due to its functional similarity (Jingyi, Z., 2018). Valuable materials can be presumed to be recovered after the system collection and disassembly, recycling, or reuse. The dismantled materials that can be recovered are metals (aluminum foil and titanium wire), membranes, plastics (polyvinyl chloride), carbon fibers (polyacrylonitrile), etc. Some of the disassembled materials from the MDC system, such as the mixture of activated carbon, polytetrafluoroethylene binder, super glue, and plastic welder, can be sent to landfill facilities for disposal.

Once the goal and scope with the system boundary are decided, the MDC materials and manufacturing and process description need to be used to prepare the LCA inventory that accounts for material and energy consumption and environmental outputs per unit. Finally, the life cycle impact assessment is used to quantify the magnitude and significance of a product's potential environmental impacts through its whole life cycle. Life cycle environmental impact assessment for each phase in MDC system from cradle to grave can include acidification, global warming potential, ecotoxicity freshwater potential, eutrophication freshwater potential, human toxicity potential with cancer effects, ozone depletion potential, respiratory inorganic, photochemical ozone creating potential, and abiotic depletion potential, which can be further subdivided to each anodic, cathodic and desalination chamber. In MDC, ion-exchange membrane fabrication, disposal technology, power density, and volume ratio between the desalination and anode chambers are important in the life cycle analysis. Thereby alternative scenarios can be performed for ion-exchange membrane fabrication and disposal technology while sensitive analysis can be carried for power density and volume ratio between desalination and anode chambers to reduce notable environmental impact. Furthermore, to check the environmental impact variations within different disposal scenarios, disposal methods such as recycling, incineration with energy recovery, and direct landfills can be considered. The selection of the disposal method is based on its environmental benefits to offset the impacts generated from the process. Additionally, in conducting LCA for the MDC system, there are a lot of uncertainties caused by the variability of foreground data from the system manufacturer, which need to be analyzed through uncertainty analysis using software tools such as Monte Carlo. A variation of ±10% in material and ±20% energy inputs can be considered to analyze the environmental impact in the aforementioned phase.

9.6.1 Major Challenges and Future Scope

The major challenges that impacted the MDC performance and limited the scalability are its low desalination rate and organic matter degradation. The effectiveness of microbes in the salt solution with their proper nutrient ratio are prime factors that impact the desalination rate. The microbial activity may get disturbed with the degradation of the organic matter that affects the active microbial colonies by their toxic degradation behavior or disturbs the C:N ratio of the MDCs. Another challenge in MDC is poor electrochemical performance, mainly due to the high cathodic activation with low anodic kinetics. The removal of salt in the anolyte solution also affects the MDC performance by making the solution hypertonic, which further deprives the microbial activity. Membrane fouling and durability is important challenges that need to be addressed. The pH of cathodic and anodic fluids is an important parameter for MDC performance since an imbalance of pH results in the ions at anode and cathode chamber aggregation. Finally, the cost-effective, durable alternative material for electrodes and ion exchange membrane and optimization of stack MDC for large scale studies with real wastewater and development of continuous flow reactors are the future opportunities in MDC.

9.7 CONCLUSIONS

MDC is an upcoming technology with the potential to treat saltwater/wastewater and simultaneously generate electricity through microbial activity. Different configurations of MDC such as capacitive MDC, upflow MDC, and osmotic MDC have been developed with their advantages and specific applications. The valuable chemical production, harmful substance removal, water softening, nutrient removal, and biosensors are additional applications for desalination. Different operational parameters (pH, COD, bacterial culture, etc.) significantly affect the MDC and need to be studied to attain higher performance. The automation of MDC needs to be carried out through a robust mathematical model that describes analyzing the impact of the different parameters. For the sustainability of MDC and its commercialization, life cycle assessment is an important study that needs to be estimated for the environmental impacts during its whole life cycle.

REFERENCES

A. Al-Mamun, O. Lefebvre, M.S. Baawain, H.Y. Ng. 2016. A sandwiched denitrifying biocathode in a microbial fuel cell for electricity generation and waste minimization. *International Journal of Environmental Science & Technology* 13: 1055–1064.

F. Alvarez, R. Alvarez, J. Coca, J. Sandeaux, R. Sandeaux, C. Gavach. 1997. Salicyclic acid production by electrodialysis with bipolar membranes. *Journal of Membrane Science* 123: 61–69.

Z. An, H. Zhang, Q. Wen, Z. Chen, M. Du. 2014. Desalination combined with copper (II) removal in a novel microbial desalination cell. *Desalination* 346: 115–121.

G. Anusha, M.T. Noori, M.M. Ghangrekar. 2018. Application of silver-tin dioxide composite cathode catalyst for enhancing performance of microbial desalination cell. *Materials Science for Energy Technologies* 1: 188–195.

J. Baek, Y.S. Cho, and W.W. Koo. 2009. The environmental consequences of globalization: A country specific time-series analysis. *Ecological Economics* 68: 2255–2264.

K.E. Bouhidel, A. Lakehal. 2006. Influence of voltage and flow rate on electrodeionization (EDI) process efficiency. *Desalination* 193: 411–421.

R.K. Brown, F. Harnisch, T. Dockhorn, U. Schröder. 2015. Examining sludge production in bioelectrochemical systems treating domestic wastewater, *Bioresource Technology* 198: 913–917.

X.X. Cao, X. Huang, P. Liang et al. 2009a. A completely anoxicmicrobial fuel cell using a photo-biocathode for cathodic carbondioxide reduction. *Energy and Environmental Science* 2: 498–501.

S. Chen, G. Liu, R. Zhang, B. Qin, Y. Luo, & Y. Hou. 2012a. Improved performance of the microbial electrolysis desalination and chemical-production cell using the stack structure. *Bioresource Technology* 118: 507–511.

X. Chen, H. Zhou, K. Zuo, Y. Zhou, Q. Wang, D. Sun, Y. Gao, P. Liang, X. Zhang, Z.J. Ren, X. Huang. 2017. Self-sustaining advanced wastewater purification and simultaneous in situ nutrient recovery in a novel bioelectrochemical system. *Chemical Engineering Journal* 330: 692–697.

X. Chen, X. Xia, P. Liang, X. Cao, H. Sun, X. Huang. 2011. Stacked microbial desalination cells to enhance water desalination efficiency. *Environmental Science and Technology* 45 (2011) 2465–2470.

Y.M. Chih,, C.H. Hou. 2019. Enhancing the water desalination and electricity generation of a microbial desalination cell with a three-dimensional macroporous carbon nanotube-chitosan sponge anode. *Science of the Total Environment* 675, 20: 41–50.

J. Choi, Y. Ahn. 2013. Continuous electricity generation in stacked air cathode microbial fuel cell treating domestic wastewater. *Journal of Environmental Management* 130:146–152.

Y. Dong, J. Liu, M. Sui, Y. Qu, J.J. Ambuchi, H. Wang, Y. Feng. 2017. A combined microbial desalination cell and electrodialysis system for copper-containing wastewater treatment and high-salinity-water desalination. *Journal of Hazardous Material* 321: 307–315.

Y. Euntae, K.J. Chae, M.J. Choi, Z. He, I.S. Kim. 2019. Critical review of bioelectrochemical systems integrated with membrane-based technologies for desalination, energy self-sufficiency, and high efficiency water and wastewater treatment. *Desalination* 452: 40–67.

L.M. Francisco, J.E. Rubio, C. Santoro, P. Atanassov, J.M. Cerrato, C.G.Arges. 2018. Investigation of patterned and non-patterned poly(2,6-dimethyl 1,4-phenylene) oxide based anion exchange membranes for enhanced desalination and power generation in a microbial desalination cell. *Solid State Ionics* 314: 141–148.

V.G. Gude, B. Kokabian, V. Gadhamshetty. 2013. Beneficial bioelectrochemical systems for energy, water, and biomass production. *Journal of Microbial and Biochemical Technology* S6: 005 1–14.

H. Gui, H, Wang, H. Zhao, P. Wu, Q. Yan. 2018. Application of polypyrrole modified cathode in bio-electro-Fenton coupled with microbial desalination cell (MDC) for enhanced degradation of methylene blue. *Journal of Power Sources* 400: 350–359.

H. Havan, S. Zain, Z. Ismail. 2020. Desalination of actual wetland saline water associated with biotreatment of real sewage and bioenergy production in microbial desalination cell. *Separation and Purification Technology*, https://doi.org/10.1016/j.seppur.2020.117110

K.S. Jacobson, D.M. Drew, Z. He. 2011b. Use of a liter-scale microbial desalination cell as a platform to study bioelectrochemical desalination with salt solution or artificial seawater. *Environmental Science & Technology* 45: 4652–4657.

K.S. Jacobson, D.M. Drew, Z. He. 2011. Efficient salt removal in a continuously operated upflow microbial desalination cell with air cathode. *Bioresource Technology* 102: 376–380.

X. Jin, I. Angelidaki, Y. Zhang. 2016. Microbial electrochemical monitoring of volatile fatty acids during anaerobic digestion. *Environmental Science & Technology* 50: 4422–4429.

Z. Jingyi, H. Yuan, Y. Deng, Y. Zha, I.M.A. Reesh, Z. He, C. Yuan. 2018. Life cycle assessment of a microbial desalination cell for sustainable wastewater treatment and saline water desalination. *Journal of Cleaner Production* 200: 900–910.

K.Y. Kim, K.J. Chae, M.J. Choi, E.T. Yang, M.H. Hwang, I.S. Kim. 2013. High-quality effluent and electricity production from non-CEM based flow-through type microbial fuel cell. *Chemical Engineering Journal* 218:19–23.

R.M. Marina, P. Rodenas, M. Aliaguilla, P.B. Jimenez, E. Borràs, P. Zamor, V. Monsalvo, F. Rogalla, J.M. Ortiz, A.E. Núñez. 2019. Comparative performance of microbial desalination cells using air diffusion and liquid cathode reactions: study of the salt removal and desalination efficiency. *Frontiers in Energy Research,* https://doi.org/10.3389/fenrg.2019.00135.

A. Morel, K. Zuo, X. Xia, J.Wei, X. Luo, P. Liang, X. Huang. 2012. Microbial desalination cells packed with ion-exchange resin to enhance water desalination rate, *Bioresource Technology* 118: 43–48.

R. Mostafa, A. Elaww, H.A. Halim. 2019. Evaluating the performance of microbial desalination cells subjected to different operating temperatures. *Desalination* 462: 56–66

S.E. Oh, B.E. Logan. 2006. Proton exchange membrane and electrode surface areas as factors that affect power generation in MFCs. *Applied Microbiology and Biotechnology* 70:182–189.

Q. Ping, Z. He. 2013 Effects of inter-membrane distance and hydraulic retention time on the desalination performance of microbial desalination cells. *Desalination Water Treatment* 52: 1324–1331.

P. Qingyun, C. Zhang, X. Chen, B. Zhang, Z. Huang, Z. He. 2014. Mathematical model of dynamic behavior of microbial desalination cells for simultaneous wastewater treatment and water desalination. *Environmental Science & Technology* 48: 13010–13019.

K. Sabina, F.M. Abbas, A. Ganesan, S. Meenatchisundaram, B. Srinivasan, A.S.B. Packirisamy, R.K. Kesavana, S. Muthusamya. 2014. Biodegradation and bioelectricity generation by microbial desalination cell, *International Biodeterioration & Biodegradation* 92: 20–25.

P. Soumya, S. Sarode, D. Das. 2018. Microbial Fuel Cell Chapter 18: Fundamentals of Microbial Desalination Cell, DOI: 10.1007/978-3-319-66793-5_18.

B. Zhang, Z. He. 2013. Improving water desalination by hydraulically coupling an osmotic microbial fuel cell with a microbial desalination cell. *Journal of Membrane Science* 441:18–24.

B. Zhang, Z. He. 2012. Energy production, use and saving in a bioelectrochemical desalination system. *RSC Advances* 28: 10673–10679.

10 Advanced Oxidation Techniques for Wastewater Treatment

M. S. Priyanka and George K. Varghese*

National Institute of Technology Calicut, India

CONTENTS

10.1 INTRODUCTION

Globally four trillion cubic meters of freshwater are consumed annually for various purposes. A major portion is discharged as wastewater, a mixture of several organic and inorganic contaminants. These compounds have to be removed by municipal or industrial sectors in order to meet legal and statutory requirements. The conventional wastewater treatment units are highly effective in removing components that can be biologically degraded. But not all wastewater constituents are amenable to biological degradation. Components like nutrients, heavy metals, salts, and ionic substances co-exist with organic matter and are resistant to biodegradation. Apart from the above-mentioned pollutants, there is a section of compounds termed "emerging contaminants (ECs)," such as pharmaceuticals, personal care products, dyes, industrial additives, and plasticizers, whose degradation and mineralization is incomplete in biological systems. ECs are generated in large quantities and discharged in their native or degraded form into water and soil. In wastewater, these components may not be present in large concentrations. However, because of their persistence in the environment and possible impacts on the receiving ecosystem and human health, these components are gaining importance in wastewater treatment. But, the discharge regulations of many nations do not address these constituents in wastewater because many of these compounds only recently started gaining attention as constituents of concern. Moreover, there is still uncertainty associated with its fate and toxicity. Because of their recalcitrant nature, special treatment methods are required for their removal.

10.2   ADVANCED OXIDATION PROCESS

AOP, first employed to treat potable drinking water in the 1980s (Deng and Zhao, 2015), started receiving renewed attention when it was found effective for the treatment of emerging contaminants. AOPs are physicochemical processes that involve the generation of highly reactive free radical species. The free radicals are termed so, as they are atoms or molecules that are capable of independent existence and have one or more unpaired electrons (Halliwel, 1991). The free radicals are highly reactive and are capable of oxidizing even those "not so readily degradable" emerging pollutants. But, because of their extremely short life/unstable nature, they need to be generated within the treatment system. These radicals include superoxide radical (O_2^{-}), hydroperoxyl radical ($HO_2\cdot$), hydroxyl radical ($\cdot OH$), alkoxyl radical ($RO\cdot$), and many other short half-life radicals. Electrons are more stable when paired together in orbitals; the two electrons in a pair have different directions of spin. Radicals, with their unpaired electrons, are generally less stable than nonradicals, although their reactivity varies. Among these radicals, $\cdot OH$ has specific advantages when used for the degradation of components in the aqueous matrices. These advantages include (i) high standard oxidation-reduction potential (2.8 eV), (ii) high bimolecular reaction rate constants (10^8–10^{10} 1/Ms) (iii) nonselective destruction of compounds, (iv) no residues in aqueous phase with a short-life of 10^{-9} s, (v) generation of hydroxylated derivatives followed by subsequent mineralization to CO_2, and (vi) generation of hydrogen and nontoxic by-products. Because of the advantages, most AOPs have $\cdot OH$ radical as primary oxidant species. The $\cdot OH$ can interact with organic/inorganic compounds through any of the three mechanisms: (i) hydroxyl addition, (ii) hydrogen abstraction, or (iii) electron transfer before the oxidization/mineralization happens. Hydroxyl addition occurs with organic compounds having an aromatic system or carbon-carbon multiple bonds. Hydrogen abstraction occurs with unsaturated organic compounds whereas electron transfer occurs when $\cdot OH$ interacts with inorganic ions (Bello et al., 2019a). The reaction mechanisms, in the same order, are given in the equations (10.1) to (10.3) below:

$$\cdot OH + R \rightarrow R(OH)\cdot \tag{10.1}$$

$$\cdot OH + RH \rightarrow R + H_2O \tag{10.2}$$

$$\cdot OH + Fe^{2+} \rightarrow Fe^{3+} + OH^- \tag{10.3}$$

As the radicals have to be generated within the treatment system, the classification of AOP is primarily based on the mechanism adopted for the generation of the radical. Some AOP systems use chemical catalysts for the generation of $\cdot OH$ radicals. There are others that use light energy, ionizing radiations, sound energy, etc. for generating

•OH radicals. But, most systems adopt a combination of mechanisms to exploit the advantages of different mechanisms. In addition to •OH radical-based systems, others use different radicals like Persulfate (PS) radical as the oxidant. Some of the common AOP systems are explained in the following sections.

10.3   FENTON-BASED AOPs

Among the AOPs, an important category is the Fenton-based systems that use a dark reaction of Fe (II) ions and H_2O_2 oxidant for the generation of •OH radicals. In the 1890s, Henry John Horstman Fenton prepared the $Fe(II)/H_2O_2$ combination (Fenton reagent) for the oxidation of an organic substance (tartaric acid) (Fenton, 1894). The Fenton oxidation can occur in homogeneous or heterogeneous systems, although the first one is more common. In the homogeneous Fenton process, ferrous salt is used as a catalyst. It dissolves in water and forms a homogeneous solution. In the latter case, solid iron (for example, zero-valent iron, magnetite, hematite, etc.) is used as a catalyst, resulting in a two-phase (liquid-solid) or heterogeneous system. The Fenton reactions are carried out under acidic conditions (pH 3–4) to avoid precipitation of iron as hydroxide (Prato-Garcia et al., 2009). The Fenton system is dependent on operation parameters like pH, catalyst concentration, and H_2O_2 concentration for the higher removal of organic pollutants.

From the Fenton reagent, hydroxyl radicals are formed through a radical mechanism as follows (Equations (10.4) to (10.9)):

$$Fe^{2+} + H_2O_2 \rightarrow Fe^{3+} + \bullet OH + OH^- \tag{10.4}$$

$$Fe^{3+} + H_2O_2 \rightarrow Fe^{2+} + HO_2 \cdot + H^+ \tag{10.5}$$

$$\bullet OH + H_2O_2 \rightarrow HO_2 \cdot + H_2O \tag{10.6}$$

$$\bullet OH + Fe^{2+} \rightarrow Fe^{3+} + OH^- \tag{10.7}$$

$$Fe^{3+} + HO_2 \cdot \rightarrow Fe^{2+} + O_2H^+ \tag{10.8}$$

$$\bullet OH + \text{Organic substrate} \rightarrow H_2O + \text{degradation products} + \bullet OH \rightarrow CO_2 + H_2O \tag{10.9}$$

In a Fenton catalytic reaction, Fe^{2+} is oxidized to Fe^{3+} along with the generation of •OH radicals. Later, Fe^{3+} reduces to Fe^{2+}, but the reaction rates are much lower than oxidation. The oxidation reaction rate is 6,000 times greater than reduction, which causes ineffective conversion of Fe^{3+} to Fe^{2+} and causes its accumulation in the reaction solution. At pH>3, Fe^{3+} precipitates and accumulates as oxyhydroxide, which is known as iron sludge. Separation and recovery of iron sludge is difficult and causes the loss of iron species and catalytic activity (Zhang et al., 2019). The rate-determining step of the overall reaction is the regeneration of Fe^{2+} from Fe^{3+}. This results in a higher initial rate of degradation, which slows down in the later stages due to reduced Fe^{2+} concentration and its slow rate of regeneration.

The main drawbacks of the Fenton process when treating emerging contaminants like pharmaceuticals, antibiotics, and other organic pollutants in wastewater is the requirement of low pH. Higher pH leads to complex reactions and precipitation of iron oxides, which leads to the production of excessive sludge. Generated sludge can cause the risk of secondary pollution and disposal concerns. Additional drawbacks of Fenton oxidation include high chemical consumption, instability of the Fenton's reagent, parasitic reactions and loss of oxidant, difficulty in optimizing the reagent concentrations, and the necessity to neutralize the treated wastewater before disposal (Bello et al., 2019). To overcome the drawbacks of the classical Fenton process, these systems are coupled with other systems. Photoelectro-Fenton, heterogeneous electro-Fenton, heterogeneous photoelectro-Fenton, and three-dimensional electro-Fenton are examples of coupled systems.

10.4 OZONE BASED AOPs

Ozone was discovered by Schönbein in 1840 and in 1872 its structure was confirmed as a triatomic oxygen molecule. In 1906, it was used as a drinking water disinfectant. By 1977, across the world, there were 1,039 drinking water treatment plants using ozone as a disinfectant. Later, it was used for the oxidation of sulfides, nitrites, and organic pollutants (Wang and Chen, 2020). Ozone is a very selective oxidizer with standard potentials of 2.07 V in acidic solution and 1.25 V in basic solution and is currently applied in water treatment. Degradation of target compounds occurs by two mechanisms: (i) direct electrophilic attack by molecular ozone, and (ii) indirect attack by ·OH radicals generated as a result of decomposition of aqueous ozone (Staehelln and Hoigne, 1985). The later mechanism involves the ·OH radicals and proceeds in three steps: initiation, propagation, and termination. Equations (10.10) and (10.11) show the initiation step where one superoxide anion and one hydroperoxyl radicals are generated. Equations (10.12) to (10.20) explain the propagation step with the generation of ozonide anion, hydroperoxyl radical, and ·OH radicals. In this step, organic radicals are generated by the reaction of ·OH radicals with organic molecules. Another possible reaction is the production of organic peroxy radicals in the presence of O_2. The termination step is generally initiated due to the lack of production of superoxides responsible for the chain reaction that produce ·OH radicals. Equations (10.21) to (10.23) show the scavenging of ·OH radical, which leads to the termination of the degradation reaction (Malik et al., 2020).

Step 1: Initiation

$$O_3 + OH^- \rightarrow O_2 \cdot^- + HO_2 \cdot \tag{10.10}$$

$$HO_2 \cdot^- \leftrightarrow O_2 \cdot^- + H^+ \tag{10.11}$$

Step 2: Propagation

$$O_3 + O_2 \cdot^- \rightarrow O_3 \cdot^- + O_2 \tag{10.12}$$

$$HO_3 \cdot \rightarrow O_3 \cdot^- + H^+ \tag{10.13}$$

$$HO_3 \cdot \rightarrow \bullet OH + O_2 \tag{10.14}$$

$$\bullet OH + O_3 \rightarrow HO_4 \cdot \tag{10.15}$$

$$HO_4 \cdot \rightarrow O_2 + HO_2 \cdot \tag{10.16}$$

$$H_2R + \bullet OH \rightarrow HR \cdot + H_2O \tag{10.17}$$

$$HR \cdot + O_2 \rightarrow HRO_2 \cdot \tag{10.18}$$

$$HRO_2 \cdot \rightarrow R + HO_2 \cdot \tag{10.19}$$

$$HRO_2 \cdot \rightarrow RO + \bullet OH \tag{10.20}$$

Step 3: Termination

$$\bullet OH + CO_3^{2-} \rightarrow CO_3 \cdot^- + OH^- \tag{10.21}$$

$$\bullet OH + HCO_3^- \rightarrow HCO_3 \cdot + OH^- \tag{10.22}$$

$$\bullet OH + \cdot OH_2 \rightarrow O_2 + H_2O \tag{10.23}$$

In the direct mechanism, molecular ozone reacts with selective organic compounds at a reaction rate much lower than the indirect mechanism. Due to its dipolar structure, the ozone reacts with the unsaturated bond and splits the bond. Depending on the degree of nucleophilicity or electron density, ozone reacts with the molecules (Malik et al., 2020). In direct reactions, three ozone molecules are required to produce 2 ·OH radicals as shown in Equation (10.24).

Direct Mechanism

$$3O_3 + OH^- + H^+ \rightarrow 2 \bullet OH + 4O_2 \tag{10.24}$$

Greater performance of ozone is obtained at high pH values (>11.0) as it acts non-selective for the oxidation of any organic and inorganic constituents present in the reacting medium. At low pH decomposition of ozone occurs at a very slow rate, favoring the direct reaction between ozone and the target compounds. Though ozonation can occur at room temperature and pressure, the process has many drawbacks. The slow reaction rate, less mass transfer, and high energy consumption for the generation of ozone are a few. To enhance the efficiency of ozonation systems, it is coupled with other systems based on UV, H_2O_2, catalysts, electrocoagulation, sonolysis, metal oxides, etc. Among the coupled systems, the most popular is the O_3/UV and O_3/H_2O_2 systems. In the O_3/UV process, ozone in an aqueous medium forms H_2O_2 and $\cdot$OH radical. Further, H_2O_2 undergoes photolysis and produces two $\cdot$OH radicals. H_2O_2 can also produce H_3O^+ and peroxyl anion HO_2^-. To propagate the reaction, O_3 and H_2O_2 react with $\cdot$OH radicals to generate superoxide radicals. These superoxides react with $\cdot$OH radicals and produce hydroxyl ions. These ions react with $\cdot$OH radicals to produce $O.^-$. Superoxides also react with O_3 to form $O_3^{\cdot-}$, which further generates $\cdot$OH radicals.

10.5 ELECTROCHEMICAL ADVANCED OXIDATION PROCESS (EAOP)

Electricity was first suggested for water treatment in 1889. Subsequently, many electrochemical technologies were developed for the treatment of wastewaters (Chen, 2004). EAOPs are recognized as one of the novel processes for the treatment of industrial effluents. They are also considered one of the effective pre-treatment methods for organic pollutants before the biological degradation process (Li et al., 2019). Electrochemical processes used in wastewater treatment can be classified into two categories: separation and degradation processes. Separation processes include membrane-based electrochemical technology, electrocoagulation, and internal microelectrolysis. In these processes, pollutants are removed from the aqueous medium without altering their chemical structure. In the case of degradation processes—the focus of our discussion—the parent compounds are degraded causing the chemical bonds to break, resulting in the formation of intermediate compounds and subsequently leading to mineralization. The fundamental process in this category is anodic oxidation (AO). In this, the oxidation of pollutants is achieved in three steps: (i) direct electron transfer at anode surface (M), (ii) generation of reactive oxygen species (ROS), $\cdot$OH radicals, and H_2O_2, and (iii) production of other weaker oxidant species electrochemically generated from ions existing in the medium (Moreira et al., 2017). See Equations (10.25) to (10.28).

$$M + H_2O \rightarrow M(\bullet OH) + H^+ + e^- \qquad (10.25)$$

$$2M(\cdot OH) \rightarrow 2MO + H_2O_2 \qquad (10.26)$$

$$3H_2O \rightarrow O_3 + 6H^+ + 6e^- \qquad (10.27)$$

$$M(\bullet OH) \rightarrow MO + H^+ + e^- \qquad (10.28)$$

The electrochemical degradation is coupled with many other AOPs, such as Fenton systems, persulfate (PS)-based systems, and O_3-based system. The most studied combinations are anodic Fenton, electro-Fenton, photoelectro-Fenton, solar pho-toelectro-Fenton, photoelectrocatalysis, EAOP-PS, and EAOP-Peroxone. There are many anode materials like RuO_2, IrO_2, Pt, graphite, Ebonex® (Ti_4O_7), PbO_2, SnO_2, and boron-doped diamond (BDD) used for O_2 generation in the process. Based on the O_2 evolution potential, these anode materials are categorized into active and nonac-tive electrodes. RuO_2, IrO_2, Pt, graphite, and other SP^2 carbon-based electrodes are active anodes whose O_2 evolution potential is low (1.8 V/SHE). On the other hand, PbO_2, SnO_2, BDD, and sub-stoichiometric TiO_2 electrodes are nonactive extruders as their O_2 evolution potential is 1.7 to 2.6 V/SHE (Moradi et al., 2020). These are clean technologies as the reagents involved in the reactions are ·OH radicals and electrons, which do not cause any sludge or secondary pollutants. Additional advan-tages are strong catalytic ability, low-volume application, mild operating conditions, and easy operation.

10.6 IONIZING RADIATION

Scientific research on the application of ionizing radiations to treat water and waste-water started in the 1950s. This category of AOPs was the most researched one in the 1990s and the first decade of the 2000s (Trojanowicz, 2020). Ionizing radiation can decompose water molecules (water radiolysis) and result in the generation of radical species like ·OH, ·H, and ·HO_2 and hydrated electrons. This phenomenon happens in three stages with different rates: physical, physicochemical (involving several processes), and the chemical stage (generation of main products of water radiolysis). The products of radiolysis of water are shown in Equation (10.29). Values in the brackets indicate the average radiation chemical yield (G-value in mol J^{-1}). As the radicals formed are highly reactive and unstable and undergo protonation, the effective G-value of a solution depends upon the pH, dissolved gases, and other scavengers of the radicals (Trojanowicz, 2020).

$$H_2O \rightarrow \bullet OH \ (2.8), \ \cdot H(0.6), \ e_{aq}^{\ -}(2.7), \ H_2(0.45), \ H_2O_2 \ (0.72), \ H_3O^+(2.6) \quad (10.29)$$

·OH with 2.72V oxidation potential and $e_{aq}-$ with 2.9 V reduction potential are two major reactive species involved in the degradation of organic pollutants present in the reaction medium via oxidation and reduction, respectively (Wang and Chu, 2016). Apart from hydrated electrons, hydrogen radicals play a predominant role in the reductive conditions. It is also important to understand the radiolytic process in deaerated and aerated solutions because the O_2 in the solution scavenges both the reducing species—electrons and hydrogen radicals—as given in Equations (10.30) and (10.31). ·OH, ·H, and $e_{aq}-$ can be scavenged by CO_3^{2-}, HCO_3^-, NO_3^- and NO_2^- (Wang et al., 2019).

$$\cdot H + O_2 \rightarrow HO_2. \quad (10.30)$$

$$e^- + O_2 \rightarrow O_2^{\cdot -} \quad (10.31)$$

Equations (10.32) to (10.35) are some of the other typical radiolytic reactions in the aqueous media.

$$HO_2 \cdot \leftrightarrow H^+ + O_2 \cdot^- \quad pK = 4.8 \tag{10.32}$$

$$HO_2 \cdot + O_2 \cdot^- \rightarrow H_2O_2 + O_2 \ (pH < 7) \tag{10.33}$$

$$e_{aq}^- + H_2O_2 \rightarrow \cdot OH + OH^- \tag{10.34}$$

$$\cdot H + H_2O_2 \rightarrow \cdot OH + H_2O \tag{10.35}$$

Acidic conditions are more favorable for the radiolytic degradation reactions than alkaline conditions. In alkaline conditions, the OH^- ions react with $\cdot OH$ to produce O^- and H_2O, which leads to a decrease in the concentration of $\cdot OH$-reducing treatment efficiency. See Equations (10.36) to (10.38).

$$e^- + H^+ \rightarrow \cdot H \tag{10.36}$$

$$e^- + \cdot OH \rightarrow OH^- \tag{10.37}$$

$$\cdot OH + OH^- \rightarrow H_2O + O^- \tag{10.38}$$

Water radiolysis is initiated by a wide variety of ionization radiations like high energy electromagnetic radiations and particle-based radiations. High energy electron beam irradiation is carried out under an electron accelerator. Whereas, γ-rays are typically produced from isotopic sources ^{60}Co and ^{137}Cs. Cobalt isotopes are the most popular, have a 1,925-days half-life time, and emit photons of energy 1.17 and 1.33 MeV (Trojanowicz, 2020). Radiation-based AOP systems are also coupled with other AOPs and biological processes to enhance the removal efficiencies of desired organic components.

10.7 SONOLYSIS

In recent years, sonochemical oxidations have emerged as one of the competitive and potential AOPs for the degradation of various recalcitrant components. Sonolysis reactions are achieved using ultrasound, which are sound waves with frequencies above the hearing range of the human ear, i.e., above 20 kHz. According to the frequency, ultrasound is divided into three categories: low, high, and very high. In the oxidation/degradation process, low- and high-frequency waves are used rather than very high-frequency waves. This technique principally works on acoustic cavitation that includes the formation, growth, and implosive collapse of bubbles in a liquid. When high-frequency ultrasound waves are passed through an aqueous medium, the acoustic bubble formed is known as acoustic cavitation. Microbubbles generated during this process increase or decrease in size. This change in size continues till they reach the resonance size to undergo the explosion. During the process of cavitation, the extremely high temperature (~5000°C) and high pressure (~500 bar) of imploding cavitation bubbles lead to the thermal dissociation of the water molecule (pyrolytic cleavage), which in turn results in the generation of various free radicals like ·OH and ·H at the bubble interface or in the liquid bulk (Chakma et al., 2015). Water sonolysis and sonochemical oxidation and reduction reactions can proceed in the presence of any gas.

The following reaction (Equations (10.39) to (10.44)) shows the generation of radicals in the sonolysis mechanism.

$$H_2O + Sonication \rightarrow \cdot OH + \cdot H \qquad (10.39)$$

$$\cdot H + O_2 \rightarrow HO_2 \cdot \qquad (10.40)$$

$$2HOO \cdot \rightarrow H_2O_2 + O_2 \qquad (10.41)$$

$$\bullet OH + \bullet OH \rightarrow H_2O_2 \qquad (10.42)$$

$$O_2 \rightarrow 2 \cdot O \qquad (10.43)$$

$$\cdot O + H_2O \rightarrow 2 \bullet OH \qquad (10.44)$$

Since bubbles in this process play a key role by acting as a center for chemical reactions, they are considered hot-spots or microreactors. Lights emitted during the bubble collapse are known as sonoluminescence and sonochemiluminescence.

Sonolytic reactions (cavitation process) can be explained in three zones: bulk solution medium, the interface of a cavitation bubble, and the innermost central cavity of the bubble. Depending on the nature of pollutants such as hydrophilic, hydrophobic, and volatile substances, their degradation takes place in any of the three zones. Hydrophilic components get placed in a bulk solution medium and get oxidized by ·OH radicals produced after the collapse of the bubble. Hydrophobic nonvolatile compounds accumulate in the interface zone and undergo degradation either by radicals or thermal reactions. Finally, volatile substances enter the innermost zone of the bubble and get pyrolyzed. These reactions in ultrasound follow pseudo-first-order kinetics (Adewuyi, 2001). Reactors for sonolysis are two types (bath and probe) for the generation of microbubbles. Based on the placement of the source of the ultrasound, sonication is classified as direct (probe) and indirect (bath). In the former class, the ultrasound source is in direct contact with the liquid medium, and in the latter one, a vessel is dipped into the medium (Fındık, 2018). In both the reactors, ultrasound waves are produced by the transducer. This part of the reactor converts electrical power to waves and is attached to one end of the probe and the vibrating plate. Apart from the transducer, a cooler/heater device regulates the temperature of the bulk medium. However, these systems depend on both ultrasonic (frequency and power) and experimental conditions (pH, temperature, dissolved gas). Apart from these parameters, sonolysis also depends on substrate characteristics like vapor pressure, density, surface tension, etc. Complications in the design of the reactor, cost of the treatment, low mineralization efficiency, and high energy consumption are the main drawbacks when sonolysis is used as a single treatment due to the poor conversion efficiency of electrical-to-sound-to-thermal energies. To overcome the drawbacks, similar to other AOPs, ultrasound reactors are coupled with photocatalysis (sonophotpocatalysis), Fenton-based systems (Sonofenton oxidation), ozonation (sonozonation), metal oxides (sonocatalysis), and oxidants like H_2O_2 and persulfate (Wu et al., 2020).

10.8   PERSULFATE OXIDATION

AOPs are gaining acceptance as an alternative to the conventional treatment methods for the removal of emerging pollutants like pharmaceuticals, pesticides, and many other aromatic molecules. Though most AOP systems have $\cdot$OH radicals as the prominent oxidant, recent studies have demonstrated the effectiveness of other species like persulfate (PS).

PS is a strong oxidizing agent with redox potential $E^0(S_2O_8^{2-}/SO_4^{2-}) = 2.05$ V but with a limited rate of reactions in the oxidation process. To enhance the efficiency and oxidization rates, PS is activated by heat, electrons, UV, metal ions, metal oxides, or bases to sulfate radicals ($SO_4\cdot^-$), which is a powerful oxidant (Yadav et al., 2018). The following (Equations (10.45) to (10.48)) are the reaction mechanisms for the generation of $SO_4\cdot^-$ in different activation methods.

$$S_2O_8^{2-} + \text{heat} \rightarrow 2SO_4\cdot^- \tag{10.45}$$

$$S_2O_8^{2-} + h\nu \rightarrow 2SO_4\cdot^- \tag{10.46}$$

$$S_2O_8^{2-} + e^- \rightarrow SO_4^{2-} + SO_4\cdot^- \tag{10.47}$$

$$S_2O_8^{2-} + Mn^+ \rightarrow SO_4\cdot^- + SO_4^{2-} + M^{(n+1)+} \tag{10.48}$$

Sulfate radicals have high redox potential (2.65–3.1 V), even greater than that of $\cdot$OH radicals. They have high solubility and undergo relatively mild reactions during the process. Though only $\cdot SO_4^-$ radicals are produced when persulphate ions are irradiated with light, $\cdot$OH can form when the sulfate radicals react with water as shown in Equation (10.50). The reaction of sulfate radicals with OH^- under alkaline conditions can also result in the formation of $\cdot$OH, as shown in Equation (10.51). The generation of sulfate and $\cdot$OH radicals accelerates the oxidation process in the reaction matrix. Once $SO_4\cdot^-$ radicals are generated, they are involved in reactions such as H_2 abstraction from saturated carbon, addition on a double bond, and electron transfer, with the target compound. Being an uncharged species, the later behavior of sulfate radicals is not possible by ·OH radicals. Apart from the mentioned activators, TiO_2 in the photocatalysis system can activate the PS with the photon generated electrons in the conduction band as in Equation (10.47). Due to this, photocatalytic efficiency can be increased by limiting the e^-/h^+ recombination (Ismail et al., 2017). Once the PS is irradiated using UV, it generates a chain of reactions leading to the production of reactive species (Ye et al., 2016) as follows (Equations (10.49) to (10.54)):

$$S_2O_8^{2-} + h\nu \rightarrow 2 \cdot SO_4^- \qquad (10.49)$$

$$\cdot SO_4^- + H_2O \rightarrow HSO_4^{2-} + \cdot OH \qquad (10.50)$$

$$\cdot SO_4^- + OH^- \rightarrow SO_4^{2-} + \cdot OH \qquad (10.51)$$

$$\cdot OH + \cdot OH \rightarrow H_2O_2 \qquad (10.52)$$

$$\cdot OH + \cdot SO_4^- \rightarrow HSO_5^- \qquad (10.53)$$

$$HSO_5^- + \cdot OH \rightarrow \cdot SO_5^- + H_2O \qquad (10.54)$$

The pH plays an important role in the generation and persistence of species in the reaction medium for the oxidation and reduction process. At pH<7, $\cdot SO_4^-$ is the predominant radical when pH=9, since both ·OH and $\cdot SO_4^-$ are present, whereas at pH>9, ·OH is the predominant radical species. PS oxidation system efficiency also depends upon the type of activator, time taken for the activation, and target compound in the aqueous medium. Many transition metal activators such as Co, Fe^{2+}, FeO, Fe_3O_4, and Cu are used in the PS systems. But their stability in the reaction medium has to be investigated. Ling et al. has reported that, though Cu was stable, it took a long time for the activation of PS. It was also reported that high catalytic activity for PS is observed with iron oxidase rather than with copper oxides (Liang et al., 2013).

10.9 PHOTOLYSIS AND PHOTOCATALYSIS

Photolysis is a mild chemical reaction triggered by artificial or natural light resulting in the degradation of a target molecule. Photochemical degradation mechanisms involve direct and indirect photolysis (Jiao et al., 2008). Direct photolysis relies on the absorption of light and consequent decomposition. Indirect photolysis comprises reactions with transient exciting species such as singlet oxygen ($1O_2$), hydroxyl radical (·OH), triplet excited-state dissolved organic matter (3DOM), or other radical species. A commonly used source of energy is ultraviolet (UV) and visible light. UV irradiations can be categorized based on their wavelength: (i) UV-A (λ = 320–400nm), (ii) UV-B (λ = 280–320 nm), and (iii) UV-C (λ< 280 nm). The higher the energy of the UV, the greater its ability for photolysis. As photolysis depends upon the sensitivity of the compound and the nature of the medium, not all organic compounds can be treated with photolysis alone. A photolysis-assisted oxidant, catalyst, etc., accelerate the rate of degradation.

Heterogeneous photocatalysis can be understood as an association of semiconductor solid catalyst with energy (light) of appropriate wavelength for degradation of molecules. At ambient temperatures, photocatalyst contains electrons in its valance band. When the suitable light (photon) with appropriate wavelength (<380 nm) and energy greater than the bandgap of semiconductor material excites electrons from the valance band to the conduction band, electron-hole pairs (e^-/h^+) are generated (Pelaez et al., 2012). In a semiconductor, the bandgap is an energy gap that spreads from the top of the filled valence band to the bottom of the empty conduction band. Wide varieties of materials have been documented as a semiconductor-nano photocatalyst and each material has a certain bandgap (Shavisi et al., 2016). Some of the catalysts and their band gaps are as follows: ZrO_2 (5 eV), ZnS (3.8 eV), TiO_2 (3.2 eV), ZnO (3.2 eV), SiC (3.0 eV), WO_3 (2.7 eV), CdS (2.8 eV), MoS_2(1.8 eV), and Si (1.1 eV) (Byrne et al., 2018). Among the different semiconductor photocatalysts, TiO_2 was widely used due to its high excitation binding energy, crystalline structure (anatase, rutile, and brookite), and high photocatalytic activity. In 1972, Fujishima and Honda discovered water splitting using TiO_2 (Zangeneh et al., 2015). Since then it is being celebrated across the world as a photocatalyst especially in the fields of energy and environment. Many studies were conducted to improve its performance and to understand its advantages over other materials (Lee et al., 2016). The basic mechanism of photocatalysis by TiO_2 occurs in five steps, given below (Equations (10.55) to (10.59)):

$$\text{Step 1: Photo reaction, } TiO_2 + h\nu \rightarrow e^- + h^+ \quad (10.55)$$

$$\text{Step 2: Electron-hole recombination, } e^- \text{ (CB) + } h^+ \text{ (VB)} \rightarrow \text{recombination + heat} \quad (10.56)$$

$$\text{Step 3: } OH^- \text{ oxidation, } OH^- \text{ (surface) + } h^+ \rightarrow \bullet OH \quad (10.57)$$

$$\text{Step 4: } O_2 \text{ reaction, } O_2 + e^- \rightarrow \cdot O_2^- \quad (10.58)$$

$$\text{Step 5: Pollutant degradation, } \bullet OH + RH \rightarrow \cdot R + H_2O \quad (10.59)$$

The detailed photocatalytic mechanism for the degradation of organic pollutants can be explained this way; when the UV light photons are absorbed by TiO_2, the charge carriers (electrons and holes) are developed, which initiates the chain of redox reactions at the surface of the catalyst with adsorbed pollutant or tends to recombine, by dispersing energy as shown in Equation (10.60). Rapid reactions of molecular O_2 and the electrons from the conduction band produce superoxides ($O_2^{\cdot-}$) as shown in Equation (10.61) and (10.62) (Nuengmatcha et al., 2016). On the other hand, positive electron vacancies interact with water molecules to generate $\cdot OH$ and hydrogen ions

H^+ as given in Equation (10.63). $\cdot OH$, which plays a pivotal role in the photocatalysis, is also generated by the reactions of superoxides, hydrogen ions, and other species present in the aqueous matrix as shown in Equations (10.40) to (10.44) and (10.67). Equations (10.64) to (10.70) explains the reactions that occurs in the photocatalytic system for the degrdation of pollutants. (Nakata and Fujishima, 2012).

$$e^- \text{ (CB)} + h^+ \text{ (VB)} \rightarrow \text{Energy} \tag{10.60}$$

$$e^-_{cb} \text{ (TiO}_2) + O_2 \rightarrow HO_2\cdot \tag{10.61}$$

$$HO_2 + TiO_2 \rightarrow H^+ + O_2\cdot^- + TiO_2 \tag{10.62}$$

$$h^+_{vb} \text{ (TiO}_2) + H_2O \rightarrow \cdot OH + TiO_2 + H^+ \tag{10.63}$$

$$O_2\cdot^- + H^+ \rightarrow HO_2\cdot \tag{10.64}$$

$$HO_2\cdot + HO_2\cdot \rightarrow H_2O_2 + O_2 \tag{10.65}$$

$$e^-_{cb} \text{ (TiO}_2) + H_2O_2 \rightarrow \cdot OH + OH^- \tag{10.66}$$

$$O_2\cdot^- + H_2O_2 \rightarrow \cdot OH + OH^- + O_2 \tag{10.67}$$

$$h^+_{vb} \text{ (TiO}_2) + OH^- \rightarrow \cdot OH + TiO_2 \tag{10.68}$$

$$\cdot OH + \text{Pollutants} \rightarrow H_2O + CO_2 \tag{10.69}$$

$$\text{Other radicals} + \text{Pollutants} \rightarrow H_2O + CO_2 \tag{10.70}$$

10.10 SUSTAINABILITY OF AOP SYSTEMS

One of the latest challenges in sewage treatment is the increasing concentration of recalcitrant compounds in the sewage. The traditional biological treatment methods are found to be incapable of addressing this challenge. With the ever-increasing use of personal care products and pharmaceuticals that contribute to the recalcitrant compounds in sewage, this challenge is only bound to intensify. So far, only AOPs have the proven ability to tackle the issue effectively. The commercialization of AOPs is relatively slow, mainly because of the associated cost. Nevertheless, many commercial installations exist and many more are coming up in different parts of the world, mostly in Europe, the United States, and China. With more stringent regulations in place regarding the discharge of the "emerging pollutants," AOP systems are expected to become much more common.

The sustainability of AOP systems depends on how green the oxidant and the reaction products are and how sustainable the oxidant generation processes are. ·OH radical as an oxidant is green as it does not generate any toxic reaction products. But, some of the processes used for the generation of ·OH have issues like high sludge generation. Incomplete mineralization of the pollutant is another concern, as there are instances where the degradation products of AOPs were found to be more toxic than the pollutants themselves. But, ensuring complete mineralization and adoption of sustainable methods for the generation of oxidants can make AOPs more sustainable. More research on process modifications, and use of less costly reagents, shall also have a positive impact on the cost aspect. In any case, as long as the removal of recalcitrant compounds from sewage remains a requirement, AOPs shall remain relevant.

REFERENCES

Adewuyi, Y.G., 2001. Sonochemistry: Environmental science and engineering applications. Ind. Eng. Chem. Res. 40, 4681–4715. https://doi.org/10.1021/ie010096l

Bello, M.M., Abdul Raman, A.A., Asghar, A., 2019a. A review on approaches for addressing the limitations of Fenton oxidation for recalcitrant wastewater treatment. Process Saf. Environ. Prot. 126, 119–140. https://doi.org/10.1016/j.psep.2019.03.028

Byrne, C., Subramanian, G., Pillai, S.C., 2018. Recent advances in photocatalysis for environmental applications. J. Environ. Chem. Eng. 6, 3531–3555. https://doi.org/10.1016/j.jece.2017.07.080

Chakma, S., Das, L., Moholkar, V.S., 2015. Dye decolorization with hybrid advanced oxidation processes comprising Sonolysis/Fenton–like/photo–ferrioxalate systems: A mechanistic investigation. Seperation Purif. Technol. 156, 596–607. https://doi.org/10.1016/j.seppur.2015.10.055

Chen, G., 2004. Electrochemical technologies in wastewater treatment. Sep. Purif. Technol. 38, 11–41. https://doi.org/10.1016/j.seppur.2003.10.006

Deng, Y., Zhao, R., 2015. Advanced oxidation processes (AOPs) in wastewater treatment. Curr. Pollut. Reports 1, 167–176. https://doi.org/10.1007/s40726-015-0015-z

Fenton, H.J.H., 1894. Oxidation of tartaric acid in the presence of iron. J. Chem. Soc. 65, 899–910.

Fındık, S., 2018. Treatment of petroleum refinery effluent using ultrasonic irradiation. Polish J. Chem. Technol. 20, 20–25.

Halliwel, B., 1991. Drug antioxidant effects: A basis for drug selection? Drugs 42, 569–605.

Ismail, L., Ferronato, C., Fine, L., Jaber, F., Chovelon, J.-M., 2017. Elimination of sulfaclozine from water with SO4– radicals: Evaluation of different persulfate activation methods. Appl. Catal. B Environ. 201, 573–581. https://doi.org/10.1016/j.apcatb.2016.08.046

Jiao, S., Zheng, S., Yin, D., Wang, L., Chen, L., 2008. Aqueous photolysis of tetracycline and toxicity of photolytic products to luminescent bacteria. Chemosphere 73, 377–382. https://doi.org/10.1016/j.chemosphere.2008.05.042

Lee, K.M., Lai, C.W., Ngai, K.S., Juan, J.C., 2016. Recent developments of zinc oxide based photocatalyst in water treatment technology: A review. Water Res. 88, 428–448. https://doi.org/10.1016/j.watres.2015.09.045

Li, J., Li, Y., Xiong, Z., Yao, G., Lai, B., 2019. The electrochemical advanced oxidation processes coupling of oxidants for organic pollutants degradation: A mini-review. Chinese Chem. Lett. 30, 2139–2146. https://doi.org/10.1016/j.cclet.2019.04.057

Liang, H.Y., Zhang, Y.Q., Huang, S. Bin, Hussain, I., 2013. Oxidative degradation of p-chloroaniline by copper oxidate activated persulfate. Chem. Eng. J. 218, 384–391. https://doi.org/10.1016/j.cej.2012.11.093

Malik, S.N., Ghosh, P.C., Vaidya, A.N., Mudliar, S.N., 2020. Hybrid ozonation process for industrial wastewater treatment: Principles and applications: A review. J. Water Process Eng. 35, 101193. https://doi.org/10.1016/j.jwpe.2020.101193

Moradi, M., Vasseghian, Y., Khataee, A., Kobya, M., Arabzade, H., Dragoi, E.N., 2020. Service life and stability of electrodes applied in electrochemical advanced oxidation processes: A comprehensive review. J. Ind. Eng. Chem. 87, 18–39. https://doi.org/10.1016/j.jiec.2020.03.038

Moreira, F.C., Boaventura, R.A.R., Brillas, E., Vilar, V.J.P., 2017. Electrochemical advanced oxidation processes: A review on their application to synthetic and real wastewaters. Appl. Catal. B Environ. 202, 217–261. https://doi.org/10.1016/j.apcatb.2016.08.037

Nakata, K., Fujishima, A., 2012. TiO 2 photocatalysis: Design and applications. J. Photochem. Photobiol. C Photochem. Rev. 13, 169–189. https://doi.org/10.1016/j.jphotochemrev.2012.06.001

Nuengmatcha, P., Chanthai, S., Mahachai, R., Oh, W., 2016. Visible light-driven photocatalytic degradation of rhodamine B and industrial dyes (texbrite BAC-L and texbrite NFW-L) by ZnO-graphene-TiO2 composite. J. Environ. Chem. Eng. 4, 2170–2177. https://doi.org/10.1016/j.jece.2016.03.045

Pelaez, M., Nolan, N.T., Pillai, S.C., Seery, M.K., Falaras, P., Kontos, A.G., Dunlop, P.S.M.M., Hamilton, J.W.J.J., Byrne, J.A.A., Shea, K.O., Entezari, M.H., Dionysiou, D.D., O'Shea, K., Entezari, M.H., Dionysiou, D.D., 2012. A review on the visible light active titanium dioxide photocatalysts for environmental applications. Appl. Catal. B Environ. 125, 331–349. https://doi.org/10.1016/j.apcatb.2012.05.036

Prato-Garcia, D., Vasquez-Medrano, R., Hernandez-Esparza, M., 2009. Solar photoassisted advanced oxidation of synthetic phenolic wastewaters using ferrioxalate complexes. Sol. Energy 83, 306–315. https://doi.org/10.1016/j.solener.2008.08.005

Shavisi, Y., Sharifnia, S., Mohamadi, Z., 2016. Solar-light-harvesting degradation of aqueous ammonia by CuO/ZnO immobilized on pottery plate: Linear kinetic modeling for adsorption and photocatalysis process. J. Environ. Chem. Eng. 4, 2736–2744. https://doi.org/10.1016/j.jece.2016.04.035

Staehelln, J., Hoigne, J., 1985. Decomposition of ozone in water in the presence of organic solutes acting as promoters and inhibitors of radical chain reactions. Environ. Sci. Technol. 19, 1206–1213. https://doi.org/10.1021/es00142a012

Trojanowicz, M., 2020. Removal of persistent organic pollutants (POPs) from waters and wastewaters by the use of ionizing radiation. Sci. Total Environ. 718, 134425. https://doi.org/10.1016/j.scitotenv.2019.134425

Wang, J., Chen, H., 2020. Catalytic ozonation for water and wastewater treatment: Recent advances and perspective. Sci. Total Environ. 704, 135249. https://doi.org/10.1016/j.scitotenv.2019.135249

Wang, J., Zhuan, R., Chu, L., 2019. The occurrence, distribution and degradation of antibiotics by ionizing radiation: An overview. Sci. Total Environ. 646, 1385–1397. https://doi.org/10.1016/j.scitotenv.2018.07.415

Wu, Z., Abramova, A., Nikonov, R., Cravotto, G., 2020. Sonozonation (sonication/ozonation) for the degradation of organic contaminants – A review. Ultrason. Sonochem. 68, 105195. https://doi.org/10.1016/j.ultsonch.2020.105195

Yadav, M.S.P., Neghi, N., Kumar, M., Varghese, G.K., 2018. Photocatalytic-oxidation and photo-persulfate-oxidation of sulfadiazine in a laboratory-scale reactor: Analysis of catalyst support, oxidant dosage, removal-rate and degradation pathway. J. Environ. Manage. 222, 164–173. https://doi.org/10.1016/j.jenvman.2018.05.052

Ye, J., Liu, J., Ou, H., Wang, L., 2016. Degradation of ciprofloxacin by 280 nm ultraviolet-activated persulfate: Degradation pathway and intermediate impact on proteome of Escherichia coli. Chemosphere 165, 311–319. https://doi.org/10.1016/j.chemosphere.2016.09.031

Zangeneh, H., Zinatizadeh, A.A.L., Habibi, M., Akia, M., Hasnain Isa, M., 2015. Photocatalytic oxidation of organic dyes and pollutants in wastewater using different modified titanium dioxides: A comparative review. J. Ind. Eng. Chem. 26, 1–36. https://doi.org/10.1016/j.jiec.2014.10.043

Zhang, M. Hui, Dong, H., Zhao, L., Wang, D. Xi, Meng, D., 2019. A review on Fenton process for organic wastewater treatment based on optimization perspective. Sci. Total Environ. 670, 110–121. https://doi.org/10.1016/j.scitotenv.2019.03.180

Part IV

Nascent Technologies

11 Hydrodynamic Cavitation: A Novel Energy Efficient Technology for Degradation of Organic Contaminants in Wastewater

Gautham B. Jegadeesan* and N. Santosh Srinivas
School of Chemical & Biotechnology,
SASTRA Deemed University, Thanjavur, India

CONTENTS

11.1 INTRODUCTION

Over the years, an increase in supply and demand for chemicals has led to unintended consequences—increased water use in the industries and thereby generation of highly contaminated industrial effluents (UNEP 2016). The World Health Organization (WHO) reports that one-third of rivers suffer from pathogenic contamination, one-seventh from organic contamination, and ten percent of all rivers from salinity (Kim and Zoh 2016; UNEP 2016). Over the years, chemical and biological methods of treatment for the removal of the recalcitrant organic chemicals of concern (COCs) have been found wanting due to their inability to degrade them completely, high sludge generation, cost-ineffectiveness, and sustainability (Ranade and Bhandari 2014; Rajoriya et al. 2016). Given the need to provide the public with clean, healthy water free of recalcitrant compounds of concern, it is therefore essential to develop a radically different approach towards water/wastewater treatment to eliminate the limitations listed above.

In the last two decades, AOPs have been the focus for their effectiveness in degrading organic COCs such as textile dyes, aromatic compounds, pharmaceuticals and personal care products (PPCPs), and chlorinated and phenolic compounds, among others (Ranade and Bhandari 2014; Gągol, Przyjazny, and Boczkaj 2018b). Table 11.1 presents the principal reactions with each different type of AOP. The use of highly reactive radicals such as hydroxyl radicals (OH$^\bullet$) and sulfate radicals (SO$_4^\bullet$) for the degradation of organics has been a fast-emerging area of study because of its inherent benefits:(1) higher efficiencies; (2) destruction of the COCs and low levels of toxic by-products; (3) faster kinetics; and (4) wider applicability for the diverse group of organic contaminants. The first reported AOP was based on Fenton's chemistry, in which a hydrogen peroxide (H$_2$O$_2$)/Fe^{2+} ion mixture was used to generate OH$^\bullet$ and destroy persistent organic pollutants (Gągol, Przyjazny, and Boczkaj 2018b; Oturan and Aaron 2014). Due to the large-sludge formation with excessive use of Fe^{2+}, peroxonation, a method of generating the hydroxyl radicals using ozone (O$_3$) and H$_2$O$_2$, was adopted (Gągol, Przyjazny, and Boczkaj 2018b; Oturan and Aaron 2014). Photocatalytic-based AOPs gained prominence given their high efficiencies, relatively low costs, ease in operation, and simplicity. When UV light is irradiated on an aqueous stream containing H$_2$O$_2$, hydroxyl radicals are generated resulting in degradation of the COCs (Badmus et al. 2018; Oturan and Aaron 2014). The rate of radical generation is dependent on several parameters including intensity of UV irradiated, pH of the medium, and amount of H$_2$O$_2$ used. Given the instability of H$_2$O$_2$ when used in large quantities, O$_3$ based UV technologies were adopted, resulting in the use of combination processes such as H$_2$O$_2$/UV, O$_3$/UV, H$_2$O$_2$/Fe^{2+}/UV (Oturan and Aaron 2014). To reduce the energy demand of the photocatalytic process owing to the large amount of UV irradiated on the semiconductor, considerable research has been conducted on reducing the band-gap energy of energetic materials (e.g., nano TiO$_2$, ZnO, CeO$_2$) to less than 2.7 eV so that the photocatalytic processes can be conducted in visible light (Ghanbari and Moradi 2017). In recent years, hydroxyl radical-based methodologies are being replaced by sulfate radical (SO$_4^\bullet$)- based methodologies due to their higher solubility, reactivity, and oxidation potential (2.5–3.1 V) (Ghanbari and Moradi 2017). Sulfate-based AOPs include persulfate (PS) and peroxymonosulfate (PMS) in combination with UV (PS/UV; PMS/UV), thermal activation of PS/PMS, and heterogeneous and homogeneous activation of PS/PMS. PS and PMS can generate both SO$_4^\bullet$ and OH$^\bullet$ radicals, which react synergistically to degrade organic COCs. The activators can be catalytic materials such as transition metals and metal oxides (e.g., Fe^0, Ni^0, Fe-oxides, etc.), photocatalytic materials, or cavitation based devices (Yang et al. 2019). However, major disadvantages of most

TABLE 11.1

Mechanisms of various advanced oxidation processes (G. Boczkaj and Fernandes 2017; Oturan and Aaron 2014)

Process	Mechanism of action	Primary Oxidant
Fenton Process	$Fe^{2+} + H_2O_2 \rightarrow Fe^{3+} + OH^{\bullet} + OH-$	$OH^{\bullet}$
	$Fe^{3+} + H_2O_2 \rightarrow Fe^{2+} + HO^{\bullet}_2 + H^+$	
	$OH^{\bullet} + H_2O_2 \rightarrow HO^{\bullet}_2 + H_2O$	
	$OH^{\bullet} + Fe^{2+} \rightarrow Fe^{3+} + OH^-$	
	$Fe^{3+} + HO^{\bullet}_2 \rightarrow Fe^{2+} + O_2H^+$	
	$Fe^{2+} + HO^{\bullet}_2 + H^+ \rightarrow Fe^{3+} + H_2O_2$	
	$2HO^{\bullet}_2 \rightarrow H_2O_2 + O_2$	
Ozone	$3O_3 + H_2O \rightarrow 2OH^{\bullet} + 4O_2$	$OH^{\bullet}$
Ozone-H_2O_2	$H_2O_2 \rightarrow HO_2^- + H^+$	$OH^{\bullet}$
	$HO_2^- + O3 \rightarrow OH^{\bullet} + O_2^- + O_2$	
Ozone-UV	$3O_3 + H_2O \rightarrow 2OH^{\bullet} + 4O_2$ (Ozonation)	$OH^{\bullet}$
	$OH^{\bullet} + O_3 \rightarrow HO_2^{\bullet} + O_2$	
	$O_3 + H_2O + hv \rightarrow H_2O_2 + O_2$ (Photolysis)	
	$H_2O_2 + hv \rightarrow 2OH^{\bullet}$ (Peroxide oxidation)	
Photocatalysis	$TiO_2 + hv \rightarrow e^-_{cb} + hv^+_{vb}$	$OH^{\bullet},\ O_2^{\bullet-},\ \bullet OOH$
	$e^-_{CB} + h^+_{VB} \rightarrow$ energy	
	$O_2 + e^-_{cb} \rightarrow O_2^{\bullet}$	
	$O_2^{\bullet} + H^+ \rightarrow \bullet OOH$	
	$\bullet OOH + \bullet OOH \rightarrow H_2O_2 + O_2$	
	$O_2^{\bullet-} + pollutant \rightarrow \rightarrow CO_2 + H_2O$	
	$\bullet OOH + pollutant \rightarrow CO_2 + H_2O$	
	Net Reaction: $\bullet OH + pollutant \rightarrow H_2O + CO_2$	
	$hv^+_{vb} + OH^-_{(surface)} \rightarrow OH\cdot + H^+$	
	$hv^+_{vb} + H_2O_{(absorbed)} \rightarrow OH\cdot + H^+$	
	$e^-_{cb} + O_2$ (absorbed) $\rightarrow O_2^-$	
Sulfate-based AOPs	$S_2O_8^{2u} (PS) + activator OPs + \left[SO_4^{\bullet} \right]$	$SO_4^{\bullet}$
	$HSO_5^- (PMS) + activator \left[SO_4^{\bullet-} \right] + OH^- + 2e^-$	$OH^{\bullet}$
	$\left[SO_4^{\bullet-} \right] + H_2O + e^- \rightarrow SO_4^{2S} + H^+ + \left[OH^{\bullet} \right]$	
	$\left[SO_4^{\bullet-} \right] + 2OH^- \rightarrow SO_4^{2S} + 2\left[OH^{\bullet} \right]$	
Hydrodynamic cavitation (HC)	$H_2O_2 + HC \rightarrow 2OH^{\bullet}$	$OH^{\bullet}$
	$OH^{\bullet} + OH^{\bullet} \rightarrow H_2O_2$	
	$H_2O_2 + OH^{\bullet} \rightarrow H_2O + HO_2$	
	$HO_2 + OH^{\bullet} \rightarrow H_2O + O_2$	
	$HO_2 + H_2O_2 \rightarrow OH^{\bullet} + H_2O + O_2$	
	$OH^{\bullet} + O_2 \rightarrow HO_2 + O^{\bullet}$	
Ultrasonic cavitation [)))))]	$H_2O +)))))) \rightarrow H + OH^{\bullet}$	$OH^{\bullet}$
	$O_2 \rightarrow 2O$	
	$O + H_2O \rightarrow 2OH^{\bullet}$	
	$H + O_2 \rightarrow OH^{\bullet} + O$	

existing AOPs are the handling of unstable chemicals, such as H_2O_2 and O_3, post-treatment to remove excess H_2O_2 and O_3, O_3 diffusion limitations, sulfate-rich treated waters, high operating costs (e.g., UV and O_3-based technologies), and formation of toxic organic by-products. Given the impetus on developing alternate methods, new technologies such as hydrodynamic and acoustic cavitation have gained prominence in recent years, with special emphasis on incorporating green chemistry principles for sustainability (Gągol, Przyjazny, and Boczkaj 2018b). In this chapter, the basic principles of cavitation for degradation of COCs are discussed, with specific focus on hydrodynamic cavitation.

11.2 CAVITATION

Cavitation is a fluid phenomenon wherein cavities or bubbles are formed within a liquid at low-pressure regions when the liquid is accelerated at high flow velocities (Gągol, Przyjazny, and Boczkaj 2018b; Bagal and Gogate 2014; Badmus et al. 2018; Suryawanshi et al. 2018; Rajoriya et al. 2016). A liquid is said to cavitate when vapor bubbles are formed and grow as a consequence of pressure reduction in a highly accelerated fluid flow and is often classified as a two-phase flow. Cavitation can be achieved via (a) acoustic or ultrasonic waves (acoustic cavitation), (b) fluid and velocity fluctuation (hydrodynamic cavitation), (c) photons (optical cavitation), and; (d) high-intensity laser (particle cavitation). Acoustic or ultrasonic cavitation occurs due to pressure variations effected by sound waves at high frequencies (20kHz–100 MHz) (Gogate 2008), hydrodynamic cavitation via fluid flow through constrictions (Rajoriya et al. 2016), optical cavitation via the impingement of high-intensity photons on the liquid, and particle cavitation due to rupturing of liquid when irradiated with a neutron beam (Gogate 2008). However, studies have shown that acoustic and hydrodynamic devices are the most efficient of all (Rajoriya et al. 2016; Gągol, Przyjazny, and Boczkaj 2018b), hence ensuing sections focus on acoustic and hydrodynamic cavitation.

11.2.1 PRINCIPLES AND MECHANISMS

As the pressure of the flowing fluid in the localized region falls below its vapor pressure, cavitation occurs in the fluid (Tao et al. 2016; Bagal and Gogate 2014). In a homogeneous liquid system, the in situ generated cavity does not form a void; rather, it contains the gaseous form of the water or pollutant (Gogate 2008). The cavities then go through the growth phase and finally collapse in the collapse region, resulting in the release of massive amounts of energy in the vicinity of the cavity (Tao et al. 2016; Bagal and Gogate 2014). It has been noted that the temperature inside the collapsing bubble can reach as much as 5000°C and the pressure close to 10,000 atm (Gogate 2008; Bagal and Gogate 2014; Tao et al. 2016). Under such extreme conditions, water molecules inside the cavity dissociate to form OH• and H• radicals, which are effective in the oxidation of the organic pollutants. In addition to water dissociation, cleavage of dissolved oxygen can also occur as shown in Table 11.1. Degradation rates of the pollutants differ based on the location of the reaction. Studies have shown that degradation of the organic COCs can occur faster near the vicinity of the cavity (at the interface) when compared to that in the bulk fluid (Gogate 2008; Bagal and Gogate 2014; Tao et al. 2016; Gągol, Przyjazny, and Boczkaj 2018b). This is most likely due to multiple reasons: (1) increased concentration of the radicals at the interface of the cavity and (2) depletion in cavity formation as the fluid exits the cavitation zone (wherein the pressure of the fluid is equal to that in the pre-cavitation zone). Degradation in cavitation-based processes can also occur due to mechanical and thermal effects. It is possible that the large mechanical forces that are used to create high-velocity force fields can induce molecular cleavage of large molecular weight organic compounds. Studies have also reported the vaporization of the organic COCs due to the high temperature in the vicinity of the cavity (Ranade and Bhandari 2014; Rajoriya et al. 2016). Nevertheless, it is generally accepted that the rate determining the mechanism for COC degradation using cavitation is largely due to the chemical effect and that mechanical and thermal effects are insignificant (Suryawanshi et al. 2018; Ranade and Bhandari 2014; Tao et al. 2016).

11.2.1.1 Acoustic or Ultrasonic Cavitation

When ultrasound waves at frequencies ranging from 20 kHz to 2 MHz impinge the flowing fluid, localized pressure variation occurs (Tao et al. 2016; Gogate 2008). When the localized pressure difference approaches negative values (*i.e.*, vapor pressure < saturated pressure), cavities are generated. This entire phenomenon due to induced ultrasound waves is also called acoustic cavitation, and the chemical change induced due to the cavitation is known as sonochemistry. The collapse of cavities and the massive release in energy results in the formation of the hydroxyl radicals: a sonochemical reaction. Ultrasonic or acoustic cavitation-assisted AOPs haves been successfully demonstrated for the degradation of several organic COCs. Gągol, Przyjazny, and Boczkaj (2018b) published an extensive review of extant studies on wastewater treatment using an ultrasonic cavitation process. In their review, it was noted that the ultrasonic cavitation (UC) process alone could disinfect bacteria by 85%, degraded trichloroethylene completely and chlorobenzene by 50%, and removed other chlorinated compounds and dioxanes by 75–100%. The operational parameters in most studies were: (1) process temperature = 20–30°C; (2) solution pH = 3–7; (3) pressure = atmospheric pressure; (4) ultrasound properties = 20–380 kHz, 40–200 W; and (5) time for reaction = 30–180 minutes. However, the formation of toxic by-products was also observed in some studies. For example, the degradation of chloroform using UC alone resulted in the formation of C_2Cl_4 and C_2Cl_6. Similarly, ethylene glycol, formaldehyde, and formic acid were some of the intermediate daughter products formed when 1,4–dioxane was treated using UC. To minimize the formation of unwanted by-products, UC has been coupled with other AOPs such as UV, Fenton's reagent, H_2O_2, PS/PMS, and O_3, among others (Park et al. 2017; Yang et al. 2019). There was a synergistic effect of other AOPs with UC-enhanced degradation of organic COCs, with efficiencies increasing by 20–50%, depending on the contaminant (Gągol, Przyjazny, and Boczkaj 2018b). Despite the promising results obtained with the ultrasonic cavitation process, their use is largely limited to laboratory scale and/or in the developing stage. This is largely due to (1) high energy costs; (2) complex scale-up design; (3) need for multiple transducers with multiple frequencies that can provide sufficient cavitation in large-scale operations; and (4) mass transfer limitations and proper distribution of cavitational activity in the sonochemical reactor. For all these reasons alone, hydrodynamic cavitation has been the preferred choice of the cavitation reactor for wastewater treatment.

11.2.1.2 Hydrodynamic Cavitation

Hydrodynamic cavitation (HC) is more favored over acoustic cavitation due to its low operating costs, simple design, and the possibility of treating large volumes of wastewater without extensive capital investments. HC occurs when fluids at accelerated velocities pass through either a constriction such as a venture, orifice, throttling valve, or via the generation of a vortex (Rajoriya et al. 2016; Suryawanshi et al. 2018; Simpson and Ranade 2018). As the liquid passes through the geometries, the change in pressure is translated to an increase in kinetic energy, in accordance with Bernoulli's principle (Suryawanshi et al. 2018; Rajoriya et al. 2016). The pressure at the throat of the constriction decreases below the vapor pressure of the flowing liquid resulting in vaporization of the liquid and generation of a large amount of cavities. With pressure recovery downstream of the constriction, the cavities collapse, releasing large amounts of energy (up to 1000 atm and 10,000 K). These so-called "hot spots" are largely responsible for the cleavage of H_2O molecules and the generation of hydroxyl radicals. As noted earlier, the high temperatures will also result in the vaporization of the organic compounds—dependent on the boiling point and solubility of the organic compound. The effectiveness of hydrodynamic cavitation is solely dependent on the number of cavities generated and their intensities (Ranade and Bhandari 2014; Rajoriya et al. 2016). It is therefore essential to design constrictions such that fluid velocity at the throat is high enough to generate cavities. To ensure efficiently, cavitation reactors (discussed in detail in the ensuing section) are often operated based on a dimensionless parameter known as cavitation number (C_v) – as given in equation (11.1) below.

$$C_v = \frac{P_2 - P_v}{\frac{1}{2}\rho V^2}$$
(11.1)

Here P_2 is the exit pressure, P_2 is the vapor pressure of the fluid, ρ is the fluid density and V is the fluid velocity. Ideally, cavitation occurs when C_v is ≤ 1. Increased inlet pressure should be beneficial for cavitation. But, it has been postulated that cavitation is efficient in a narrow inlet pressure range because a greater number of cavities tend to coalesce forming a so-called cavity cloud that collapses at a lowered intensity than individual cavities. Therefore, cavitation processes should be operated accounting for the physiochemical properties of the fluid, chemical substance to be degraded and geometry of the cavitation device (Rajoriya et al. 2016; Gągol, Przyjazny, and Boczkaj 2018b). Regardless of the mechanisms involved, hydrodynamic cavitation-based processes have shown promise with good wastewater treatment efficiencies at low operational costs and ease in scale-up.

11.2.2 CAVITATION-ASSISTED ADVANCED OXIDATION PROCESSES

Over the last decade, several investigators have reported the effectiveness of HC in the treatment of aromatic/aliphatic compounds, pharmaceutical residues, and chlorinated compounds among others (Li, Song, and Yu 2014; Tao et al. 2016; Badmus et al. 2018). An exhaustive review of extant literature has been published by Rajoriya et al. (2016), Gagol et al. (2018b), and Badmus et al. (2018). In the last couple of years, research has been focused on optimizing the parameters for wastewater treatment using HC, in conjunction with other AOPs, and determining which cavitating device provides the best efficiencies. Table 11.2 summarizes some of the key studies reported in the literature using hydrodynamic cavitation. The cavitating devices used in most studies were venture (circular or slit), orifice plate and vortex diode (Rajoriya et al. 2016; Suryawanshi et al. 2018; Badmus et al. 2018). The operational parameters in most studies were (1) inlet fluid pressure = 2–15 bar (depending on the type of device), (2) cavitation number = 0.067–0.4, (3) solution pH = 2–12 (though most studies were performed at a pH range of 2–6.5), (4) fluid temperature = 25–50°C, and (5) pump power = 0.75–3.5 kW. It has been commonly reported that the use of HC alone (either using orifice or venture as the cavitating device) resulted in less than 40% organic COCs degradation, irrespective of the operating conditions. As seen in Table 11.2, the degradation observed were as follows: 2, 4-dinitrophenol (12%); 2,4,6- trichloro phenol (28%), phenol from bitumen effluent (25%); carbamazepine (27–38.7%); Rhodamine B (22%); Reactive Red 180 and Direct Orange 46 (4.6%); imidaclorprid (26.24%); methyl parathion (45%); dimethyl hydrazine (37%); and diclofenac (26.85%). However, better efficiencies were reported for bacteria and single industrial dyes. In the case of bacteria, HC process alone can disrupt *E. coli* by 90%, and *Microsystis aeruginosa* by 94% (Li, Song, and Yu 2014). Multiple studies observed 37.23–92% dye decolorization (Acid Red 88, Reactive Red 120, Rhodamine B, etc.) with 22–35% total organic carbon (TOC) removal with the use of venturi-based HC process (Rajoriya, Bargole, and Saharan 2017a; Rajoriya et al. 2016; Saxena, Saharan, and George 2018). A slit-venturi-based HC process provided better decolorization for Orange G dye when compared to a circular venture or an orifice plate. Badmus et al. (2020) used a venture-based jet loop HC process to degrade the simulated Orange-G solution and observed 70% decolorization within 60 minutes. However, similar results were not obtained with a mixture of dyes or industrial dye effluent. Kumar et al. (2018) reported 28.23% decolorization of a ternary mixture of dyes (methylene blue, methyl orange, and Rhodamine B) using an orifice based HC process. When an industrial textile dyeing effluent was treated with a slit-venturi-based HC process, only 17% TOC, 12% COD, and 25% color removal were observed at an inlet pressure of 5 bar (Rajoriya et al. 2018). An increase in pressure reduced TOC removal to less than 10%. Suryawanshi et al. (2018) investigated the degradation of common solvents, i.e., toluene, acetone, and methyl ethyl ketone (MEK) using a patented vortex diode HC reactor (VoDCa™). At a low-pressure drop (ΔP) of 0.5 bar, 78% degradation of toluene was observed, while acetone and MEK degradation was 25–27%, respectively. An increase in pressure drop across the cavitating unit reduced percent degradation by two to three times, while an increase in solvent concentration also reduced % degradation.

TABLE 11.2

Summary of different studies using hydrodynamic cavitation (HC) for wastewater treatment

Process	Contaminant (Wastewater)	Process Conditions	Efficiency	Proposed Mechanism	Comments	References
Hydrodynamic cavitation, assisted with Ozone and H_2O_2	Industrial pesticide effluent (34,000 mg/L COD)	Circular venturi (2mm throat diameter)	TOC: *HC+ O₃:* 25.69% degradation *HC+H₂O₂:* 60.29%	OH•, O• radicals.	Optimal conditions: Dilution of effluent: 50% Ozone loading: $0.75gH^{-1}$ H_2O_2 loading 2 g L^{-1}; Biodegradability index raised to 0.325 post HC/O_3 (120min).	(Raut-Jadhav et al. 2016)
Hydrodynamic cavitation, assisted with H_2O_2, Fenton, Photo-Fenton, photolytic process	Ternary dye mixture (Methylene blue, Methyl orange and Rhodamine-B) –each 10ppm.	Orifice (2mm throat dia)	Decolorisation *HC:* 28.23% (180min) *HC+ H₂O₂:* 100% (40min) *HC+fenton :* 100% (40min) *HC+Photo-fenton:* 98.86% (20min) *HC+Photolytic:* 74.53% (120min) *HC+Photocatalytic:* 82.13% (120min)	OH•, Fe^{2+}, Photocatalysis based radicals.	Optimal conditions: *Pressure:* 6 Bar, *pH* 3 *Dye: H₂O₂*(1:40) *Fenton:H₂O₂:* 1:10 TiO_2 as photocatalyst. Improved mass transfer rates and better contact b/w photocatalyst and contaminant in HC assisted PC process. Higher synergetic coeff. Was obtained for $HC+H_2O_2$	(Kumar et al. 2018)
Hydrodynamic cavitation assisted with air, oxygen, Ozone and Fenton's reagent	Textile dyeing industry effluent (Reactive, direct and acid dyes)	Slit venturi	TOC: *HC:* 17.27% *HC+O₂:* 48.07% *HC+O₃:* 48%	OH•, HO_2•, O• radicals.	Optimal conditions: No dilution. Pressure: 5bar; pH 6.8 O_2: 2 L min^{-1} HC assisted with ozone showed 98% decolorization. Turbulence in HC improved O_3 contact time and reactivity with contaminant.	(Rajoriya et al. 2018)

Process	Pollutant	Reactor/Conditions	Results	Radicals	Optimal conditions	Reference
Hydrodynamic cavitation assisted with heterogeneous Fenton process (HFP) (Fe⁰/H₂O₂)	Orange G (10–100 ppm)	Orifice plate	Decolorization: HC/Fe⁰: 74.6% (30min) HC/ H₂O₂: 93.7% HC/Fe⁰(fixed): 99.8% (60min)	$OH^•$, $O^•$ and Fe^0 derived radicals.	Optimal conditions: pH: Dye: 20 mg L^{-1}. Fe^0: 0.7 g L^{-1}. H_2O_2: 20mg L^{-1}. The location of Fe^0 was critical to the HFP assisted HC due to the liquid velocity beyond orifice and its reach of cavitation bubbles. The decolorization was imminent in first 20min whereas the intermediates did not form.	(Cai et al. 2016)
Hydrodynamic cavitation assisted with acoustic cavitation	Rhodamine B (20–40 $\mu molL^{-1}$)	HC: 3 layered wherein each layer consists of 6 rotating venturi tubes (30°). AC: 27.90 kHz	HC: ~22% AC: ~12% HC/AC synergistic effect: 119% Time dependent Synergistic factor: 40%	$OH^•$ radicals	Optimal conditions: Pressure: 0.4 MPa. Rhodamine conc.: 30 μmol L^{-1}. AC power: 176watt. HC/AC prevents bubble cluster at high velocities, higher rate of radical generation.	(Yi et al. 2018)
Hydrodynamic cavitation with H_2O_2 as oxidant	Dyes such as Acid red 88, Reactive red 120 and Reactive Orange 4	Venturi; Inlet pressure of 5 bar, cavitation number = 0.15	60–92% decolorization occurred with HC alone, 100% decolorization with HC/ H_2O_2: 60% maximum TOC removal	$OH^•$ radicals	Hydrodynamic cavitation (HC) better than acoustic cavitation. pH for max. degradation < 3. Hybrid processes such as HC/ H_2O_2 or HC/O₃ better than HC alone.	(Rajoriya, Bargole, and Saharan 2017b)
Hydrodynamic cavitation	Simulated Orange II sodium salt	Jet-loop cavitation; inlet pressure=400kPa, pH=2, volume=10L	HC alone = 70% decolorization; H_2O_2/Fe (II) – Ineffective; HC/ H_2O_2/nZVI = 99% decolorization; 74% TOC removal	$OH^•$ radicals	Hybrid combination of H_2O_2 and nano-zerovalent iron (nZVI) – best efficiency; Maximum decolorization at 400 kPa, and pH = 2.	(Kassim O Badmus et al. 2020)

(Continued)

TABLE 11.2

Summary of different studies using hydrodynamic cavitation (HC) for wastewater treatment (Cont.)

Process	Contaminant (Wastewater)	Process Conditions	Efficiency	Proposed Mechanism	Comments	References
Hydrodynamic cavitation	*Microcystis aeruginosa*	Orifice	Reduction: 94% HC alone Air flow enhanced degradation by 12% Photosynthetic activity reduced by 90%	OH• radicals	Pressure: 0.4MPa; airflow rate: 0.5 L min⁻¹; The algae were grown post treatment to study the effect of cavitation (Optical density and photosynthetic activity).	(Li, Song, and Yu 2014)
Hydrodynamic cavitation, assisted with Ozone and H_2O_2	2,4,6- Trichloro phenol (20 mg/L)	Slit-Venturi cavitating device equipped with ozone generator, operated at 2-5 bar.	Degradation: *HC:* 24.36% Optimal C_v: 0.23 $HC+H_2O_2$: 62.07% $HC+ O_3$: 94.4% $HC+ O_3+ H_2O_2$: Complete degradation.	Oxidation by OH• radicals.	Choked cavitation at higher pressures (>4bar). Optimal degradation at pH 7. Optimal quantity TCP: H_2O_2 (1:5). Operating ozone at higher flow rate is preferable. Slit venture better than orifice.	(Barik and Gogate 2018)
Hydrodynamic and acoustic cavitation assisted with hydrogen peroxide, ozone and peroxone.	Effluents from the production of bitumens (Contains 15 compounds each 40ppm)	*HC:* Venturi (2mm throat dia) *Acoustic:* 1000 W, 25 kHz, 16 A	Acoustic: 90% (360min) HC: 25% (phenol removal) HC+ Fenton: 60% (phenol removal) $HC+H_2O_2$: 100% $AC/HC+O_3$: ~80% Nitrobenzene HC+ peroxone: 100% (60min)	OH• radicals from C=C in benzyl rinds, or from its alkyl substitutes,	AC alone removed organo-sulfurs completely (post 6hrs treatment). Nitro compounds were difficult to degrade. Electron donor group were readily targeted. Compound emulsification was observed. HC/AC Peroxone degraded ~100% of all compounds. HC: 8 bar; C_v: 0.14; pH 10.5; Peroxone: 12.32 g h⁻¹; H_2O_2: 3 ml/min	(Gągol, Przyjazny, and Boczkaj 2018a)
Hydrodynamic cavitation assisted with UV, H_2O_2, O_3 and H_2O_2/O_3	Carbamazepine (10ppm)	Slit venturi	Degradation: HC: 38.7% HC/UV: 52.9% (16W lamp); HC/ H_2O_2: 58.3% HC/O_3: 91.3%; HC/ H_2O_2/ O_3: 100%	OH•, HO_2•, O• radicals.	Optimal conditions: Pressure: 4 bar; pH 4; CBZ-H_2O_2 ratio: (1:5); O_3:400mg h⁻¹	(Thanekar, Panda, and Gogate 2018)

Process	Pollutant/Material	Reactor/Conditions	Radicals	Results	Findings	Reference
Hydrodynamic cavitation	Waste activated sludge (1% thickened)	Rotational generator	OH• radicals, shear stress between rotor and stator.	SCOD improvements: Increase from 45 ppm to 602 ppm in 20 passes.	Cavitational aggressiveness is formed due to the pressure fluctuations between the stator and the rotor.	(Petkovšek et al. 2015)
Hydrodynamic cavitation assisted with iron metal blades (IMBs)	Unsymmetrical dimethyl hydrazine (UMDH) (10 ppm)	Orifice plate (2mm dia, 1mm thickness)	OH• radicals and Fe^{2+} radicals	Degradation: HC: 37%; HC/IMBs: 71% (120mins)	Optimal conditions: Pressure: 7.8 bar; Operating at pH 5 or lower is preferred due to the formation of partial carboxylic group which is readily degraded. IMBs blades (Fe^{0}) facilitated double the degradation of UMDH.	(Torabi Angaji and Ghiaee 2015)
Hydrodynamic cavitation assisted with electric field	Waste activated sludge	Orifice (single hole, 2mm dia) Modified with Criss-cross section (M-HC). Electric field equipped orifice (EFM)	OH• radicals	Disintegration degree: HC: 17.2% M-HC: 33.5% EFM-HC: 47% (180min)	Combining electrolysis and cavitation. Storage tank equipped with graphite electrodes, 80V DC power supply. Criss-cross section enhances solubilizing efficiency. No chemical addition. Better methane production was achieved	(Jung et al. 2014)
Hydrodynamic cavitation with aeration	Acetone, toluene and methyl-ethyl ketone (MEK)	Vortex diode, Pressure drop = 0.5-2 bar, temperature = 30°C; Orifice plate = 3 mm diameter hole	OH• radicals	80% degradation of toluene, 20-60% degradation of acetone and MEK	Chemical oxidation as a predominant mechanism. Increase in pressure drop reduces % degradation. Higher pressure drop required for orifice for similar degradation. Energy requirement lower for HC than orifice plate	(Suryawanshi et al. 2018)
Hydrodynamic cavitation with chlorine dioxide	2,4,6-Triamino-1,3,5-Trinitrobenzene Explosive	Orifice plate = 22 holes, 1.5 mm hole; Pressure = 0.4 MPa; temperature = 30°C	OH•; ClO_2• radicals	TOC reduction: HC alone = 13%; ClO_2 alone = 28%; HC+ClO_2 = 65%	HC/ClO_2 has a higher energy efficiency compared to HC, and ClO_2 alone.	(K. Wang et al. 2020)

Given the lower degradation efficiency using HC alone, most literature studies recommend the use of a hybrid AOP process—a combination of HC with H_2O_2, O_3, aeration, UV, H_2O_2/UV, O_3/UV, H_2O_2/Fe^{2+}/UV, or photocatalysis. Combination of H_2O_2/HC or Fenton/HC increased decolorization of dyes (Acid Red 88, Reactive Red 120, Rhodamine B, etc) by two to three times, with a similar increase in TOC reduction, when compared to HC alone (Rajoriya, Bargole, and Saharan 2017a; Rajoriya et al. 2016; Saxena, Saharan, and George 2018). It has been observed that complete decolorization was observed within 20 minutes in most studies, indicating the higher reactivity of Fenton/HC and H_2O_2/HC. Similar treatment efficiencies were reported upon the use of Fenton/HC, H_2O_2/HC, Photo-Fenton/HC for insecticides, pharmaceutical drugs, and chlorinated compounds. Using HC with H_2O_2, Fenton's reagent, air, and oxygen for the treatment of industrial wastewater resulted in up to 63% COD removal, with a combination of oxygen + Fenton + HC providing the best results (Joshi and Gogate 2019). In recent years, Gogate and his co-workers have reported the use of several hybrid technologies for the degradation of organic contaminants including HC + H_2O_2 + O_3 + CuO (potassium thiocyanate); HC + O_3 (industrial textile effluent); HC + Fenton (municipal wastewater) and HC+ H_2O_2 and O_3 (naproxen) with reasonable success in a pilot-scale set-up, with reactor sizes ranging from 40 liters to 100 liters (Jawale and Gogate 2019; J. Wang et al. 2020; Gogate, Thanekar, and Oke 2020; Thanekar, Garg, and Gogate 2020). The operational cost for the treatment in these studies was observed to be the lowest for HC + H_2O_2-based hybrid AOPs, and significantly lower than acoustic cavitation-based AOPs. From the above discussion, it is fairly evident that the HC process in synergy with other AOPs, particularly H_2O_2, has tremendous potential in reducing the adverse effects of organic COCs in wastewater and is extremely viable for an easy scale-up.

11.2.2.1   Effect of Operating Parameters in Cavitation Process Efficiency

The key to obtaining near complete efficiencies at low costs is the optimization of the operational parameters such as fluid pressure, temperature, and solution pH, oxidant, and contaminant concentration among others. Table 11.3 presents the optimum operating conditions for hydrodynamic cavitation.

Irrespective of the design of the HC reactor, inlet pressure and C_V (equation 11.1) are considered to be the most important parameters. An increase in the inlet pressure of the medium increases the fluid velocity and decreases C_v. The lower the C_v, the higher the number of cavitation bubbles generated, the higher the generation of hydroxyl radicals formed, and the greater the degradation efficiency. However, any increase in pressure beyond optimum value will result in an extremely high density of cavities, which will coalesce to form larger cavitation bubbles resulting in a phenomenon called choked cavitation (Tao et al. 2016). These large bubbles are unlikely to completely collapse, resulting in a lower rate of hydroxyl ion generation and therefore lower degradation efficiency. In most studies, efficient COC degradation was observed at an inlet pressure range of 2–15 bar, corresponding to a cavitation number of 0.15–0.5, though C_v in the range of 0.15–0.29 is preferred. However, cavitational effects have been observed at values greater than 1 due to the large number of cavities generated, small amounts of dissolved gases, and suspended particles (Rajoriya

TABLE 11.3
Optimum process conditions for operating HC reactors

Parameter	Favorable condition	Optimized value	Reference
Inlet pressure	Increased pressure or pump speed, but within an optimum value	2–15 bar	(Rajoriya et al. 2016; Badmus et al. 2018; Joshi and Gogate 2019)
Cavitation number (C_v)	Depends on inlet pressure, $C_v < 1$ preferred	0.15–0.29	(Rajoriya et al. 2016; Badmus et al. 2018)
Solution pH	Degradation preferable in acidic pH; increase in reactivity of OH˙ radical	2.5–4.5	(Gągol, Przyjazny, and Boczkaj 2018b)
Temperature	Lower temperature	30–50°C	(Tao et al. 2016)
Initial contaminant concentration	Lower concentration preferred; issues during treatment of industrial wastewater observed	Depends on the contaminant and degradation rate	(Rajoriya et al. 2016; Badmus et al. 2018; Joshi and Gogate 2019)
Oxidant concentration	Depends on the contaminant concentration and type of oxidant	No specific range	-
Physiochemical properties of liquid (viscosity, surface tension, etc.)	Achieve lower size of nuclei	No specific range	(Gogate 2008)

et al. 2016; Suryawanshi et al. 2018). Thus, it can be said that a lower cavitation number is optimal and that care should be taken to avoid choked cavitation.

Acidic conditions are preferred for HC-based treatment because of the faster generation of hydroxyl ions (from H_2O_2 decomposition). Most studies have reported >90% treatment efficiency at pH <3.5, because recombination of OH˙ radicals is prevented and oxidation potential increased (Rajoriya et al. 2016; Gągol, Przyjazny, and Boczkaj 2018b). As noted in Table 11.3, a solution pH of 2.5–4.5 is optimal. In addition to pH, the solution temperature is also a key parameter. Water vapor pressure depends on the temperature of the treated medium. It has been observed that increasing solution temperature increased the intensity of cavity generation, which results in reduced degradation. Gogate (2008) opined that for a chemical reaction wherein cavitational collapse is required (i.e., resulting in the formation of oxidative radicals), low temperature operating conditions are preferred. The reason is the formation of more water vapors, leading to choked cavitation. Tao et al. (2016) noted that an increase in liquid temperature, physiochemical parameters such as viscosity, surface tension, and gas solubility reduce, resulting in a reduction in cavitation intensity and cavity nuclei. Because of a certain cushioning effect with an increase in vapor pressure inside the bubble, the collapse of the bubble is inefficient, and hence degradation suffers. Hence, most studies favored lower temperatures in the range of 30–50°C. Gągol et al. (2018b) noted increased temperature led to cavity suppression and formation of cloud cavitation (when concentration and intensity of cavity bubbles increase beyond a critical value). Additionally, organic compounds tend to volatilize at higher temperatures and are therefore not subjected to degradation.

Most bimolecular chemical reactions—one the contaminant and the other the oxidant—either follow second-order or pseudo-first-order kinetics. The higher the pollutant concentration, the lower the degradation efficiency since the number of hydroxyl radicals generated may be insufficient or there is limited diffusion of the radical to the interface of the collapsing bubbles. It should be noted that if the system contains higher amounts of the contaminant, the rate of degradation or consumption of hydroxyl radicals is higher. Studies have reported percent degradation decrease almost 1.5–2 times at large contaminant concentration, irrespective of the contaminant present (Rajoriya et al. 2016, 2018; Gągol, Przyjazny, and Boczkaj 2018b). However, as in any bimolecular reaction, the second reactant also plays an important role, both in terms of type and concentration. As evident in many studies, H_2O_2 is the preferred oxidant (Table 11.2). Because the dissociation energy of the O-O bond, H_2O_2 is -213kJ/mol and lower compared to dissociation energy for the O-H bond in water (-418 kJ/mol), and cavitation can easily break H_2O_2 into OH˙ radicals. As noted in Table 11.2 and elsewhere in the literature, the HC+H_2O_2 hybrid process increased degradation efficiency two to three times, compared to HC alone. Other oxidants that are considered viable in an HC-based hybrid process include O_3, Fenton's reagent, TiO_2 photocatalysis, PS/PMS, and aeration. Suryawanshi et al. (2018) reported a 10% increase in solvent degradation upon the use of aeration in an HC process when compared to HC alone. Irrespective of the oxidant used, the nature and amount of COCs present in the contaminated water determine the amount of oxidant to be used. The more OH˙ radicals present, the easier it is to completely degrade the complex pollutants. The molar ratio of the pollutant: H_2O_2 for efficient degradation has been observed to range from 1:5 to 1:60 (H_2O_2). The amount of O_3 for efficient degradation used has been 0.6–8 g/hr; while the amount of TiO_2 used for an HC+ photocatalysis process has been 25–500 mg/L (Rajoriya et al. 2016; Gągol, Przyjazny, and Boczkaj 2018b; Tao et al. 2016; Barik and Gogate 2018). The optimization of oxidant concentration should consider the fact that the entirety of the OH˙ radicals is used in the degradation process, and scavenging of the radicals is prevented. Overloading the system with excess oxidant can result in secondary reactions that produce radicals (e.g., $OH_2˙$, O˙) having lower oxidation potential than OH˙ radicals (Gągol, Przyjazny, and Boczkaj 2018b). The other method to minimize oxidant use is the stepwise injection of the oxidant, as it prevents the unnecessary decomposition of the oxidant, thereby reducing efficiencies. In all, the complexity of the pollutant, solution pH, cavitating device, and kinetics of the degradation process should determine the optimum amount of oxidant to be used. Other physicochemical parameters of importance are the viscosity and surface tension of the liquid. The favorable conditions for the inception of cavitation are low liquid vapor pressure, low viscosity, and high surface tension. The lower the vapor pressure, the higher the vapor content in the cavity resulting in an increase in the intensity of bubble collapse. Lower viscosities of liquid are preferred for multiple reasons: (1) it is easier to transport low viscous fluids across the constriction and (2) natural cohesive forces are lower in a low viscous fluid, thereby reducing the threshold pressure for cavitation (Tao et al. 2016). It has also been reported that higher surface tension reduces the radius of the nucleus of the bubble, enhancing collapse intensity.

11.3 REACTOR DESIGN FOR HYDRODYNAMIC CAVITATION REACTORS

The geometry of the cavitating device affects the intensity of cavitation and, therefore, the efficiency of the HC process. A proper design for the fluid flow through the constriction is required as it influences pressure conditions during flow; thus, the cavitation number and the intensity of the cavities are generated. Additionally, constrictions should be located at appropriate positions such that enough cavities are generated for efficient degradation of the pollutant. The cavitating devices reported in various studies include throttling valves, venture, orifice plates, high-speed rotors, homogenizers, and vortex-diodes, among others (Rajoriya et al. 2016; Suryawanshi et al. 2018; Gągol, Przyjazny, and Boczkaj 2018b; Jawale and Gogate 2019). Figure 11.1 provides a snapshot of some of the common cavitation devices reported in the literature. Schematics for other less commonly used cavitating devices (e.g., swirling – jet, high-speed rotor, etc.) are illustrated in other studies (Rajoriya et al. 2016; Tao et al. 2016; and Gągol, Przyjazny, and Boczkaj 2018b). Venturi and orifice-based HC reactors are the most effective because (1) cavitation occurring in an orifice is transient, generating a high-intensity cavity collapse; and (2) smooth recovery of pressure occurs in the downstream section, as well as stable formation of cavities in venturi. It has been reported that the highest cavitational intensity is observed when the lowest free area of the orifice is used (Mancuso, Langone, and Andreottola 2020). If the free area is kept constant, larger diameter orifice plates with a smaller number of holes generated low-intensity cavities. The cavitational intensity increased when multiple-hole orifices are used (Rajoriya et al. 2016). For a venturi-based HC system, the size and shape of the throat and convergent section determines the number of cavities generated, while the divergent section determines the maximum size of cavities. The ratio of the perimeter of the throat to its cross-sectional area (α) of an orifice and ratio of throat area to pipe cross-sectional area (β) are key to active cavitation and magnitude collapse (Rajoriya

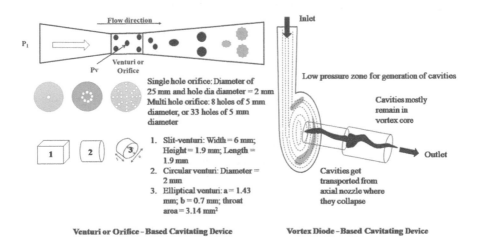

FIGURE 11.1 Illustration of commonly used cavitating devices [Adapted from (Sarvothaman, Nagarajan, and Ranade 2018; Rajoriya et al. 2016)].

et al. 2016; Tao et al. 2016). It has been reported that an increase in α and a decrease in β show higher percent degradation. An increase in orifice plate thickness reduced percent degradation, while an increase in the hole diameter increased degradation. Typical values of α reported are 0.4 to 1.33; values of β reported are 0.023–0.1 (Jiang et al. 2016; Boczkaj et al. 2018). Saxena, Saharan, and George (2018) reported that slit and circular venturi devices offered better dye degradation than orifices, largely because the cavitation number was lower.

The geometry of the venturi is designed such that the ratio of outlet to inlet liquid pressure is 0.8, the optimum ratio of diameter to length of the throat is 1:1, and the angle of the divergent section is 7°. It has been reported that for a given cross-section area, a higher perimeter of throat generated more cavities; hence, rectangular and elliptical shapes are preferred over circular shapes. In recent years, vortex-based HC units have also shown promise in wastewater treatment applications. They are sometimes called low pressure swirling cavitation or vortex cavitation. As seen in Figure 11.1, the fluid enters the vortex chamber tangentially, and cavities are generated in the swirl chamber (Suryawanshi et al. 2018; Mancuso, Langone, and Andreottola 2020). The cavities are generated due to the vortex flow and collapse at the exit outlet. It was observed that cavitation was observed at a pressure drop of 0.3–0.5 bar, which was two to three times lower than the pressure drop required when orifices and ventures are used (Suryawanshi et al. 2018; Jain et al. 2019). Jain et al. (2019) reported 99% *E. coli* disinfection at 0.5 bar pressure drop using vortex-based HC; however, the disinfection rate was higher when orifices were used compared to a vortex-based device. Eight times better cavitational yields were reported when toluene was the COC used with the vortex diode, compared to orifice-based HC (Suryawanshi et al. 2018). Given that various studies have reported diverse geometries and configurations of cavitation reactors, it should be noted that the optimum values presented here are subject to change, depending on COCs and their physiochemical properties.

11.4 SUSTAINABILITY—ROAD TO THE FUTURE

Sustainability is defined by three major factors: environmental, social, and economic. Any sustainable water treatment process should ensure that (1) contaminants are completely removed before being discharged to an aquatic source, (2) it is energy efficient and green, and (3) it is cost effective. The environmental advantages of the HC-based process—either alone or in synergy with other AOPs—are obvious; hybrid technologies offer more degradation of the organic pollutants than single AOPs. In most cases, the use of a hybrid AOP ensures that toxic by-products are not formed, as seen in most studies. It has been widely documented that using HC alone may result in the formation of by-products, but the use of HC + Fenton or HC + H_2O_2 ensures complete mineralization of the contaminant (conversion to CO_2). HC-based reactors, in addition to offering a wide variety of setups, also present a greater superiority over acoustic and optical cavitation in terms of less complexity of scale and expend less energy (therefore reducing operational costs), making them more suitable for industrial scale-up. However, this statement is based mostly on laboratory-based and pilot-scale investigations that are limited to treating volumes less than 100 L. Joshi and Gogate (2019) determined that cavitational yield (ratio of the amount treated to the energy supplied) was the lowest for HC + aeration process, and the total treatment cost was the lowest ($398/$m^3$ of treated water) for an HC + Fenton + aeration process when compared to other HC-based hybrid process. Similarly, when naproxen-laden wastewater was treated using an HC-hybrid process, the cost was Rs. 0.12/L for HC + H_2O_2, compared to other HC processes and biological treatments (Thanekar, Garg, and Gogate 2020). While biological-based treatments alone may cost significantly less ($7–10/$m^3$), they are inefficient. A comparison of other AOPs for different wastewater treatment effluents indicated that ozone-based treatment could cost $250–300/$m^3$, while HC+ H_2O_2 and H_2O_2-based processes would cost 30 times less (G. Boczkaj and Fernandes 2017). The third wheel of sustainability is the process of being green. Most other AOPs use significant amounts of chemicals such as H_2O_2, PS/PMS, and O_3, which are not exactly green technologies. Since the HC process minimizes the use of such chemicals even in a hybrid-based process, it can be considered a fairly green technology. In conclusion, while the use of HC as an effective wastewater treatment technology is still in its nascent stage, it should be explored for industrial scale-up for its sheer potential. Hence, more studies need to be performed to optimize the reactor design for industrial-scale operations.

ACKNOWLEDGEMENTS

The authors are grateful to Department of Science & Technology, New Delhi for support on this project (DST/TMD(EWO)/OWUIS-2018/RS-17 (G)). We also acknowledge SASTRA Deemed University, Thanjavur for extending infrastructure support to pursue this work.

REFERENCES

Badmus, Kassim O, Ninette Irakoze, Olushola Rotimi Adeniyi, and Leslie Petrik. 2020. "Synergistic Advance Fenton Oxidation and Hydrodynamic Cavitation Treatment of Persistent Organic Dyes in Textile Wastewater." *Journal of Environmental Chemical Engineering* 8 (2): 103521. doi:10.1016/j.jece.2019.103521.

Badmus, Kassim Olasunkanmi, Jimoh Oladejo Tijani, Emile Massima, and Leslie Petrik. 2018. "Treatment of Persistent Organic Pollutants in Wastewater Using Hydrodynamic Cavitation in Synergy with Advanced Oxidation Process." *Environmental Science and Pollution Research* 25 (8):7299–7314. doi:10.1007/s11356-017-1171-z.

Bagal, Manisha V., and Parag R. Gogate. 2014. "Wastewater Treatment Using Hybrid Treatment Schemes Based on Cavitation and Fenton Chemistry: A Review." *Ultrasonics Sonochemistry* 21 (1): 1–14. doi:10.1016/j.ultsonch.2013.07.009.

Barik, Arati J, and Parag R. Gogate. 2018. "Hybrid Treatment Strategies for 2,4,6-Trichlorophenol Degradation Based on Combination of Hydrodynamic Cavitation and AOPs." *Ultrasonics Sonochemistry* 40: 383–94. doi:10.1016/j.ultsonch.2017.07.029.

Boczkaj, G., and A. Fernandes. 2017. "Wastewater Treatment by Means of Advanced Oxidation Processes at Basic PH Conditions: A Review." *Chemical Engineering Journal* 320: 608–33. doi:10.1016/j.cej.2017.03.084.

Boczkaj, Grzegorz, Michał Gągol, Marek Klein, and Andrzej Przyjazny. 2018. "Effective Method of Treatment of Effluents from Production of Bitumens Under Basic PH Conditions Using Hydrodynamic Cavitation Aided by External Oxidants." *Ultrasonics Sonochemistry* 40 (May 2017): 969–79. doi:10.1016/j.ultsonch.2017.08.032.

Cai, Meiqiang, Jie Su, Yizu Zhu, Xiaoqing Wei, Micong Jin, Haojie Zhang, Chunying Dong, and Zongsu Wei. 2016. *Decolorization of Azo Dyes Orange G Using Hydrodynamic Cavitation Coupled with Heterogeneous Fenton Process. Ultrasonics Sonochemistry* 28. doi:10.1016/j.ultsonch.2015.08.001.

Gągol, Michał, Andrzej Przyjazny, and Grzegorz Boczkaj. 2018a. "Highly Effective Degradation of Selected Groups of Organic Compounds by Cavitation Based AOPs Under Basic PH Conditions." *Ultrasonics Sonochemistry* 45: 257–66. doi:10.1016/j. ultsonch.2018.03.013.

Gągol, Michał, Andrzej Przyjazny, and Grzegorz Boczkaj. 2018b. "Wastewater Treatment by Means of Advanced Oxidation Processes Based on Cavitation – A Review." *Chemical Engineering Journal* 338: 599–627. doi:10.1016/j.cej.2018.01.049.

Ghanbari, Farshid, and Mahsa Moradi. 2017. "Application of Peroxymonosulfate and Its Activation Methods for Degradation of Environmental Organic Pollutants: Review." *Chemical Engineering Journal.* doi:10.1016/j.cej.2016.10.064.

Gogate, P. R., P. D. Thanekar, and A. P. Oke. 2020. "Strategies to Improve Biological Oxidation of Real Wastewater Using Cavitation Based Pre-Treatment Approaches." *Ultrasonics Sonochemistry* 64: 105016. doi:10.1016/j.ultsonch.2020.105016.

Gogate, Parag R. 2008. "Cavitational Reactors for Process Intensification of Chemical Processing Applications: A Critical Review." *Chemical Engineering and Processing: Process Intensification* 47 (4): 515–27. doi:10.1016/j.cep.2007.09.014.

Jain, Pooja, Vinay M. Bhandari, Kshama Balapure, Jyotsnarani Jena, Vivek V. Ranade, and Deepak J. Killedar. 2019. "Hydrodynamic Cavitation Using Vortex Diode: An Efficient Approach for Elimination of Pathogenic Bacteria from Water." *Journal of Environmental Management* 242: 210–19. doi:10.1016/j.jenvman.2019.04.057.

Jawale, Rajashree H., and Parag R. Gogate. 2019. "Novel Approaches Based on Hydrodynamic Cavitation for Treatment of Wastewater Containing Potassium Thiocyanate." *Ultrasonics Sonochemistry* 52: 214–23. doi:10.1016/j.ultsonch.2018.11.019.

Jiang, Tao, Danni Kong, Kun Xu, and Fahai Cao. 2016. "Hydrogenolysis of Glycerol Aqueous Solution to Glycols over Ni–Co Bimetallic Catalyst: Effect of Ceria Promoting." *Applied Petrochemical Research* 6 (2): 135–44. doi:10.1007/s13203-015-0128-8.

Joshi, Saurabh M., and Parag R. Gogate. 2019. "Intensification of Industrial Wastewater Treatment Using Hydrodynamic Cavitation Combined with Advanced Oxidation at Operating Capacity of 70 L." *Ultrasonics Sonochemistry* 52: 375–81. doi:10.1016/j. ultsonch.2018.12.016.

Jung, Kyung Won, Min Jin Hwang, Yeo Myeong Yun, Min Jung Cha, and Kyu Hong Ahn. 2014. "Development of a Novel Electric Field-Assisted Modified Hydrodynamic Cavitation System for Disintegration of Waste Activated Sludge." *Ultrasonics Sonochemistry* 21 (5): 1635–40. doi:10.1016/j.ultsonch.2014.04.008.

Kim, Moon Kyung, and Kyung Duk Zoh. 2016. "Occurrence and Removals of Micropollutants in Water Environment." *Environmental Engineering Research* 21 (4): 319–32. doi: 10.4491/eer.2016.115.

Kumar, M. Suresh, S. H. Sonawane, B. A. Bhanvase, and Bhaskar Bethi. 2018. "Treatment of Ternary Dye Wastewater by Hydrodynamic Cavitation Combined with Other Advanced Oxidation Processes (AOP's)." *Journal of Water Process Engineering* 23 (January): 250–56. doi:10.1016/j.jwpe.2018.04.004.

Li, Pan, Yuan Song, and Shuili Yu. 2014. "Removal of Microcystis Aeruginosa Using Hydrodynamic Cavitation: Performance and Mechanisms." *Water Research* 62: 241–48. doi:doi.org/10.1016/j.watres.2014.05.052.

G. Mancuso, M. Langone, G. Andreottola, A critical review of the current technologies in wastewater treatment plants by using hydrodynamic cavitation process: principles and applications, *J. Environ. Heal. Sci. Eng.* 18 (2020) 311–333. https://doi.org/10.1007/ s40201-020-00444-5.

Oturan, Mehmet A., and Jean Jacques Aaron. 2014. "Advanced Oxidation Processes in Water/ Wastewater Treatment: Principles and Applications. A Review." *Critical Reviews in Environmental Science and Technology* 44 (23): 2577–2641. doi:10.1080/10643389. 2013.829765.

Park, Jungsu, Jared Church, Younggyu Son, Keug-Tae Kim, and Woo Hyoung Lee. 2017. "Recent Advances in Ultrasonic Treatment: Challenges and Field Applications for Controlling Harmful Algal Blooms (HABs)." *Ultrasonics Sonochemistry* 38: 326–34. doi:10.1016/j.ultsonch.2017.03.003.

Petkovšek, Martin, Matej Mlakar, Marjetka Levstek, Marjeta Stražar, Brane Širok, and Matevž Dular. 2015. "A Novel Rotation Generator of Hydrodynamic Cavitation for Waste-Activated Sludge Disintegration." *Ultrasonics Sonochemistry* 26: 408–14. doi: 10.1016/j.ultsonch.2015.01.006.

Rajoriya, Sunil, Swapnil Bargole, Suja George, and Virendra Kumar Saharan. 2018. "Treatment of Textile Dyeing Industry Effluent Using Hydrodynamic Cavitation in Combination with Advanced Oxidation Reagents." *Journal of Hazardous Materials* 344: 1109–15. doi:10.1016/j.jhazmat.2017.12.005.

S. Rajoriya, S. Bargole, V.K. Saharan, Degradation of reactive blue 13 using hydrodynamic cavitation: Effect of geometrical parameters and different oxidizing additives, Ultrason. *Sonochem.* 37 (2017) 192–202. https://doi.org/10.1016/j.ultsonch.2017.01.005.

———. 2017b. "Degradation of a Cationic Dye (Rhodamine 6G) Using Hydrodynamic Cavitation Coupled with Other Oxidative Agents: Reaction Mechanism and Pathway." *Ultrasonics Sonochemistry* 34 (January): 183–94. doi:10.1016/j.ultsonch.2016.05.028.

Rajoriya, Sunil, Jitendra Carpenter, Virendra Kumar Saharan, and Aniruddha B. Pandit. 2016. "Hydrodynamic Cavitation: An Advanced Oxidation Process for the Degradation of Bio-Refractory Pollutants." *Reviews in Chemical Engineering* 32 (4): 379–411. doi:10.1515/revce-2015-0075.

Ranade, Vivek V., and Vinay M. Bhandari. 2014. "Industrial Wastewater Treatment, Recycling, and Reuse: An Overview." In *Industrial Wastewater Treatment, Recycling and Reuse*, 1–80. doi:10.1016/B978-0-08-099968-5.00001-5.

Raut-Jadhav, Sunita, Mandar P Badve, Dipak V Pinjari, Daulat R Saini, Shirish H Sonawane, and Aniruddha B Pandit. 2016. "Treatment of the Pesticide Industry Effluent Using Hydrodynamic Cavitation and Its Combination with Process Intensifying Additives (H2O2 and Ozone)." *Chemical Engineering Journal* 295: 326–35. doi:10.1016/j. cej.2016.03.019.

Sarvothaman, Varaha Prasad, Sanjay Nagarajan, and Vivek V. Ranade. 2018. "Treatment of Solvent-Contaminated Water Using Vortex-Based Cavitation: Influence of Operating Pressure Drop, Temperature, Aeration, and Reactor Scale." *Industrial and Engineering Chemistry Research* 57 (28): 9292–9304. doi:10.1021/acs.iecr.8b01688.

Saxena, S., Saharan, V. K., & George, S. (2018). Enhanced synergistic degradation efficiency using hybrid hydrodynamic cavitation for treatment of tannery waste effluent. *Journal of Cleaner Production*, 198, 1406–1421. https://doi.org/10.1016/j.jclepro.2018.07.135

Simpson, Alister, and Vivek V. Ranade. 2018. "Modelling of Hydrodynamic Cavitation with Orifice: Influence of Different Orifice Designs." *Chemical Engineering Research and Design* 136: 698–711. doi:10.1016/j.cherd.2018.06.014.

Suryawanshi, Pravin G., Vinay M Bhandari, Laxmi Gayatri Sorokhaibam, Jayesh P. Ruparelia, and Vivek V. Ranade. 2018. "Solvent Degradation Studies Using Hydrodynamic Cavitation." *Environmental Progress & Sustainable Energy* 37 (1): 295–304. doi:10.1002/ep.12674.

Tao, Yuequn, Jun Cai, Bin Liu, Xiulan Huai, and Zhixiong Guo. 2016. "Application of Hydrodynamic Cavitation to Wastewater Treatment Hydrodynamic Cavitation in Wastewater Treatment: A Review." *Chemical Engineering Technology* 39 (8): 1363–76. doi:10.1002/ceat.201500362.

Thanekar, Pooja, Sakshi Garg, and Parag R. Gogate. 2020. "Hybrid Treatment Strategies Based on Hydrodynamic Cavitation, Advanced Oxidation Processes, and Aerobic Oxidation for Efficient Removal of Naproxen." *Industrial & Engineering Chemistry Research* 59 (9): 4058–70. doi:10.1021/acs.iecr.9b01395.

Thanekar, Pooja, Mihir Panda, and Parag R. Gogate. 2018. "Degradation of Carbamazepine Using Hydrodynamic Cavitation Combined with Advanced Oxidation Processes." *Ultrasonics Sonochemistry* 40: 567–76. doi:10.1016/j.ultsonch.2017.08.001.

Torabi Angaji, Mahmood, and Reza Ghiaee. 2015. "Decontamination of Unsymmetrical Dimethylhydrazine Waste Water by Hydrodynamic Cavitation-Induced Advanced Fenton Process." *Ultrasonics Sonochemistry* 23: 257–65. doi:10.1016/j.ultsonch.2014.09.007.

UNEP, A Snapshot of the World's Water Quality: Towards a global assessment Table of Contents A Snapshot of the World's Water Quality: Towards a global assessment, United Nations Environ. Program. (2016) 162. https://doi.org/978-92-807-3555-0.

Wang, Jihong, Huilun Chen, Rongfang Yuan, Fei Wang, Fangshu Ma, and Beihai Zhou. 2020. "Intensified Degradation of Textile Wastewater Using a Novel Treatment of Hydrodynamic Cavitation with the Combination of Ozone." *Journal of Environmental Chemical Engineering* 8 (4): 103959. doi:10.1016/j.jece.2020.103959.

K. Wang, R. Jin, Y. Qiao, Z. He, X. Wang, C. Wang, Y. Lu, 2,4,6-Triamino-1,3,5-Trinitrobenzene Explosive Wastewater Treatment by Hydrodynamic Cavitation Combined with Chlorine Dioxide, *Propellants, Explos. Pyrotech.* 45 (2020) 1243–1249. https://doi.org/10.1002/prep.201900356.

Yang, Lie, Jianming Xue, Liuyang He, Li Wu, Yongfei Ma, Huan Chen, Hong Li, Pai Peng, and Z. Zhang. 2019. "Review on Ultrasound Assisted Persulfate Degradation of Organic Contaminants in Wastewater: Influences, Mechanisms and Prospective." *Chemical Engineering Journal* 378 (June). doi:10.1016/j.cej.2019.122146.

Yi, Chunhai, Qianqian Lu, Yun Wang, Yixuan Wang, and Bolun Yang. 2018. "Degradation of Organic Wastewater by Hydrodynamic Cavitation Combined with Acoustic Cavitation." *Ultrasonics Sonochemistry* 43: 156–65. doi:10.1016/j.ultsonch.2018.01.013.

Part V

Bio and Bio-Inspired Materials
for Water Reclamation

12 Bioremediation and Phytoremediation as the Environmentally Sustainable Approach for the Elimination of Toxic Heavy Metals

Joyabrata Mal
Department of Biotechnology, Motilal Nehru National
Institute of Technology Allahabad, Uttar Pradesh, India

S. Rangabhashiyam*
Department of Biotechnology, School of Chemical
and Biotechnology, SASTRA Deemed University,
Thanjavur, India

CONTENTS

12.1 INTRODUCTION

Pollution due to toxic heavy metals (HMs) causes hazard effects to living organisms and damages environmental quality. The results of different industrial activities involve the discharge of HMs into bodies of water, which is the major increasing public concern. To overcome the problems linked with these issues, the World Health Organization (WHO) and the United States' Environmental Protection Agency (EPA) imposed standard permissible levels for the environmental release of heavy metal (Rangabhashiyam and Selvaraju, 2015; Cheng et al., 2019). Heavy metal represents metals and metalloids having a density of greater than 5 g/cm^3 (Duruibe et al., 2007). HMs are associated with the property of non-biodegradability and tend to accumulate in vegetation and animals; human beings consume them and the metals are absorbed indirectly through food. Heavy metal contaminations caused a considerable menace in food safety due to proven heavy metal accumulation in rice, fish, vegetables, and water (Singh and Kumar, 2017; Ramona et al., 2019).

Anthropogenic activities contribute as an important factor that affects the geochemical metals cycle. Compared to other environmental pollutants including radioactive and organic compounds, the sum of yearly toxicity effects of human activity-based mobilized metals was found to be excessive (Sadar et al., 2020). The massive rise of soil contamination with metals over the years is mainly because of different industrial development, waste discharges, waste from metropolitan cities, air deposition, and more usage of pesticides and fertilizers in the soil for the sake of agricultural practices. Such activities result in the loss of soil potential, affect the soil morphology, interrupt biological systems distributed in soil, reduce the yield of crops, and pollute the groundwater (Dedeke et al., 2016). According to the EPA, soil contamination due to heavy metal ions has affected about 10 million humans worldwide (EPA, 2016).

HMs are broadly classified into three categories: toxic metals, precious metals, and radionuclides. The toxic heavy metal ions include arsenic, chromium, nickel, lead, cadmium, cobalt, zinc, and iron; they impose major hazardous effects on the living system. The environmental contamination due to HMs is mainly from industrial activities: electroplating, electrolysis, tannery, fertilizer, pesticide, textile, surface finishing, paper and pulp, energy sector, iron and steel, photography, electrical equipment manufacturing, wood processing, aerospace, petroleum refining, alloy industries, atomic energy installation, and metal processing industries (Wang and Chen, 2006; Zhou et al., 2010; Gautam et al., 2014). Apart from anthropogenic activities linked to metal pollution on the environment, another source of metal pollution includes leaching, settling from air, corrosion, and evaporation (Weerasundara et al., 2017).

12.2   HARMFUL EFFECTS OF HMs

HMs can be grouped into essential and nonessential groups. The essential HMs include Co, Cr, Cu, Fe, Mn, Ni and Zn and other include Pb, Cd, and Hg. Figure 12.1 illustrates the different anthropogenic sources of toxic HMs and their maximum contaminant levels (MCL) (Umar et al., 2010). Heavy metal ions are otherwise referred to as cumulative poison due to their property of persistence nature; their accumulation tends to damage the deoxyribonucleic acid and they act as effective carcinogens (Knasmuller et al., 1998). Nickel is often detected in the industrial effluents of paint manufacturing, battery manufacturing, and electroplating. The permissible concentration intake of nickel ions involves enzyme functions associated with the metabolic activities of ureolysis, hydrogen, and acidogenesis. Nevertheless, the nickel intake at higher concentrations

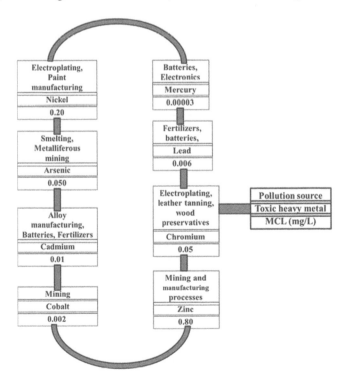

FIGURE 12.1   Anthropogenic sources of HMs and maximum contaminant levels.

consequently leads to lung disease, skin problems, renal edema, and gastrointestinal problems such as vomiting, diarrhea, and nausea (Borba et al., 2006).

Exposure to lead create smore risks for children's health. The higher concentration of lead contamination affects the central nervous system, which can lead to coma and even death. The effects of lead-poisoned children who survive are intellectual disability and behavioral disorder. Even a low concentration of lead exposure to children influences normal brain growth and reduces performance in education. The impacts on human health due to lead poisoning are mostly permanent. Apart from children, lead poison in adults also leads to kidney damage and causes high blood pressure (Dapul and Laraque, 2014). A trace concentration of zinc is useful as an essential element for human health, contributes to normal physiological functions, and maintains metabolic activities. However, zinc contamination at higher concentrations causes anemia, nausea, vomiting, and skin problems. Arsenic pollution in water bodies is considered harmful even at very low concentrations. According to pH and redox potential, arsenic occurs in the form of either trivalent arsenite or pentavalent arsenate. Arsenic ingestion in human beings causes cancers in the skin, liver, lung, bladder, and affects human organs. It also disturbs the function of the cardiovascular system, nervous system, and causes bone marrow disorder. Chronic exposure to arsenic causes the destruction of deoxyribonucleic acid, enzyme inhibition, and synthesis of more reactive oxygen species, which in turn promotes tumor formation (Sang et al., 2019; Sherlala et al., 2019).

Chromium in the aquatic environment mostly exists in the form of trivalent and hexavalent. Trivalent chromium is involved in metabolic activity at a lower concentration. Hexavalent chromium exhibits relatively highly toxic and mutagenic effects. Hexavalent chromium with the property of a strong oxidizing agent causes carcinogenic diseases; it also affects human health with harmful effects including allergic skin reactions, diarrhea, nasal septum perforation, chronic ulcers, pulmonary congestion, and liver inflammation (Rangabhashiyam and Selvaraju 2015; Rangabhashiyam et al., 2016).

Mercury contamination in pregnant patients harms the fetus's central nervous system and mental growth. It affects neurodevelopment and lowers the IQ in children. Mercury poisoning includes the symptoms of tremors, depression, loss of memory, vision and hearing problems, and renal injury (Yu et al., 2017; Siyu et al., 2019). Similarly, high-level exposure to cadmium affects endocrine and cardiovascular and systems. Bone and renal diseases are the predominant symptoms due to chronic cadmium toxicity (Jarup and Akesson, 2009). Cobalt at permissible concentration plays a vital role as a metal component in vitamin B12; nevertheless, higher exposure of cobalt brings various health effects. The cobalt exposure symptoms are neurological disorders affecting vision and hearing; they can damage cardiovascular and endocrine systems and cause hematological dysfunctions (Laura et al., 2017).

12.3 DIFFERENT METHODS FOR HMs REMOVAL

HMs discharged from industrial wastewater cause a negative impact on the ecosystem. Therefore, a suitable method must be employed in a systematic approach to eliminate or minimize the concentration of metal ions in industrial waste streams.

TABLE 12.1
Treatment methods for heavy metal removal from wastewaters

Treatment methods	Advantages	Disadvantages
Adsorption	Effective even at low concentration	Higher cost of activated carbon
	Good performance	Regeneration problem
		Performance variation with adsorbents
Membrane filtration	Higher efficiency	Higher cost investment
	Less generation of solid wastes	Operates at low flow rates
	Lower chemical utilization	
Flotation	High selective process	Requires more capital, maintenance,
	Fewer retention times	and function costs
	Separate even small particles	
Coagulation–flocculation	Efficient settling and dewatering	Requirement of more chemicals
	characteristics	Higher generation of chemical sludge
Ion-exchange	Highly selective process	Requirement of high initial
	Provision for regeneration	investment and maintenance cost
Chemical precipitation	Simple, economical process	Generation of secondary sludge
	Removes different metals	Cost investment for disposal of
		sludge wastes

The conventional methods based on the physicochemical approach for HMs elimination include precipitation (Quanyuan et al., 2018), coagulation, flocculation (Yongjun et al., 2020), ion-exchange (Kuljit and Rajeev 2018), oxidation, reduction (photocatalysis (Al-Sayed et al., 2019), adsorption (Rangabhashiyam et al., 2019), and membrane filtration (Miaolu et al., 2020). The comparisons of the different methods towards heavy metal removal are mentioned in Table 12.1.

12.4 BIOREMEDIATION

Bioremediation represents the use of enzymes or microbes for the toxic contaminant sequestration. The microorganism related decontamination occurs through the biological process of the metabolic pathway (Plan and Van den Eede, 2010). The merits of bioremediation compared to the conventional physicochemical methods include an economical and eco-friendly approach without leaving hazardous wastes. Bioremediation technology occurs through the various methods of biosorption, bioaccumulation, bioventing, bioleaching, biotransformation, biomineralization, biostimulation, and bioaugmentation. The potential of microbes explored towards the elimination of toxic HMs through bioremediation offers a safe method compared to conventional treatment methods. Various strategies have developed over the different in-situ and ex-situ bioremediation for the decontamination of the toxic HMs (Malik, 2004; Rangabhashiyam et al., 2014; Lee et al., 2015; Yongsong et al., 2018).

HMs are associated with the characteristics of non-degradability and different oxidation states. Microbial remediation of HMs depends on mobility, metal speciation, and bioavailability. The interaction of the heavy metal with microbes involves complex phenomenon including physical adsorption, precipitation, ion-exchange complexation, oxidation-reduction reactions, intracellular accumulation, and extracellular complexation (Yao et al., 2012; Rangabhashiyam and Balasubramanian, 2019). The process of heavy metal bioremediation is influenced by the factors of the microorganisms' population, chemicals, and environment (Chang-Ho et al., 2016; Samakshi and Arindam, 2019).

12.4.1 MECHANISM OF BIOREMEDIATION OF HEAVY METAL

Microorganisms are essential in the remediation of HMs as they are able to adopt different mechanisms to interact and endure the metal toxicity. Various mechanisms used by microbes—biotransformation, sequestration, precipitation, and production of exopolysaccharide (EPS)—have been widely studied (Wu et al., 2010; Dixit et al., 2015; Igiri et al., 2018). In presence of toxic metals in the environment, microorganisms develop ingenious mechanisms involving several procedures such as ion exchange, electrostatic interaction, redox process, precipitation, and surface complexation (Yang et al., 2015). The major mechanisms microbes used to detoxify HMs are methylation, demethylation, metal oxidation, metal-organic complexion, intracellular and extracellular metal sequestration, metal efflux pumps, and production of metal chelators like metallothioneins and biosurfactants (Ramasamy et al., 2006). The following mechanisms are used for microbial remediation of HMs (Figure 12.2).

12.4.1.1 Biosorption and Bioaccumulation

Biosorption is a very simple physicochemical process resembling conventional adsorption or ion exchange except that the sorbent is of biological origin. The uptake

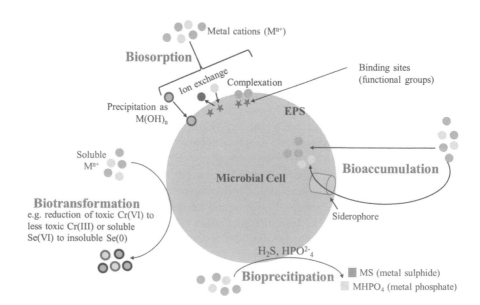

FIGURE 12.2 Bioremediation mechanisms used by microorganisms for removal of heavy metals.

of HMs by microbial cells through biosorption is a metabolically passive process that mostly happens on the cell's exterior. Biosorption can be carried out by dead biomass also as it occurs through ion exchange, surface complexation, and precipitation onto the cell wall and surface layers. The factors that influence biosorption of HMs are i) the type of the biomass and the functional groups present on the EPS, ii) pH, iii) temperature, and iv) the presence of other cations and anions (Purchase et al., 2009; Chojnacka, 2010; Li and Yu, 2014). During the complexation, anionic functional groups like carboxyl and hydroxyl interact electrostatically and covalently with the aqueous metal cations for the binding of metal. The speciation and solubility of metal ions and the properties of the biomass largely depend on pH (Xu et al., 2006). Generally, fewer protons in solution at a higher pH compete with the target metal ion for the binding sites in EPS, causinge an increase in higher metal adsorption. Similarly, the competition between positively charged metal ions and protons increases at lower pH and leads to lesser metal adsorption (Li and Yu, 2014). As biosorption is a reversible process, lowering the pH can be used for the desorption of metals from the biosorbents and in the reuse of biosorbents (Chojnacka, 2010).

On the contrary, bioaccumulation is a more complex metabolism-dependent process than biosorption and occurs in twostages. The first stage is biosorption of the metal and is followed by the accumulation of the metals inside the cell by an active transport system (Chojnacka k 2010; Igiri et al., 2018). In bioaccumulation, the selection of organisms plays an important role. Bioaccumulating organisms should possess features including strong resistance to high loads of toxic metals and the ability to release proteins (e.g. metallothioneins, phytochelatins) containing thiol groups for intracellular binding with toxic metal ions. Although it is usually slower than biosorption, it is possible to achieve higher metals removal via bioaccumulation as cells offer more binding sites both on the surface and inside the cell. In addition, it is a nonequilibrium process because the binding of metals keeps increasing due to the continuous growth of the biomass (Chojnacka, 2010; Igiri et al., 2018).

12.4.1.2   Biotransformation

The biotransformation of HM ions to a less harmful form by microorganisms plays a significant role to detoxify heavy metal ions (Liu et al., 2017). The change in the redox state of HM ions through either reduction or oxidation can effectively reduce theirtoxicity (Giovanella et al., 2016). Bacteria can utilize HMs as electron donors or acceptors for their energy generation (Mal et al., 2016; Yin et al., 2019). *Bacillus arseniciselenatis*, selenium-reducing bacteria, use soluble toxic selenate or selenite as a terminal electron acceptor under anaerobic conditions and reduce it to insoluble and less toxic elemental selenium via dissimilatory reduction (Nancharaiah and Lens, 2015). The reduction of metal ions via enzymatic activity can also lead to the detoxification of metals (Viti et al., 2003). Bacteria like *Bacillus* sp. executes the reduction of mercuric ion to less toxic metallic mercury through mercuric ion reductase (Noroozi et al., 2017). The toxic As(III) can be oxidized by using *Micrococcus* sp. and *Acinetobacter* sp. into less soluble and less toxic As(V) as part of its detoxification mechanisms (Nagvenkar and Ramaiah 2010). Methylation of metals is also considered an important microbial detoxification strategy in metal remediation. Toxic Hg(II) can be biomethylated to gaseous methyl mercury by bacteria e.g.

Escherichia sp., *Bacillus* spp., *Pseudomonas* sp., *Clostridium* sp., etc. as part of their defense mechanism (Igiri et al., 2018). Biomethylation of selenium (Se) and arsenic (As) to gaseous dimethyl selenide and arsines, respectively, also plays a significant role for Se/As detoxification via volatilization (Ramasamy et al., 2006).

12.4.1.3   Intracellular and Extracellular Sequestration

Intracellular sequestration is the complexation of metal ions within cytoplasm once they cross the cell wall and enter microorganisms; it protects the cellular components from exposure to the metal ions (Igiri et al., 2018; Yin et al., 2019) and prevents the metal ions from reaching atoxic level. Several studies have been doneon the intracellular accumulation of metals, predominantly in the treatment of effluent treatment. Microorganisms can decontaminate metal ions by using sulfides or cytosolic polyphosphates or cysteine-rich proteins to form insoluble metal precipitates (Yin et al., 2019). Some bacteria and cyanobacteria possess the ability to deposit heavy metal ions via cytosolic polyphosphates (Remonsellez et al., 2006). On the other hand, when exposed to heavy metal ions, metallothionein, a cysteine-rich protein, can be over-expressed to overcome this stress when it acts as a sink for the accumulation of excess toxic HMs ions. For example, *Synechococcus* sp. releases metallothionein to bind Cd(II) and Zn(II) as part of the detoxification mechanism (Blindauer et al., 2008). *Pseudomonas putida* utilizes cysteine-rich low molecular weight proteins for the intracellular sequestration of Cd, Cu, and Zn ions (Igiri et al., 2018). Thiol-mediated metal reduction in the cytoplasm is also widely recognized as part of microbial detoxification mechanisms against HMs.

Extracellular sequestration of HM ions is the accumulation and complexation of metal ions as insoluble compounds by cellular components in the periplasm. Cu-resistant *Pseudomonas syringae* secrets copper-inducible proteins CopA, CopB in the periplasm, and CopC in the outer membrane, which binds copper ions and decreases the metal toxicity (Igiri et al., 2018). Precipitation of metal ions is also a part of the extracellular sequestration strategy. *Geobacter metallireducens* are capable of reducing toxic Mn(IV) to insoluble Mn(II) and U(VI) to insoluble U(IV) (Gavrilescu, 2004). Similarly, *Geobacter sulfurreducens* and *Geobacter metallireducens* can decrease chromium toxicity by reducing the toxic Cr(VI) to less toxic Cr(III) (Bruschi and Florence, 2006). *Desulfovibrio desulfuricans* and *Klebsiella planticola* can generate large amounts of hydrogen sulfide that fosters metal precipitation and protects the cells from the toxicity of HMs like cadmium (Yin et al., 2019). Detoxification of metals can be achieved by siderophores-mediated reduction, too (Yin et al., 2019). Reduction of selenite to insoluble elemental selenium through iron siderophore, produced by *Pseudomonas stutzeri*, has been reported for the detoxification of selenite (Nancharaiah and Lens, 2015). Siderophores are capable of binding and accumulating not only ferric iron but other metals as well and help in protecting the microorganisms from HMs toxicity (Sharma et al., 2018; Yin et al., 2019).

12.4.2   Microbes Involved in Bioremediation

Nowadays, various bioremediation technologies have been employed to remove toxic HMs from polluted soil/water for on-site remediation. Metal transformation

TABLE 12.2
Bioremediation of toxic heavy metal using microorganisms

Microorganism	Heavy metal	Reference
Bacteria		
Bacillus, Shewanella, Lysinibacillus, Acinetobacter	Manganese	Xunchao et al., 2019
Bacillus subtilis	Chromium	Kim et al., 2015
Burkholderia sp.	Zinc	Yan et al., 2018
Serratia marcescens	Manganese	Pollyana et al., 2018
Pseudomonas azotoformans	Cadmium, Copper, Lead	Anna et al., 2018
Fungi		
Beauveria bassiana	Zinc, Copper, Cadmium, Chromium, Nickel	Deepak et al., 2016
Sterigmatomyces halophilus	Iron, Zinc	Amna et al., 2018
Alternaria alternata	Lead, Cadmium	Preeti et al., 2019
Talaromyces islandicus	Lead	Rohit et al., 2020
Bacillus marisflavi	Copper, Lead	Kayalvizhi and Kathiresan 2019
Penicillium citrinum	Manganese	Kayalvizhi and Kathiresan 2019
Phanerochaete chrysosporium	Lead, Cadmium	Ningqin et al., 2020
Algae		
Chlorella coloniales	Chromium, Cadmium, Cobalt, Iron, Arsenic	Jalil and Kamyar 2019
Microcystis aeruginosa	Zinc, Cadmium	Jiancai et al., 2020
Ascophyllum nodosum, Fucus spiralis, Laminaria hyperborea, Pelvetia canaliculata,	Copper, Zinc, Nickel	Cechinel et al., 2016
Enteromorpha sp.	Hexavalent chromium	Rangabhashiyam et al., 2016
Cystosiera compressa, Sargassum vulgare, Turbinaria sp.	Copper, Lead	Nabel et al., 2018
Cladophora glomerata	Hexavalent chromium	Ali et al., 2018
Spirulina platensis, Chlorella vulgaris	Cadmium, Lead, Copper	Mohammad et al., 2019

by microbes is considered an important strategy in the bioremediation of HMs due to its wide applicability. However, choosing suitable biomaterial is a major challenge in the bioremediation of heavy metals. At present, living/dead microorganisms and surface modified biomass is the most widely utilized strategy due to its outstanding advantages including simple-to-use, low cost, high adsorption capacity, and wide availability. Among these microorganisms, bacteria, fungi, and algae are most widely used for bioremediation of HMs (Table 12.2).

12.4.2.1 Algae

Microalgae have potential in high biomass production and are very efficient in the removal of HMs through biosorption. Microalgae growth can be controlled by important factors such as sunlight and carbon dioxide. The suitability of microalgae in heavy metal bioremediation is linked with the mechanisms of extracellular and intracellular interactions (Kumar et al. 2015). Metal ions interacting with the cell wall of microalgae act as the initial bioaccumulation step. Various binding groups such as hydroxyl, carboxyl, nitrate, amine, phosphate, and carbohydrate act as a binding group in the cell surface (Talebi et al., 2013). *Enteromorpha* sp. showed the removal of hexavalent chromium and maximum removal performance reported at pH 2.0. The kinetic model of pseudo-second-order well fitted the biosorption experimental data (Rangabhashiyam et al., 2016). The removal of multimetallic solutions consisting of arsenic, boron, copper, manganese, and zinc performed using *Chlorella vulgaris* and *Scenedesmus almeriensis*. The influence of organic matter and carbon dioxide was examined; these affected the growth and metal removal (Ricardo et al., 2019). *Desmodesmus* sp. and *Heterochlorella* sp. are the acid-tolerant microalgae for the simultaneous HMs removal and biodiesel production. The metals of iron, manganese, and zinc, at the solution pH 3.5, best supported the growth of microalgae and metals removal. Intracellular mechanisms have removal efficiency of 40–80% and 40–60% of iron and manganese, respectively (Sudharsanam et al., 2019).

The removal of copper was reported using *Cystoseira crinitophylla* with a biosorption capacity of 160 mg/g. The batch equilibrium data was well described with the Langmuir and Freundlich models. The continuous biosorption experimental data was best described using the column models of Thomas and Clark (Christoforidis et al., 2015). Good removal of cadmium was demonstrated using flocculating microalgae, *Scenedesmus obliquus* in the acidic conditions. The removal efficiency achieved 93.39% within the contact time of 20 min. The regeneration of the biosorbent was assessed using the 0.1 M HCl and the biosorbent illustrated with good biosorption capacity during the entire five recycles (Xiaoyue et al., 2016). Biomass of *Durvillaea antarctica* investigated for the biosorption of copper. Hydroxyl groups distributed in the cell wall are weaker compared to carboxyl groups and undergo interaction with copper ions at the initial solution pH greater than 10. The native form of biomass is amorphous and has a smooth surface with few pore structure. The acid wash of biomass results in cavity structure and removes the intrinsic cations. The porosity structure was induced with the acid treatment of biomass because of crosslink existed biomass surface polysaccharides and intrinsic cations (Hector et al., 2015).

12.4.2.2 Bacteria

The cell wall of bacteria contributes an important role to undergo interaction with metal ions and deposition on the cellular organization. The biomass sorbed the metal ions on the extracellular cell wall and cell wall functional group. The cell wall functional groups of bacteria include hydroxyl, carboxyl, phosphate, and amine. The occurrence of such functional groups including carboxyl groups favors the biosorption of metal ions (Vijayaraghavan and Yun, 2008). Sulfate-reducing bacteria have extensively explored cadmium removal via bioprecipitation and biosorption. The bioprecipitation is connected with bacterial metabolism, while metabolism-independent represents the biosorption. Biosorption presented with 77% of the total cadmium removal compared to the bioprecipitation (Francesca et al., 2010).

Hexavalent chromium removal assessed using magnetotactic bacteria and removal performance investigated the presence of co-existing ions cobalt and copper. The higher removal efficiency of 77% towards hexavalent chromium using living cells was reported for the contact time of 10 min. The co-existing ions cobalt and copper along with the electric field application favored the hexavalent chromium removal (Yingmin et al., 2014). Microorganisms of six bacterial strains mediated biomineralization of six metals investigated. The removal rates of HMs ranged from 88% to 99%. The accumulated HMs in the cell envelope appear as rhombohedral and crystalline structures in the pH range of 8–9. (Meng et al., 2013). Two indigenous bacterial strains *Brochothrix thermosphacta* and *Vibrio alginolyticus* tested for the removal of aluminum using the bioaugmentation method. The result of bioaugmentation showed a better performance to treat aluminum in wastewater (Ipung et al., 2019). Maghemite nanoparticles coated on the bacterial strain *Bacillus subtilis* tested for cadmium removal. At the optimal process conditions, 83.5% maximum removal efficiency and 32.6 mg/g biosorption capacity attained. Biosorbent recovery of 76.4% was reported after the metal removal process (Devatha and Shivani, 2020).

Bacteria with metal tolerance has been successfully used in bioremediation. Metal tolerance by bacteria evolved through various processes including transport across the bacterial cell membrane, cell wall accumulation, entrapment, metal complex formations, oxidation, and reduction reactions. The toxic metal resistance genetic determinants were present in plasmid and chromosome. The process of heavy metal removal by bacteria occurs through ATP, both dependent and independent. (Nanda et al., 2019). The indigenous bacteria was illustrated with good potential towards the removal of HMs from synthetic acid mine drainages. Carbon supplements in continuous mode with a sufficient adaptation period for indigenous bacteria improved the removal efficiency in synthetic acid mine drainage. Minimum inhibitory concentration analysis was performed using *Bacillus cereus* for the metals of copper, chromium, lead, zinc, iron, manganese, and magnesium. The bacterial strain showed high resistance to lead and was reported with the bioremediation efficiency of 97.17% at the optimal conditions of pH 7.0 and 35°C. The distinct functional groups of lead treated *Bacillus cereus* are N–H, C–C, and N–O (Anusha and Natarajan, 2020). Polyphenol from sugarcane molasses was studied as carbon for the reduction and removal of hexavalent chromium using bacteria and showed no inhibitory effect on bacterial growth under the pH-dependent condition. The removal of hexavalent chromium showed promising results under additive or synergistic influence with the combination of molasses and hexavalent chromium reducing bacteria (Kento et al., 2020).

12.4.2.3 Fungi

Fungi with the numerous hydrolyzing enzymes including oxidoreductases, transferases, hydrolases, and lyases are involved in HMs detoxification (Rangabhashiyam et al., 2013; Shabeena et al., 2018; Rajesh et al., 2020). Fungi constitute 13% of the bioremediation agents compared to bacteria, enzymes, algae, plants, and protozoa (Cristina et al., 2019). The mycoremediation of metals was performed using the fungal consortia including *Ascomycota* and *Basidiomycota*. The bioremediation analyzed using the measurements of pH, redox potential, electrical conductivity, metal concentration, fungal biomass, and activity of the enzyme. The consortium of fungi showed efficient removal of arsenic, manganese, chromium, and copper. Iron removal was higher in the case of the *Ascomycota* consortium (Auwalu et al., 2020). Different fungi isolated from my location were assessed for the bioremediation of uranium. The results showed that the 22 fungal species were resistant to uranium, which included 10 *Penicillium* sp. The uranium bioremediation with high potential was revealed in *Gongronella* sp., *Talaromyces* sp, and *Penicillium* sp. (Ednei et al., 2020). Reduction of hexavalent chromium was carried out using *Aspergillus flavus*. The experiments of the batch investigation showed an 89.1% reduction of hexavalent chromium to trivalent chromium within 24 hours. The instrumental characterization showed that the removal of chromium was associated with the mechanisms of adsorption/precipitation on the surface of mycelia (Kumar and Dwivedi, 2019). Isolates *Talaromyces islandicus* and *Aspergillus terreus* were created using 18 S rRNA sequencing. These two strains exhibited higher resistance to lead ions. Both these strains have the potential to remove 80% of lead ions from aqueous solutions (Rohit et al., 2020). Silver bioreduction performed in a nonenzymatic approach with the *Aspergillus foetidus* biomass. The thermodynamic analysis of the biosorption experiment showed an endothermic and spontaneous nature. There was greater than 93% arsenic removal efficiency within 3.5 hours and the results were because of porous structure and silver nanocrystals (Triparna et al., 2017).

12.5 PHYTOREMEDIATION

Several conventional methods have been explored to tackle the toxicity of HMs despite phytoremediation getting more attention as a promising technology for the environmental cleanup option, using specialized plants for removing toxic metals and site restoration (Muthusaravanan et al., 2018; Patra et al., 2020). The remediation process using plants is distinct and only selective potential plants (either native or genetically modified) can be involved to remove HMs from the soil, water, and air. Different phytotechniques such as phytoextraction, phytovolatilization, phytostabilization, and phytodegradation have been studied in detail for decontaminating land and water bodies (Muthusaravanan et al., 2018; Patra et al., 2020). In general, various parts of plants play a significant role in phytoremediation such as uptake by root, bioaccumulation, and translocating them to the aerial parts through the designated tissues, transformation, and/or degradation within the various plant parts. Finally, after harvesting the plant on maturity, the biomass is processed by drying, ashing, or composting or can be reprocessed for further use.

There are several advantages of phytoremediation (Lee, 2013; Muthusaravanan et al., 2018):

- Aesthetically pleasing, publicly accepted, and environmentally friendly technology
- Relatively cheap compared to other methods
- Can be applied against a wide range of contaminants

- Minimum environmental disruption
- Reduces soil erosion due to plantation
- Produces recyclable plant residue with a high metal content with a possibility of recovering and reusing the metals. The contaminated and unusable plant biomass can also be reduced to only 20 to 30 tons of ash per 5,000 tons of soil by the thermal or microbial or chemical process

However, phytoremediation also has several limitations:

- Phytoremediation is most effective only in the contaminated soil or the groundwater (< 5 m depth) due to its shorter roots
- Takes a much longer time clean up a site
- Climate or seasonal conditions, and organic and inorganic contaminants, may influence the growth and survival of the plant. Thus, proper care of the site is needed.
- Plants absorbing toxic HMs may pose a significant threat of transferring HMs to the food chain and result in nonedible plant products
- Possible incomplete removal of HMs from the contaminated site. Moreover, volatilization of HMs may just convert soil or groundwater contamination to air contamination instead of complete bioremediation of HMs.

Although phytoremediation has some limitations, it is increasingly recognized as a successful technology for removing HMs from contaminated soil and water that promises effective and cost-effective clean-up. The rapidly increasing market and the success of phytoremediation for HMs removal from the metal-contaminated area are evident in many places including industrial areas and agricultural fields (Emenike et al., 2018).

The best plants for phytoremediation use should have the following traits: high biomass production, high growth rate, extensive fibrous root systems, easy to cultivate and harvest, and highly tolerant to contaminants, diseases, and pests with an ability to accumulate high levels of contaminants (Patra et al., 2020). To date, only a few plants, known as hyperaccumulators, have been isolated and identified fulfilling the above criteria and explored in the phytoremediation for the removal of HMs (Lee, 2013; Souza et al., 2013). Some nonhyperaccumulator plants can also be used as an alternative to the hyperaccumulators after proper modification.

12.5.1 Mechanisms of Phytoremediation

Phytoremediation generally happens via various mechanisms (Figure 12.3) including phytosequestration, phytoextraction, phytostabilization, phytodegradation, phytovolatilization, and rhizofiltration (Figure 12.4) (Muthusaravanan et al., 2018; Yadav et al., 2018; Patra et al., 2020). In fact, the five main subgroups involved in this process function simultaneously to some extent to remove the HMs from the contaminated sites. Plants used for phytoremediation processes and their mechanisms of contaminant removal are summarized in Table 12.3.

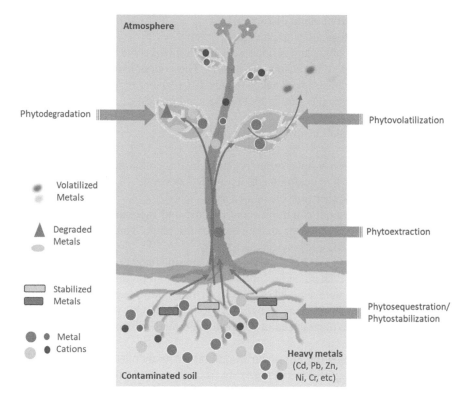

FIGURE 12.3   A pictorial depiction of different mechanisms involved in phytoremediation.

12.5.1.1 Phytosequestration

Phytosequestration is the process that reduces the contaminant's mobility and prevents its migration into the water, soil, or air (Prasad, 2011). There are three mechanisms of phytosequestration:

- Phytochemical complexation in the root zone helps in reducing the bioavailability of a fraction of the HMs. Phytochemicals released into the rhizosphere leads to the precipitation and immobilization of HMs in the root area.
- Transport proteins present on the exterior root membranes can irreversibly adhere and stabilize the HMs on the root surfaces and consequently inhibit the HMs from entering the plant.
- Vacuoles in the plant cells act as a storage and waste container for the plant. Some transport proteins associated with the root assist the transfer of HMs between cells. However, HMs can be stored in those vacuoles of the roots and prevent further translocation to the xylem.

12.5.1.2 Phytoextraction

Phytoextraction is a process used for removing HMs where plants absorb HMs from the contaminated site into their roots and translocate them to the shoots and leaves.

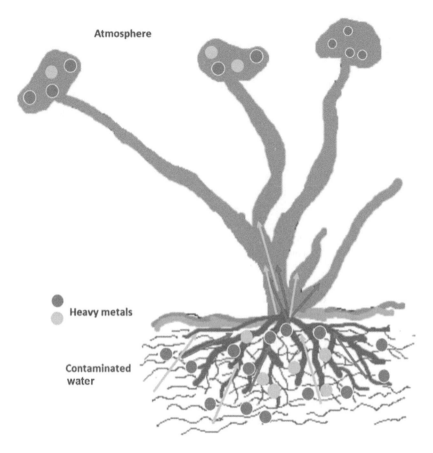

FIGURE 12.4   Rhizofiltration of metals from contaminated water.

This method is highly effective in removing HMs from polluted soils, water, and commercially feasible as well (Yadav et al., 2018; Patra et al., 2020). Plants used for this process should have the following properties: i) capability of fast growth, ii) high biomass, iii) easy to cultivate and harvest, and resistant to pathogens and pests, iv) have extended root networks, v) potential for high metal accumulation, vi) high tolerance level against HMs, vii) well adapted to prevailing climatic conditions, and viii) must be nonedible to avoid food chain contamination (Ali et al., 2013; Patra et al., 2020). Popular species for phytoextraction like *Pteris vittata, Ricinus communis, Puccinellia frigida, Helianthus annuus, Pisum sativum, Brassica napus, Jatropha curcas, Brassica juncea,* and *Stanleya pinnata* have shown the great capability to accumulate the HMs along with high tolerance of HMs in a large scale (Yadav et al., 2018).

For effective phytoextraction of HMs, the contaminants must be bioavailable in the contaminated zone and to the plant roots. Alternatively, the uptake can also happen via vapor adsorption onto the organic root membrane in the vadose zone (Prasad, 2011). After the adsorption, the contaminant may be dissolved into the water and

TABLE 12.3
Various plants reported for phytoremediation technologies

Process	Metals	Plant species	Reference
Phytoextraction	Arsenic	*Pteris vittata*	Yang et al. 2017a,b
	Cadmium	*Ricinus communis*	Yang et al. 2017a,b
	Chromium	*Helianthus annuus*	Farid et al. 2017
	Lead	*Pisum sativum*	Tariq and Ashraf 2016
	Zinc	*Brassica napus*	Dhiman et al. 2016
	Mercury	*Jatropha curcas*	Marrugo-Negrete et al. 2015
	Nickel	*Brassica juncea*	Kathal et al. 2016a,b
	Selenium	*Stanleya pinnata*	Bañuelos et al. 2015
Phytostabilisation	Arsenic	*Epilobium dodonaei* Vill.	Randelović et al. 2016
	Cadmium	*Iris sibirica*	Ma et al. 2017
	Chromium	Rose plant	Ramana et al. 2013
	Lead	*Hordeum vulgare*	Katoh et al. 2017
	Zinc	*Epilobium dodonaei* Vill.	Randelović et al. 2016
	Nickel	*Typha domingensis*	Bonanno and Vymazal 2017
Phytodegradation	Arsenic	*Pteris vittata*	Sakakibara et al. 2010
	Mercury	*Azolla caroliniana*	Bennicelli et al. 2004
Phytovolatilization	Arsenic	*Lupinus albus*	Vázquez et al. 2006
	Cadmium	*Acanthus ilicifolius* L.	Shackira and Puthur 2017
	Lead	*Athyrium wardii*	Zhao et al. 2016
	Zinc	*Agrostis castellana*	Pastor et al. 2015
	Nickel	*Typha latifolia* L.	Varun et al. 2011
Rhizofiltration	Copper	*Noccaea caerulescens*	Dinh et al. 2018
	Lead	*Pistia stratiotes*	Galal et al. 2017a,b
	Zinc	*Noccaea caerulescens*	Dinh et al. 2018
	Nickel	*Tagetes erecta*	Pal et al. 2013

taken up via transport mechanisms along with other nutrients by the plants. After that, the plant either stores it or sequesters it into the cell vacuoles of aboveground tissues or maybe metabolized via phytodegradation and/or phytovolatilized. The HMs can be removed permanently from the polluted zone after the accumulation of metals in the roots and shoots followed by removal and ultimate disposal of the plant biomass (Lee, 2013).

12.5.1.3 Phytostabilization

Phytostabilisation/phytoimmobilization is a process where the toxic compounds are immobilized or stabilized at contaminated sites (Ali et al., 2013; Yadav et al., 2018). Immobilization of HMs using certain plant species can be accomplished through either sorption or accumulation via precipitation, complexation, or reduction of metal within the rhizosphere (Ali et al., 2013). Degradation of metals is not possible, but leaching of metals into the nearby water or agricultural lands can be stopped. Hence, the phytostabilization of HM-contaminated sites is proven as the best alternative to capture them in-situ but large-scale removal is not feasible (Muthusaravanan et al., 2018; Patra et al., 2020).

The phytostabilization was applied to immobilize Pb, As, Cd, Cr, Cu, and Zn in the contaminated soil. For example, highly toxic chromium (VI) reduces to the less soluble chromium(III) using deep-rooting plants (Ali et al., 2013). *Agrostis* spp. and *Festuca* spp. are two of the most commonly used plants to remediate Cu, Zn, and Pb contaminatedsites in European countries (Mahar et al., 2016). Metal-tolerant species like *Epilobium dodonaei*, Rose plant, *Lupinus luteus*, *Brassica juncea*, *Vicia villosa*, and *Phragmites australis* have been employed successfully for its ability to remediate and promote stabilization and immobilization of HMs for soil restoration (Yadav et al., 2018).

During phytostabilization there is no need for disposal of hazardous biomass. It also has the following advantages: i) limits the soil erosion by wind and rain ii) provides hydraulic control, and iii) increases the chance of re-establishing vegetation at a contaminated site, where the natural vegetation is not possible. However, the process has certain restrictions as contaminants remain in the soil, so constant monitoring of the process is required (Muthusaravanan et al., 2018).

12.5.1.4   Phytodegradation

Phytodegradation is the process where contaminants taken up by the plants, after the breakdown or metabolization, are converted into simpler and less toxic forms (Prasad, 2011). The breakdown of the contaminants occurs by the plant itself through the metabolic process and various enzymatic reactions inside the plant. Since HMs are nonbiodegradable, this process of phytoremediation is mainly limited to the removal of organic contaminants (Yadav et al., 2018). Some of the enzymes that help in the process are nitroreductase (degradation of explosives and other nitrated compounds), dehalogenase (transformation of chlorinated compounds), oxygenase (conversion of hydrocarbons, e.g. aliphatic and aromatic compounds), peroxidase (transformation of phenolic compounds), nitrilase (conversion of cyanated aromatic compounds), and phosphatise (the breakdown of organophosphate pesticides) (Jabeen et al., 2009; Yadav et al., 2018). Genetically modified plant species, e.g. *Brassica juncea* and *Liliodendron tulipifera*, are currently being explored for their enhanced phytodegradation capabilities (Kärenlampi et al., 2000).

12.5.1.5   Phytovolatilization

Phytovolatilization involves the uptake of toxic metals by root, translocation to shoot, and finally release/volatilization of metals as a lesser toxic form into the atmosphere from the stem/trunk and leaf surfaces (Patra et al., 2020). It can especially be applied to the remediation of volatile organic compounds (VOCs). Similarly, certain toxic metals like Hg and Se can be uptaken from the contaminated sites and biomethylated to form volatile compounds such as mercuric oxide and dimethyl selenide, which are further evaporated or volatilized into the atmosphere (Wang et al., 2012; Patra et al., 2020).

For example, *Typha latifolia* was used for the volatilization of selenium from contaminated soil into a less toxic form (LeDuc and Terry, 2005). Similarly, mercury can be volatilized as less toxic gaseous elemental mercury by plants, but there are very few naturally occurring plants that can perform this (Ghosh and Singh, 2005). Genetically modified *Arabidopsis thaliana* and *Nicotiana tabacum* show the strong capability to uptake the highly toxic Hg(II) and methyl mercury from the soil and release them as less toxic volatile elemental mercury via mercuric reductase and bacterial organomercurial lyase enzyme (Pilon-Smits and Pilon, 2000). *Pteris vittata* were employed for the remediation of arsenic (As) from As-contaminated soil, where As was effectively volatilized into the atmosphere (Sakakibara et al., 2010).

Although phytovolatilization remains a controversial phytoremediation technique, it is considered one of the best because, once volatilized, the chance of redeposition of molecules at or near the contaminated zone are highly unlikely (Muthusaravanan et al., 2018). However, the volatile dimethyl selenide and mercuric oxide or arsenic are still toxic to the ecosystem, and the release of them in large quantities during the process may create secondary air pollution inthe surrounding area. Therefore, further risk assessment studies are required, so that phytovolatilization can become accepted by regulators and the public (Lee, 2013; Muthusaravanan et al., 2018).

12.5.1.6 Rhizofiltration

Rhizofiltration (Figure 12.4) refers to the use of plant roots to absorb, precipitate, and concentrate toxic compounds, particularly HMs or radionuclides from the aquatic ecosystem (Patra et al., 2020). Both hydrophytes and mesophytes are used in rhizofiltration. However, mesophytes are considered a perfect plant for the rhizofiltration because of their fast-growing, extensive fibrous root system and their ability to remove HMs rapidly from the solution. Rhizofiltration is considered one of the most economically attractive solutions for the treatment of surface and groundwater contaminated with HMs and radionuclides, albeit in relatively low concentrations (Prasad, 2011). For example, it was successfully employed to remove cesium and strontium in Chernobyl, Ukraine (Dushenkov and Kapulnik, 2002). The advantage of this technique is that hydroponically cultivated plants rapidly remove HMs from water and subsequently translocate them to the aerial parts of the plants.

Constructed wetlands can be created using terrestrial, aquatic and wetland plants and all of the contaminated water needs to be exposed to the roots for removing the metals (Rezania et al., 2016; Vymazal and Březinová, 2016). *Pistia stratiotes* for Cu, Zn, and Pb, *Limnocharis flava* for Cd, *Salix matsudana* for Pb, and *Typha domingensis* for Cd, Cr, and Hg have been employed successfully for rhizofiltration (Yadav et al., 2018). *Micranthemum umbrosum* showed great potential to remove As and Cd in hydroponic nutrient solutions. *Warnstorfia fluitans* can also be adsorbed and removes As efficiently from arsenite/arsenate contaminated water bodies without any phytotoxic effects. Another study suggests that metal uptake by phytofiltration can be enhanced significantly using a mixed culture of *Phragmites australis* and *Typha latifolia* for the removal of Cu, Ni, Fe, Cd, Pb, Cr, and Zn within 14 days. Finally, the harvested plants that contain the HMs accumulated from the wastewater can be disposed of or treated to recycle the metals (Yadav et al., 2018).

During rhizofiltration, the disappearance of metals from the solution follows nonlinear kinetics by suggesting that many mechanisms operate simultaneously for the removal of metals (Prasad, 2011). Surface absorption of contaminants by the roots (even by dead roots) depends on the physicochemical processes, including ion exchange and chelation, which is the fastest and the most used mechanism. Apart from that, rhizofiltration is also largely dependent on biosorption through microbial, fungal, or other living or dead biomass. In addition, other, slower mechanisms like intracellular uptake, deposition in vacuoles, and translocation to the shoot, or precipitation of the metal from solution by plant exudates may also play a significant role in rhizofiltration.

12.5.2 FACTORS AFFECTING THE PHYTOREMEDIATION OF HEAVY METALS

Several factors affect the phytoremediation of HMs including others cations present in the soil, soil pH and texture, redox potential, organic matter, root exudates and system, and plant species (Ma et al., 2016; Sheoran et al., 2016; Yadav et al., 2018; Patra et al., 2020). These factors not only influence the bioavailability of HMs but also affect their uptake from the contaminated soil and thus need to be adjusted to enhance the remediation.

12.5.2.1 Soil Properties

Soil pH is the most important factor influencing the availability of metals in the soil for plant uptake as it affects the metal solubility and availability. A lower soil pH increases the concentration of bioavailable HMs. At low pH, the increased amount of H^+ ions enhances the cation exchange capacity between HMs. As a result, the adsorption of HMs on the surface of colloids and clay mineral particles into the soil decrease and increase the availability of contaminants for plant uptake (Sheoran et al., 2016). For example, an increase in uptake of Cd and Zn by using *Thlaspi caerulescens*, an important hyperaccumulator, was observed by reducing the pH of the soil (Yadav et al., 2018). HMs can be divided into two categories based on common soil pH range: metals (e.g. Cd, Zn, Ni) with comparatively high mobility, and metals (e.g. Cu, Cr, Pb) with low mobility (Kim et al., 2015). Many studies reported that Cd, Hg, Pb, Cr, Fe, and Zn are more soluble and bioavailable at low pH and offer better phytoremediation (Ma et al., 2016; Yadav et al., 2018). In another study, lead uptake by plants was significantly enhanced by changing the pH, phosphorous content, and organic matter of the soil (Ma et al., 2016; Sheoran et al., 2016).

12.5.2.2 Plant Species Optimal for Phytoremediation

The success of the phytoremediation largely depends on the selection of suitable plant species for HMs remediation. Plants growing on contaminated sites can be divided into three categories: excluders, accumulators, and indicators. Excluders restrict the entry of metals into the cells of the roots and are used for the phytostabilization of toxic metals in contaminated soil (Patra et al., 2020). The accumulators cannot stop metals entering the roots and accumulatingin the biomass of the plants while the indicators are used to study the toxicity of the metals. The most favored plant species should have the following characteristics: the ability to adapt easily to local climates and soils, ease of planting and maintenance, fast growth rate, depth of the plant's root structure, and ability to extract or degrade the concerned contaminants to less toxic forms (Souza et al., 2013; Muthusaravanan et al., 2018).

Hyper-accumulator plants are nature's gift; they are able to grow easily on metal contaminated soils and can accumulate high concentration (100–1,000 times higher

TABLE 12.4

List of some hyperaccumulators for HMs remediation

Metal	Plant Species	Family	Metal accumulation capacity (mg kg⁻¹ DW)	Reference
As	*Tagetes minuta*	Asteraceae	381	Salazar MJ and Pignata ML 2014
Cd	*Lantana camara* L.	Verbenaceae	N/A	Liu S et al., 2019
	Cannabis sativa L.	Cannabaceae	151	Ahmad R et al., 2016
	Thlaspi caerulescens	Brassicaceae	5,000	Koptsik GN 2014
Cr	Eichhornia crassipes	Pontederiaceae	NA	Sarkar M et al., 2017
Cu	*Eichhornia crassipes*	Pontederiaceae	NA	Sarkar M et al., 2017
	Cannabis sativa L.	Cannabaceae	1,530	Ahmad R et al., 2016
Ni	*Alyssum markgrafii*	Brassicaceae	4,038	Salihaj M et al., 2016
	Thlaspi caerulescens	Brassicaceae	16,200	Koptsik GN 2014
Pb	*Noccaea caerulescens*	Brassicaceae	1,700–2,300	Dinh N et al., 2018
	Sorghum halepense L.	Poaceae	1,407	Salazar MJ and Pignata ML 2014
	Betula occidentalis	Betulaceae	1,000	Koptsik GN 2014
	Helianthus annuus	Asteraceae	5,600	Koptsik GN 2014
	Brassica nigra	Brassicaceae	9,400	Koptsik GN 2014
	Medicago sativa	Fabaceae	43,300	Koptsik GN 2014

than nonhyperaccumulating species) of metals in their shoots without showing any phytotoxic effects (Souza et al., 2013; Saxena et al., 2020). Three basic features that differentiate hyperaccumulators from nonhyperaccumulating species include a significantly higher HMs uptake rate and sequestering of them in leaves, a faster root-to-shoot translocation, and a strong ability to tolerate HMs toxicity.

Bini et al. (2017) demonstrated that *Alyssum bertolonii* has the ability to accumulate a very high concentration of Ni (i.e., 2,118 mg kg⁻¹) in its aerial parts. So far, more than 500 plant species have been isolated and identified as metal hyperaccumulators, e.g. *Asteraceae* sp., *Brassicaceae* sp., *Cyperaceaesp.*, *Fabaceae sp.* (Souza et al., 2013; Saxena et al., 2020). Table 12.4 lists some important hyperaccumulators for heavy metal remediation. These species, however, are generally slow-growing plants and generate little biomass, which makes their commercial use difficult. Nonetheless, these species are of great interest for studies to understand the mechanisms that allow them to accumulate and tolerate phytotoxicity at high concentrations and the knowledge can be used for developing genetically modified plant species with higher biomass production (Souza et al., 2013).

Although nonhyperaccumulator plant species cannot accumulate high concentrations of metal, due to their high biomass production it is possible to overcome the typical capacity hyperaccumulator. Hence, not only crop plant species, but willow, poplar, and the Brazilian leguminous tree should also be considered and investigated for phytoremediation because of their enormous biomass (Souza et al., 2013).

12.5.3 Approaches for Improving the Phytoremediation Process

In this section, we will discuss several ways to enhance the phytoremediation process.

12.5.3.1 Application of Chelating Agents

The release and mobility of metals into soil solution can be enhanced by adding synthetic chelating agents and therefore increase the potential for uptake into roots (Souza et al., 2013; Yadav et al., 2018). Several chelating agents such as EDTA (the most widely used and successfully utilized), DTPA, EGTA, and NTA have been explored successfully for their capability to mobilize metals and enhance metal accumulation in plant species (Muthusaravanan et al., 2018; Yadav et al., 2018). The addition of high levels of EDTA increases the soluble lead concentrations by forming a soluble Pb-EDTA complex. This leads to the increased movement of Pb-EDTA from roots to shoots and hence, the phytoextraction of lead from the contaminated site (Yadav et al., 2018). However, the application of such chelating agents also has certain environmental risks such as leaching of metals from contaminants to uncontaminated areas and into the groundwater (Souza et al., 2013).

In contrast, low-molecular-weight natural organic chelating agents such as citric acid, malic acid, oxalic acid, and acetic acid can also be used as an interesting alternative to induced phytoextraction of toxic metals. For example, it is reported that an increase in Cd accumulation in the *Solanum nigrum* after the addition of citrate and acetate as a chelating agent, while phytoextraction of Ni by using *Thlaspi goesingense* increased significantly after adding citrate (Souza et al., 2013). The use of organic chelating agents is also advantageous because of having lower negative impacts on the environment due to their high biodegradation rate in the soil. In addition, organic chelating agents like citric acid also impart less toxic effects on plant growth in contrast to EDTA.

12.5.3.2 Application of Plant Growth-Promoting Rhizobacteria (PGPR)

The plant-microbe interactions could be exploited to enhance the plant growth under toxic metal stress and to increase the remediation of HM-contaminated sites. PGPR has been reported to not only enhance the host plant growth in toxic metal-contaminated sites, but it also improves the HMs phytoextraction by affecting the solubility, bioavailability, and transport of HMs (Yadav et al., 2018; Saxena et al., 2020). Various beneficial hormones such as auxins, gibberellins, cytokinins, and ethylene along with various secondary metabolites produced by PGPR can improve the plant growth as well (Yadav et al., 2018). Siderophores, released by PGPR, also help in the mobilization and extraction of metals. However, plant-microbe interaction can also hinder the metal uptake by decreasing the metal bioavailability in the root. For example, it is reported that although *Burkholderia* sp. enhanced the growth of the plant, it reduces the uptake and accumulation of Ni and Cd in the roots and shoots of tomatoes (*Solanum lycopersicum*) (Madhaiyan et al., 2007). Hence, further research is necessary to explore the novel microbial diversity to isolate and characterize suitable plant-associated beneficial microbes for improving microbe-assisted phytoremediation of HM-contaminated sites.

12.5.3.3   Genetically Modified Plants for Enhanced Phytoremediation

As an advanced technique, genetic engineering is being applied to modify the low biomass and slow-growing hyperaccumulating plants for improved phytoremediation (Fasani et al., 2017). The main objective is to manipulate the plant's genes with increased metal uptake, and accumulation properties and the ability to tolerate high metal toxicity (Goel et al., 2009; Saxena et al., 2020). The general approach for developing transgenic plants include over-expression or knock-down of genes encoding metal-binding or metal-chelating, and membrane transporter proteins, which control the detoxification and phytoremediation of HMs (Goel et al., 2009; Sarwar et al., 2017). For example, over-expression of glutamyl cysteine synthetase results in enhanced accumulation of HMs by genetically modified *Populus angustifolia, Nicotiana tabacum,* and *Silene cucubalus.* Yadav et al. (2018) and Xia et al.(2018) demonstrated enhanced biomass production and Cd uptake in genetically modified *Arabidopsis* plants having over-expression of Caffeoyl-CoA O-methyltransferase.

Although the use of transgenic plants in phytoremediation of HMs has been immensely successful, several environmental risks associated with it are still there to overcome. Some of the risks are as follows: i) enhanced exposure of toxic metals to humans and wildlife due to increased metal accumulation in edible parts of plants, ii) uncontrolled spread of transgenic plants due to cross-pollination (Davison, 2005), and iii) spreading of the transgenic plants due to better fitness can increase the risk of the potential loss of diversity (Saxena et al., 2020). Societal concern opposing the use of genetically modified plants is another hurdle, which needs to be overcome (Yadav et al., 2018; Saxena et al., 2020).

12.5.4   Emerging Phytotechnologies

Phytoremediation is an emerging field for the remediation of HMs from contaminated soil and water bodies by using plants as a more efficient alternative to traditional remediation methods. This technology holds promise because of its cheap capital and maintenance cost, higher success rates, and aesthetic nature. Its successful applications provide a new outlook for engineers and scientists; some emerging phytotechniques are briefly discussed below.

12.5.4.1   Constructed Wetlands

The use of constructed wetlands (CWs), as a low-cost and eco-friendly natural wastewater treatment system has been gaining attention worldwide. CWs are constructed to remediate various wastewaters such as municipal, agricultural, and industrial wastewater (Rezania et al., 2016; Saxena et al., 2020). In CWs, wetland plants with a fast growth rate and high biomass properties that arecapable of high metal accumulation, such as *Phragmites australis, Typha latifolia,* and *Canna indica,* are used within a more controlled environment for the treatment of metal-contaminated wastewater (Bharagava et al., 2017). The removal of HMs in CWs mainly followed the biological method including uptake by both plant and microbe simultaneously. The removal efficiency of CWs mainly depends on plant species and theirgrowth rate, and concentration of the HMs in the wastewater (Rezania et al., 2016). Currently, researchers are exploring integrating CW plants with a microbial fuel cell (MFC) as an innovative approach for combining wastewater treatment and electricity generation. Habibul et al. (2016) demonstrated the simultaneous removal of chromium and electricity generation in an integrated plant-microbial fuel cell system. CWs have many economic and ecological benefits, as theyhave low costs for installation, operation, and maintenance.

12.5.4.2 Phytomining

Phytomining is an emerging technology where selected plants are used to accumulate precious metals in their shoots from mineral-rich substrates (Sheoran et al., 2009). The process is more suitable for the treatment and recovery of valuable metals like Ag, Au, Ni, Ti, due to their high price (Mahar et al., 2016; Saxena et al., 2020). For example, phytomining of Ni has been commercialized using hyperaccumulator plants such as *Alyssum murale* and *Alyssum corsicum*, which can uptake and accumulate Ni (>400 kg ha^{-1}) with a production cost of $250–$500 ha^{-1} (Ali et al., 2013). Antony et al. (2015) also demonstrated phytomining of Ni using *Berkheya coddii* for the remediation and accumulation of Ni from industrially contaminated soil in Rustenburg, South Africa.

The economics of phytomining depends on various factors including the metal content of the soil and plants, the plant's biomass production per annum, and, most importantly, the revenue generation and world price of the metals being phytomined (Saxena et al., 2020). The profitability of Ni and Au phytomining using *B. coddii* and *B. juncea*, respectively, in Australia is estimated at AU$ 11,500 and 26,000 ha^{-1} yield^{-1}, respectively (Mahar et al., 2016). However, phytomining of U is reportedly unprofitable due to its low concentration in the biomass (10 t ha^{-1}) and low plant growth rate under harsh climatic conditions in the northern regions of the world (Sheoran et al., 2009). Nevertheless, phytomining has the possibility to become economically beneficial for simultaneous metal remediation and recovery of mine sites and mine tailing waste, particularly in developing countries in tropical regions.

12.6 CONCLUSION

Contamination of toxic heavy metals in the water and soil is a major problem worldwide and requires a public health emergency. This chapter highlights that HMs are among the most decisive threats to the environment and human health and it is therefore very important to maintain the minimum concentration limit of HMs according to the environmental agencies. Conventional remediation techniques are neither cheap nor environmentally friendly, thus necessitating the implementation of cost-effective and environmentally friendly technologies to remediate HM-polluted soil and water. The feasibility of bio- and phytoremediation of HM-contaminated sites as an attractive eco-friendly and sustainable approach for the elimination of HMs has been summarized and shows much promise for metal detoxification and removal. The chapter explores the current scientific progress, sustainability issues, and the suitability of bio- and phytoremediation in metal clean-up as an interesting alternative to physicochemical methods with their benefits associated with both economic and environmental terms and with the possibilities of scaling-up from the bench level. A better understanding of the uptake of HMs and detoxification by microbes or plants is required. Exploration of better biological systems along with the introduction of advanced biotechnological techniques like genetic engineering in the future will also help the scientific community to boost the ability and affordability for HMs removal from anthropogenic activities.

REFERENCES

Ahmad R, Tehsin Z, et al. (2016) Phytoremediation potential of hemp (*Cannabis sativa* L.): identification and characterization of heavy metals responsive genes. Clean Soil Air Water, 44, 195.

Ali H, Khan E, et al. (2013) Phytoremediation of heavy metals concepts and applications. Chemosphere, 91, 7, 869.

Ali A Al-H, Hussein S Al-Qahtani, Abdullah A Al-Ghanayem, Ameen F, Ibraheem BM (2018) Potential use of green algae as a biosorbent for hexavalent chromium removal from aqueous solutions. Saudi J Biol Sci., 25, 1733.

Al-Sayed A Al-Sherbini, Ghannam HEA, El-Ghanam GEM, El-Ella AE, Youssef AM (2019) Utilization of chitosan/Ag bionanocomposites as eco-friendly photocatalytic reactor for Bactericidal effect and heavy metals removal. Heliyon, 5, e01980.

Amna B, Javaid H, Ali A, Khalid M, Muhammad A, Muhammad SH, Sami U, Sumbal S, Imran A (2018) Biosorption of heavy metals by obligate halophilic fungi. Chemosphere, 199, 218.

Anna CP, Justyna SB, Wojciech L (2018) Optimization of copper, lead and cadmium biosorption onto newly isolated bacterium using a Box-Behnken design. Ecotox Environ Safe., 149, 275.

Antony VDE, Alan JM, et al. (2015) Environ Sci Technol., 49, 4773.

Anusha, P, Natarajan D (2020) Bioremediation potency of multi metal tolerant native bacteria Bacillus cereus isolated from bauxite mines, kolli hills, Tamilnadu: A lab to land approach, Biocatal Agric Biotechnol., 25, 101581.

Auwalu H, Agamuthu P, Innocent CO, Fauziah SH (2020) Bioaugmentation assisted mycoremediation of heavy metal and/metalloid landfill contaminated soil using consortia of filamentous fungi. Biochem Eng J., 157, 107550.

Bañuelos GS, Arroyo I, et al. (2015). Selenium biofortification of broccoli and carrots grown in soil amended with Se-enriched hyperaccumulator *Stanleya pinnata*. Food Chem., 166, 603.

Bennicelli R, Stępniewska Z, et al. (2004). The ability of *Azolla caroliniana* to remove heavy metals (Hg(II), Cr(III), Cr(VI)) from municipal waste water. Chemosphere, 55, 141.

Bharagava RN, Saxena G, et al. (2017) Constructed wetlands: an emerging phytotechnology for the degradation and detoxification of industrial wastewaters. In Environmental pollutants and their bioremediation approaches. Bharagava RN (ed.), CRC Press, Taylor & Francis Group, USA: 397.

Bini C, Maleci L, et al. (2017) Potentially toxic elements in serpentine soils and plants from Tuscany (Central Italy). A proxy for soil remediation. Catena, 148, 60.

Blindauer CA, Harrison MD, et al. (2008) Isostructural replacement of zinc by cadmium in bacterial metallothionein. Metal Ions in Biology and Medicine, 10, 167.

Bonanno G and Vymazal J (2017). Compartmentalization of potentially hazardous elements in macrophytes: insights into capacity and efficiency of accumulation. J Geochem Explor., 181, 22.

Borba CE, Guirardello R, Silva, EA, Veit MT, Tavares CRG (2006) Removal of nickel(II) ions from aqueous solution by biosorption in a fixed bed column: experimental and theoretical breakthrough curves. Biochem Eng J., 30, 184.

Chang-Ho K, Yoon-Jung K, Jae-Seong S (2016) Bioremediation of heavy metals by using bacterial mixtures. Ecol Eng., 89, 64.

Cheng SY, Show PL, Lau BF, Chang JS, Ling TC (2019) New prospects for modified algae in heavy metal adsorption. Trends Biotechnol., 37, 11.

Chojnacka K (2010). Biosorption and bioaccumulation – the prospects for practical applications. Environ Int., 36, 299.

Christoforidis A, Orfanidis S, Papageorgiou S, Lazaridou A, Favvas E, Mitropoulos AC (2015) Study of Cu (II) removal by *Cystoseira crinitophylla* biomass in batch and continuous flow biosorption, Chem Eng J., 277, 334.

Claveria RJR, Pereza TR, et al. (2019). *Pteris melanocaulon* Fée is an As hyperaccumulator. Chemosphere, 236, 124380.

Cristina MQ, Ana MTM, Leandro CPL (2019) Overview of bioremediation with technology assessment and emphasis on fungal bioremediation of oil contaminated soils. J Environ Manage., 241, 156.

Dapul H., Laraque, D (2014) Lead poisoning in children. Adv Pediatr., 61, 313.

Dedeke GA, Owagboriaye FO, Adebambo AO, Ademolu KO (2016) Earthworm metallothionein production as biomarker of heavy metal pollution in abattoir soil. Applied Soil Ecology., 104, 42.

Deepak G, Priyadarshini D, Arghya B, Abhishek M, Anushree M, Maneesh N, Shaikh ZA (2016) Multiple heavy metal removal using an entomopathogenic fungi *Beauveria bassiana*. Bioresour Technol., 218, 388.

Devatha CP, Shivani S (2020) Novel application of maghemite nanoparticles coated bacteria for the removal of cadmium from aqueous solution. J Environ Manage., 258, 110038.

Dhiman SS, Selvaraj C, et al. (2016). Phytoremediation of metal-contaminated soils by the hyperaccumulator canola (*Brassica napus* L.) and the use of its biomass for ethanol production. Fuel, 183, 107.

Dinh N, Ent AVD, et al. (2018). Zinc and lead accumulation characteristics and in vivo distribution of Zn^{2+} in the hyperaccumulator *Noccaea caerulescens* elucidated with fluorescent probes and laser confocal microscopy. Environ Exp Bot., 147, 1.

Dixit R, Wasiullah, et al. (2015). Bioremediation of heavy metals from soil and aquatic environment: an overview of principles and criteria of fundamental processes. Sustainability, 7, 2, 2189.

Duruibe, JO, Ogwuegbu MOC, Egwuruhwu JN (2007) Heavy metal pollution and human biotoxic effect. Int J Phys Sci, 2, 112.

Ednei C, Tatiana AR, Marycel C, Thomas KM, Benedito C (2020) Resistant fungi isolated from contaminated uranium mine in Brazil shows a high capacity to uptake uranium from water. Chemosphere, 248, 126068.

Emenike CU, Jayanthi B, et al. (2018). Biotransformation and removal of heavy metals: a review of Phytoremediation and microbial remediation assessment on contaminated soil. Environ Rev., 26, 2, 156.

Farid M, Ali S, et al. (2017). Citric acid assisted phytoextraction of chromium by sunflower; morpho-physiological and biochemical alterations in plants. Ecotoxicol Environ Saf., 145, 90.

Fasani E, Manara A, et al. (2017). The potential of genetic engineering of plants for the remediation of soils contaminated with heavy metals. Plant Cell Environ., 41, 5, 1201.

Francesca P, Carolina CV, Luigi T (2010) Isolation and quantification of cadmium removal mechanisms in batch reactors inoculated by sulphate reducing bacteria: Biosorption versus bioprecipitation. Bioresour Technol., 101, 2981.

Galal TM, Eid EM, et al. (2017). Bioaccumulation and rhizofiltration potential of *Pistia stratiotes* L. for mitigating water pollution in the Egyptian wetlands. Int J Phytoremediation., 20, 5, 40.

Gautam RK, Mudhoo A, Lofrano G, Chattopadhyay MC (2014) Biomass-derived biosorbents for metal ions sequestration: Adsorbent modification and activation methods and adsorbent regeneration. J Environ Chem Eng., 2, 239.

Gavrilescu M (2004). Removal of heavy metals from the environment by biosorption. Engineering in Life Sci., 4, 3, 219.

Giovanella P, Cabral L, et al. (2016). Mercury (II) removal by resistant bacterial isolates and mercuric (II) reductase activity in a new strain of *Pseudomonas* sp. B50A. New Biotechnol., 33, 216.

Goel S, Malik JA, et al. (2009). Molecular approach for phytoremediation of metalcontaminated sites. Arch Agron Soil Sci, 55, 451.

Habibul N, Hu Y, et al. (2016). Bioelectrochemical chromium (VI) removal in plant-microbial fuel cells. Environ Sci Technol., 50, 3882.

Hector C, Claudia O, Jaime P, Daniel B, Ximena C, Liliana G, Juan Carlos MP (2015) Characterization of copper (II) biosorption by brown algae Durvillaea antarctica dead biomass. Adsorption, 21, 645.

Hwang, SK, Jho, EH (2018) Heavy metal and sulfate removal from sulfate-rich synthetic mine drainages using sulfate reducing bacteria. Sci Total Environ., 635, 1308.

Igiri BE, Okoduwa SIR, et al. (2018). Toxicity and bioremediation of heavy metals contaminated ecosystem from tannery wastewater: A review. J Toxicol, 2018, 2568038.

Ipung FP, Setyo BK, Nur II, Muhammad FI, Siti Rozaimah SA (2019) Aluminium removal and recovery from wastewater and soil using isolated indigenous bacteria. J Environ Manage., 249, 109412.

Jabeen R, Ahmad A, et al. (2009). Phytoremediation of heavy metals: physiological and molecular mechanisms. Bot Rev., 75, 339.

Jalil J, Kamyar Y (2019) Optimization of heavy metal biosorption onto freshwater algae (*Chlorella coloniales*) using response surface methodology (RSM). Chemosphere, 217, 447.

Jarup L, Akesson A (2009) Current status of cadmium as an environmental health problem. Toxicol Appl Pharmacol., 238, 201.

Jiancai D, Dongwang F, Weiping H, Xin L, Yonghong W, Heather B (2020) Physiological responses and accumulation ability of Microcystis aeruginosa to zinc and cadmium: Implications for bioremediation of heavy metal pollution. Bioresour Technol., 303, 122963.

Kärenlampi S, Schat H, et al. (2000). Genetic engineering in the improvement of plants for phytoremediation of metal polluted soils. Environ Pollut., 107, 225.

Kayalvizhi, K., Kathiresan, K (2019) Microbes from wastewater treated mangrove soil and their heavy metal accumulation and Zn solubilisation. Biocatal Agric Biotechnol., 22, 101379.

Kathal R, Malhotra P, et al. (2016). Phytoextraction of Pb and Ni from the polluted soil by *Brassica juncea* L. J Environ Anal Toxicol., 6, 394.

Katoh M, Risky E, et al. (2017). Immobilization of lead migrating from contaminated soil in rhizosphere soil of barley (*Hordeum vulgare* L.) and hairy vetch (*Vicia villosa*) using hydroxyapatite. Int J Environ Res Public Health, 14, 1273.

Kento I, Yuki H, Hiroaki S, Ryuto M, Tomoyasu S (2020) Role of polyphenol in sugarcane molasses as a nutrient for hexavalent chromium bioremediation using bacteria. Chemosphere, 250, 126267.

Kim RY, Yoon JK, et al. (2015). Bioavailability of heavy metals in soils: definitions and practical implementation – a critical review. Environ Geochem Health, 41, 1041.

Kim IH, Choi J-H, Joo JO, Kim Y-K, Choi J-W, Oh B-K (2015) Development of a microbe-zeolite carrier for the effective elimination of heavy metals from seawater. J Microbiol Biotechnol., 25, 1542.

Knasmuller S, Parzefall W, Sanyal R, Ecker S, Schwab C, Uhl M, Darroudi F (1998) Use of metabolically competent human hepatoma cells for the detection of mutagens and antimutagens. Mutat Res Fundam Mol Mech Mutagen., 402, 1, 185.

Koptsik GN (2014). Problems and prospects concerning the phytoremediation of heavy metal polluted soils: a review. Eurasian Soil Sci., 47, 92.

Kuljit K, Rajeev J (2018) Synergistic effect of organic-inorganic hybrid nanocomposite ion exchanger on photocatalytic degradation of Rhodamine-B dye and heavy metal ion removal from industrial effluents. J Environ Chem Eng., 6, 7091.

Kumar SK, Dahms HU, Won EJ, Lee JS, Shin KH (2015) Microalgae - A promising tool for heavy metal remediation. Ecotoxicol Environ Saf. 113, 329

Kumar V, Dwivedi SK (2019) Hexavalent chromium reduction ability and bioremediation potential of *Aspergillus flavus* CR500 isolated from electroplating wastewater. Chemosphere, 237, 124567.

Laura L, Bart V, Catherine Van DS, Floris W, Leen M (2017) Cobalt toxicity in humans: A review of the potential sources and systemic health effects. Toxicology, 387, 43.

Lee, E., Han, Y., Park, J., Hong, J., Silva, R.A., Kim, S., et al., 2015. Bioleaching of arsenic from highly contaminated mine tailings using *Acidithiobacillus thiooxidans*. J Environ Manage., 147, 124131.

Lee JH (2013). An overview of phytoremediation as a potentially promising technology for environmental pollution control. Biotechnol Bioprocess Eng., 18, 431.

Liu S, Ali S, et al. (2019). A newly discovered Cd-hyperaccumulator *Lantana camara* L. J Hazard Mater., 371, 233.

Liu SH, Zeng GM, et al. (2017). Bioremediation mechanisms of combined pollution of PAHs and heavy metals by bacteria and fungi: a mini review. Bioresour Technol., 224, 25.

Ma Y, Oliveira RS, et al. (2016). Biochemical and molecular mechanisms of plant-microbe-metal interactions: relevance for phytoremediation. Front Plant Sci., 7, 918.

Ma N, Wang W, et al. (2017). Removal of cadmium in subsurface vertical flow constructed wetlands planted with *Iris sibirica* in the low-temperature season. Ecol Eng., 109, 48.

Madhaiyan M, Poonguzhali S, et al. (2007). Metal tolerating methylotrophic bacteria reduces nickel and cadmium toxicity and promotes plant growth of tomato (*Lycopersicon esculentum* L.). Chemosphere, 69, 220.

Mahar A, Wang P, et al. (2016). Challenges and opportunities in the phytoremediation of heavy metals contaminated soils: a review. Ecotoxicol Environ Saf, 26, 111.

Mal J, Nancharaiah YV, et al. (2016). Metal chalcogenide quantum dots: biotechnological synthesis and applications. RSCAdv., 6, 41477.

Malik A (2004) Metal bioremediation through growing cells. Environ Int., 30, 261.

Cechinel MA, Mayer DA, Pozdniakova TA, Mazur LP, Boaventura RA, de Souza AAU, de Souza SMGU, Vilar VJ (2016) Removal of metal ions from a petrochemical wastewater using brown macro-algae as natural cation-exchangers. Chem Eng J., 286, 1.

Marrugo-Negrete J, Durango-Hernández J, et al. (2015). Phytoremediation of mercury-contaminated soils by *Jatropha curcas*. Chemosphere, 127, 57.

Meng Li, Xiaohui C, Hongxian G (2013) Heavy metal removal by biomineralization of urease producing bacteria isolated from soil. Int Biodeter Biodegr., 76, 81.

Miaolu H, Lei W, Yongtao L, Xudong W, Jiani Z, Yan Z, Tingting L (2020) Novel polydopamine/metal organic framework thin film nanocomposite forward osmosis membrane for salt rejection and heavy metal removal. Chem Eng J., 389, 124452.

Mohammad HS, Omolbanin R, Elham S (2019) Application of modified Spirulina platensis and Chlorella vulgaris powder on the adsorption of heavy metals from aqueous solutions, J Environ Chem Eng., 7, 103169.

Muthusaravanan S, Sivarajasekar N, et al. (2018). Phytoremediation of heavy metals: mechanisms, methods and enhancements. Environ Chem Lett., 16, 1339.

Negm NA, Abd El Wahed MG, Hassan ARA, Abou Kana MTH (2018) Feasibility of metal adsorption using brown algae and fungi: Effect of biosorbents structure on adsorption isotherm and kinetics. J MolLiq., 264, 292.

Nanda, M, Kumar, V, Sharma, D.K (2019) Multimetal tolerance mechanisms in bacteria: The resistance strategies acquired by bacteria that can be exploited to "clean-up" heavy metal contaminants from water. Aquat Toxicol., 212, 1.

Ningqin L, Tianjue H, Yunbo Z, Huaqing Q, Jamila A, Hao Z (2020) Fungal cell with artificial metal container for heavy metals biosorption: Equilibrium, kinetics study and mechanisms analysis. Environ Res., 182, 109061.

Noroozi M, Amoozegar M, et al. (2017). Isolation and characterization of mercuric reductase by newly isolated halophilic bacterium, *Bacillus firmus* MN8. Global J Environ Sci Manage., 3, 427.

Pal S, Singh HB, et al. (2013). Potential of different crop species for nickel and cadmium phytoremediation in peri-urban areas of Varanasi district (India) with more than twenty years of wastewater irrigation history. Italian J Agron., 8, 8.

Pastor J, Gutiérrez-Ginés MJ, Hernández AJ. (2015). Heavy-metal phytostabilizing potential of *Agrostis castellana* Boiss. and reuter. Int J Phytoremediat., 17, 988.

Patra DK, Pradhan C, et al. (2020). Toxic metal decontamination by phytoremediation appraoch: Concept, challenges, opportunities and future perspectives. Environ TechnolInno., 18, 100672.

Plan D., Van den Eede G., 2010. The EU legislation on GMOs: an overview. JRC Scientific and Technical Reports, Luxembourg.

Pollyana SQ, Natalia RB, Monica MC, Versiane AL, Renata GS (2018) Rich growth medium promotes an increased on Mn(II) removal and manganese oxide production by *Serratia marcescens* strains isolates from wastewater. Biochem Eng J., 140, 148.

Prasad MNV (2011). A State-of-the-Art Report on Bioremediation, Its Applications to Contaminated Sites in India, Ministry of Environment & Forests, Government of India, New Delhi.

Preeti D, Shouvik M, Antara G, Papita D, Punarbasu C (2019) Role of Manglicolous fungi isolated from Indian Sunderban mangrove forest for the treatment of metal containing solution: Batch and optimization using response surface methodology. Environ Technol Innov., 13, 166.

Purchase D, Scholes LN, et al. (2009). Effects of temperature on metal tolerance and the accumulation of Zn and Pb by metal-tolerant fungi isolated from urban runoff treatment wetlands. J Appl Microbiol., 106, 1163.

Quanyuan C, Yuan Y, Xinying L, Jun L, Juan Z, Zhaolu H (2018) Comparison of heavy metal removals from aqueous solutions by chemical precipitation and characteristics of precipitates, J Water Process Eng., 26, 289.

Rajesh KS, Ruchita T, Amit R, Akhileshwar KS (2020) Fungi as potential candidates for bioremediation. In Abatement of Environmental Pollutants, Trends and Strategies. Singh P, Kumar A, Borthaku A (eds.), 177.

Ramana S, Biswas AK, et al. (2013). Potential of rose for phytostabilization of chromium contaminated. Ind J Plant Physiol., 18, 381.

Ramona M, Mohammad RH, Pegah H, Kianoush KD (2019) Bioremediation of heavy metals in food industry: Application of Saccharomyces cerevisiae. Electron J Biotechnol., 37, 56.

Ramasamy K, Kamaludeen S, et al. (2006). Bioremediation of metals microbial processes and techniques. In: Environmental Bioremediation Technologies. Singh SN and Tripathi RD (eds.), Springer Publication, New York, NY: 173.

Rangabhashiyam S., Anu N. and Selvaraju N (2013) The significance of fungal laccase in textile dye degradation: A review. Res J Chem Environ, 17, 88.

Rangabhashiyam S, Balasubramanian P (2019) Characteristics, performances, equilibrium and kinetic modelling aspects of heavy metal removal using algae. Bioresour Technol Rep., 5, 261.

Rangabhashiyam S, Nandagopal G, Nakkeeran MS, Selvaraju E (2016) Adsorption of hexavalent chromium from synthetic and electroplating effluent on chemically modified *Swietenia mahagoni* shell in a packed bed column. Environ Monit Assess., 188, 411.

Rangabhashiyam S, Sayantani S, Balasubramanian P (2019) Assessment of hexavalent chromium biosorption using biodiesel extracted seeds of *Jatropha* sp., *Ricinus* sp. and *Pongamia* sp. Int J Environ Sci Technol., 16, 5707.

Rangabhashiyam S, Selvaraju N (2015) Evaluation of the biosorption potential of a novel *Caryota urens* inflorescence waste biomass for the removal of hexavalent chromium from aqueous solutions. J Taiwan Inst Chem Eng, 47, 59.

Rangabhashiyam S, Suganya E, Varghese L, Selvaraju N (2016) Equilibrium and kinetics studies of hexavalent chromium biosorption on a novel green macroalgae *Enteromorpha* sp. Res Chem Intermed. 42, 1275.

Remonsellez F, Orell A, et al. (2006). Copper tolerance of the thermoacidophilic archaeon *Sulfolobus metallicus*: possible role of polyphosphate metabolism. Microbiology, 152, 59.

Rezania S, Taib SM, et al. (2016). Comprehensive review on phytotechnology: heavy metals removal by diverse aquatic plants species from wastewater. J Hazard Mater., 318, 587.

Ricardo S, Raul M, Maria Elisa Taboada, Silvia B (2019) Influence of organic matter and CO_2 supply on bioremediation of heavy metals by *Chlorella vulgaris* and *Scenedesmus almeriensis* in a multimetallic matrix. Ecotoxicol Environ Saf., 182, 109393.

Rohit S, Daizee T, Shefali B, Sundeep J, Rajeev K, Raman K, et al. (2020) Bioremediation potential of novel fungal species isolated from wastewater for the removal of lead from liquid medium. Environ Technol Innov., 18, 100757.

Rohit S, Daizee T, Shefali B, Sundeep J, Rajeev K, Raman K, M.Shaheer A, Vikas B, Ahmad U (2020) Bioremediation potential of novel fungal species isolated from wastewater for the removal of lead from liquid medium. Environ Technol Innov., 18, 100757.

Sadar A, Malik Wajid HC, Ghazala S, Grzegorz B, et al. (2020) A comprehensive assessment of environmental pollution by means of heavy metal analysis for oysters' reefs at Hab River Delta, Balochistan, Pakistan. Marine Pollution Bulletin, 153, 110970.

Sakakibara M, Watanabe A, et al. (2010). Phytoextraction and phytovolatilization of arsenic from As-contaminated soils by *Pteris vittata*. Proc. of the Ann. Int. Conf. on Soils, Sediments, Water and Energy, 12, 26.

Salihaj M, Bani A, et al. (2016). Heavy metals uptake by hyperaccumulating flora in some serpentine soils of Kosovo. Global NEST J, 18, 214.

Samakshi V, Arindam K (2019) Bioremediation of heavy metals by microbial process, Environ Technol Innov., 14, 100369.

Sang HL, Young HJ, Dinh DN, Soon WC, Sung CK, Sang ML, Sung SK (2019) Adsorption properties of arsenic on sulfated TiO_2 adsorbents, J Ind Eng Chem., 80, 444.

Sarkar M, Rahman AKML, et al. (2017). Remediation of chromium and copper on water hyacinth (*E. crassipes*) shoot powder. Water Resour Ind., 17, 1.

Sarwar N, Imran M, et al. (2017). Phytoremediation strategies for soils contaminated with heavy metals: modifications and future perspectives. Chemosphere, 171, 710.

Saxena G, Purchase D, et al. (2020). Phytoremediation of heavy metal-contaminated sites: eco-environmental concerns, field studies, sustainability issues, and future prospects. Rev Environ Contam Toxicol., 249, 71.

Shabeena B, Anwar H, Muhammad H, Hazir R, Amjad I, Mohib S, Muhammad I, Muhammad Q, Badshah I (2018) Bioremediation of hexavalent chromium by endophytic fungi; safe and improved production of *Lactuca sativa* L. Chemosphere, 211, 653.

Shackira AM and Puthur JT (2017). Enhanced phytostabilization of cadmium by a halophyte - *Acanthus ilicifolius* L. Int J Phytoremediat, 19, 319.

Sharma R, Bhardwaj R, et al. (2018). Microbial Siderophores in Metal Detoxification and Therapeutics: Recent Prospective and Applications. In Plant Microbiome: Stress Response. Egamberdieva D and Ahmad P (eds.), Springer, Singapore: 337.

Sherlala AIA, Raman AAA, Bello MM, Buthiyappan, A (2019) Adsorption of arsenic using chitosan magnetic graphene oxide nanocomposite. J Environ Manage., 246, 547.

Singh UK, Kumar B (2017) Pathways of heavy metals contamination and associated human health risk in Ajay River basin, India. Chemosphere, 174, 183.

Sheoran V, Sheoran AS, et al. (2009). Phytomining: A review. Miner Eng., 22, 1007.

Sheoran V, Sheoran AS, et al. (2016). Factors affecting phytoextraction: a review. Pedosphere, 26, 148.

Siyu L, Ruiqi B, Zhanjun L, Jiayi L, Dongxu H, Wenyue Z, Lanjie Y, Ning D, Zhiyan L, Zhigang Z (2019) Exploring the kidney hazard of exposure to mercuric chloride in mice: Disorder of mitochondrial dynamics induces oxidative stress and results in apoptosis. Chemosphere, 234, 822.

Souza LA, Piotto FA, et al. (2013). Use of non-hyperaccumulator plant species for the phytoextraction of heavy metals using chelating agents. Sci Agric., 70, 4, 290.

Sudharsanam A, Suresh RS, Logeshwaran P, Kadiyala V, Mallavarapu M (2019) Potential of acid-tolerant microalgae, *Desmodesmus* sp. MAS1 and *Heterochlorella* sp. MAS3, in heavy metal removal and biodiesel production at acidic pH. Bioresour Technol., 278, 9.

Talebi AF, Tabatabaei M, Mohtashami SK, Tohidfar M, Moradi F (2013) Comparative salt stress study on intracellular ion concentration in marine and salt-adapted freshwater strains of microalgae. Not Sci Biol., 5, 309.

Tariq SR and Ashraf A (2016). Comparative evaluation of phytoremediation of metal contaminated soil of firing range of four different plant species. Arab J Chem., 9, 806.

Triparna M, Shatarupa C, Aksar AB, Tapan KD (2017) Bioremediation potential of arsenic by non-enzymatically biofabricated silver nanoparticles adhered to the mesoporous carbonized fungal cell surface of *Aspergillus foetidus* MTCC8876. J Environ Manage., 201, 435.

Umar F, Janusz AK, Misbahul AK, Makshoof A (2010) Biosorption of heavy metal ions using wheat based biosorbents: A review of the recent literature. Bioresour Technol., 1015043.

Varun M, Souza RD, et al. (2011). Evaluation of phytostabilization, a green technology to remove heavy metals from industrial sludge using *Typha latifolia* L. Experimental design. Biotechnol Bioinf Bioeng., 1, 137.

Vázquez S, Agha R, et al. (2006). Use of white lupin plant for phytostabilization of Cd and As polluted acid soil. Water Air Soil Pollut., 177, 349.

Vijayaraghavan K., Yun YS (2008) Bacterial biosorbents and biosorption. Biotechnol Adv., 26, 266.

Viti C, Pace A, et al. (2003). Characterization of Cr(VI)-resistant bacteria isolated from chromium-contaminated soil by tannery activity. Curr Microbiol., 46, 1, 1.

Wang J, Feng X, et al. (2012). Remediation of mercury contaminated sites: A review. J Hazard Mater., 221, 1.

Wang JL, Chen C (2006) Biosorption of heavy metals by *Saccharomyces cerevisiae*: a review. Biotechnol Adv., 24, 427.

Weerasundara L, Amarasekara RWK, Magana-Arachchi DN, Ziyath AM, Karunaratne DGGP, Goonetilleke A, Vithanage M (2017) Microorganisms and heavy metals associated with atmospheric deposition in a congested urban environment of a developing country: Sri Lanka. Sci Total Environ, 584.

Wu G, Kang H, et al. (2010). A critical review on the bio-removal of hazardous heavy metals from contaminated soils: Issues, progress, eco-environmental concerns and opportunities. J Hazard Mater., 174, 1.

Xia Y, Liu J, et al. (2018). Ectopic expression of *Vicia sativa* Caffeoyl-CoA O-methyltransferase (VsCCoAOMT) increases the uptake and tolerance of cadmium in Arabidopsis. Environ Exp Bot, 145, 47.

Xiaoyue Z, Xinqing Z, Chun W, Bailing C, Fengwu B (2016) Efficient biosorption of cadmium by the self-flocculating microalga *Scenedesmus obliquus* AS-6-1. Algal Research, 16, 427.

Xu H, Liu Y, et al. (2006). Effect of pH on nickel biosorption by aerobic granular sludge. Bioresour Technol., 97, 359.

Xunchao C, Xin Z, Dunnan Z, Waheed I, Changkun L, Bo Y, Xu Z, Xiaoying L, Yanping M (2019) Microbial characterization of heavy metal resistant bacterial strains isolated from an electroplating wastewater treatment plant. Ecotoxicol Environ Saf., 181, 472.

Yadav KK, Gupta N, et al. (2018). Mechanistic understanding and holistic approach of phytoremediation: A review on application and future prospects. Ecol Eng., 120, 274.

Yan L, Tao L, Wenyu X, Zhenghui L, Jianping C, Junfeng W (2018) Removal of Zn(II) from aqueous solutions by *Burkholderia* sp. TZ-1 isolated from soil of oil shale exploration area. J Environ Chem Eng, 6, 7062.

Yang T, Chen M, et al. (2015). Genetic and chemical modification of cells for selective separation and analysis of heavy metals of biological or environmental significance. TrAC Trends in Analytical Chemistry, 66, 90.

Yang J, Yang J, et al. (2017). Role of co-planting and chitosan in phytoextraction of As and heavy metals by *Pteris vittata* and castor bean – a field case. Ecol Eng., 109, 35.

Yao Z, Li J, Xie H, Yu C, 2012. Review on remediation technologies of soil contaminated by heavy metals. Procedia Environ Sci., 16, 722–729.

Yin K, Wang Q, et al. (2019). Microorganism remediation strategies towards heavy metals. Chem Eng J., 360, 1553.

Qu Y, Zhang X, Xu J, Zhang W, Guo Y (2014) Removal of hexavalent chromium from wastewater using magnetotactic bacteria. Sep Purif Technol., 136, 10.

Yongjun S, Shengbao Z, Shu-Yuan P, Sichen Z, Yang Y, Huaili Z (2020) Performance evaluation and optimization of flocculation process for removing heavy metal. Chem Eng J., 385, 123911.

Yongsong Ma, Xi Li, Hongmin Mao, Bing Wang, Peijie Wang (2018) Remediation of hydrocarbon–heavy metal co-contaminated soil by electrokinetics combined with biostimulation. Chem Eng J., 353, 410.

Zhao L, Li T, et al. (2016). Pb uptake and phytostabilization potential of the mining ecotype of *Athyrium wardii* (Hook) grown in Pb-contaminated soil. Clean Soil Air Water, 44, 1184.

Yu BH, Nor HA, Hazwanee H, Eugenie SST (2017) Mercury contamination in facial skin lightening creams and its health risks to user. Regulat Toxicol Pharma., 88, 72.

13 Biopolymers as a Sustainable Approach for Wastewater Treatment

J. Juliana
Department of Civil Engineering, National Institute
of Technology Trichy, India

K. L. Jesintha, S. Mariaamalraj and
V. C. Padmanaban*
Centre for Research, Department of Biotechnology,
Kamaraj College of Engineering & Technology, India

A. R. Neelakandan and G. K. Rajanikant
School of Biotechnology, National Institute
of Technology Calicut, India

CONTENTS

13.1 INTRODUCTION

As environmental pollution is increasing on a daily basis, both the number of pollutants and the effluent produced are also increasing. These pollutants affect the quality of water bodies and make them unfit for human use and consumption. Because modern treatment technologies require more space and are expensive to implement, the biopolymer use for wastewater treatment can be cost-effective and does not produce much sludge. The biopolymers act as excellent biosorbents by entrapping and cross-linking the micropollutants such as dyes, heavy metal ions, and many other organic pollutants within their three-dimensional structure by various mechanisms. Biopolymers are currently used as efficient coagulants and flocculants in industrial applications and can reduce the pollutant load in sedimentation tanks during effluent treatment operations. These biopolymers are long-chain molecules, such as polypeptides, polynucleotides, or polysaccharides, derived from living organisms and can be easily cast and processed into different shapes and structures. Biopolymers can also be chemically modified to improve their properties, and the demand for these biopolymers is therefore increasing. Their structural, chemical, and biological properties help them to be used in wider applications in different fields. They are renewable, biodegradable, and regenerable and are more stable in nature. The use of biopolymers is sometimes limited due to their higher cost of production, but the use of cost-effective substrates and smart strategies in large-scale production can overcome this problem. Ancient and modern approaches to the use of a wide range of biopolymers and their structural, chemical, and biological properties are discussed briefly in this chapter. A schematic outline of the types of biopolymers and their role in wastewater treatment for the future are given in Figure 13.1.

13.2 ANCIENT AND MODERN APPROACH TO THE USE OF BIOPOLYMERS IN WATER TREATMENT

Water is one of the basic human needs. Improper disposal of industrial and urban effluents and pollutants is on the rise due to rapid industrialization, making most of the water available unsuitable for human consumption and use. The World Health Organization has reported that 85% of rural populations have no safe drinking water and more than 85% of diseases in developing countries are caused by consuming contaminated drinking water. While we are now developing advanced technologies for water treatment, the idea of treating and clarifying drinking water before use began with our ancestors. Jahn (1988) has stated that the Egyptians were the first to use the kernels of genus Prunus to purify water; later this technique was also used by tribes in African regions. The sap of the species *Opuntia* has also been used as a flocculant by natives of Chile and Peru and many other countries. *Strychnos potatorum* seeds

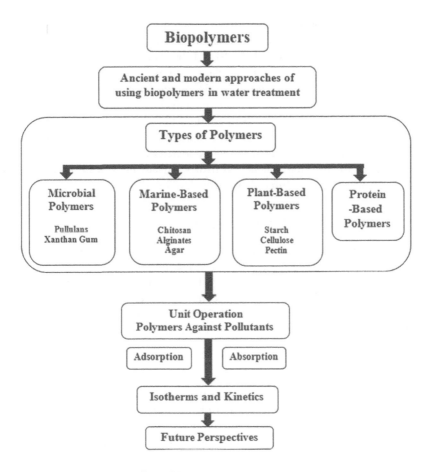

FIGURE 13.1 Schematic outline of biopolymers: their types and role in wastewater treatment and their future prospects.

were used in India four thousand years ago to detoxify river water before consumption. Two and a half thousand years ago, the ancient text of Hinduism and Buddhism also referred to the flocculating properties of various gems, pearls, and minerals, and the water being treated with them (Jahn 1999). The seeds, leaves, and barks of many plant families of *Acanthaceae*, *Moringaceae*, *Papilionaceae*, and *Capparidaceae* have also traditionally been used for domestic water treatment applications because of their excellent coagulating properties. Among them, *Moringa oleifera*, one of the most important species in the *Moringaceae* family, has been used to treat water efficiently in many countries around the world for centuries. Dimeric cationic proteins present in the moringa extract coagulate the contaminants using adsorption and neutralization and cause them to settle down in the form of flocs without affecting the pH and conductivity of the water being treated. Moringa has many advantages, as it is widely available, has high purification potentials, and costs little so that people from rural areas of extreme poverty can afford to use it as a natural water purifier.

Different species of moringa have shown differences in efficiency in the removal of contaminants, which may be due to changes in their chemical components and weight proportions (Pritchard et al. 2009). Pritchard et al. (2009) compared the shallow well-water purification potential of plant extracts of *Moringa oleifera, Jatropha curcas,* and guar gum and concluded that *Moringa oleifera* improved the quality of well water by significantly reducing turbidity and coliform counts. After the British intervention in the middle of the 20th century, the use of alum as a coagulant in industrial processes came into existence. Alum and ferric salts were then used in industry but had many disadvantages, such as toxicity at higher doses, high cost, and carcinogenicity of resulting sludge. With these disadvantages in mind, bio-based flocculants from different sources have been developed.

13.3 MICROBIAL-BASED POLYMERS

Microorganisms such as bacteria, fungi, and yeast are the major sources for the production of microbial-based polymers. These polymers are eco-friendly, readily degradable, less toxic than synthetic polymers, and have excellent flocculating properties. This provides a sustainable technology over synthetic methods for the treatment of wastewater (Pernicova et al. 2019). Some of the widely used microbial polymers employed in water treatment are discussed here.

13.3.1 PULLULANS

Pullulan is a naturally produced linear exopolysaccharide consisting of maltotriose units that are linked by α-1,6 glycosidic linkage, which makes them readily soluble in water and offers a high degree of flexibility so that they can be made into different forms depending on their application. Pullulan is produced from starch by the fungus *Aureobasidium pullulans* and can be used for the removal of micropollutants. The production cost of pullulan is slightly higher than other polysaccharides, which restricts its use in several fields. The use of cost-effective substrates, such as agro- and food-industry wastes, can thus increase pullulan production and can also be a rich source of nutrition for the growth of the microorganism that produces it (Terán Hilares et al. 2019; Rishi et al. 2020). The composites of pullulans have improved adsorptive characteristics that contribute to effective adsorption of heavy metals and micropollutants from the wastewater by cross-linking and electrostatic interactions. A novel adsorbent of magnesia/pullulan composite developed by Kang et al. (2011) had high adsorption capacities of 4535 mg/kg and remained stable over wide pH ranges of 2–11 and had good potential to perform defloridation in water treatment processes (Zeng et al. 2020). Quinkan fabricated a stable pullulan/polydopamine hydrogel for decontamination of heavy metals of Cu^{2+}, Co^{2+} and Ni^{2+} from aqueous solutions and found that they had maximum adsorption of 100.9mg/g towards *Cu²⁺* that could be effectively used for the removal of micropollutants from industrial effluents in the future (Su et al. 2019). Ting also fabricated a hybrid hydrogel using a pullulan/polydopamine cross-linked with neopentyl glycol diglycidyl ether (NGDE) and found that it exhibited a good adsorption capacity by removing 96 mg/g of crystal violet as well as 25.8 mg/g of methylene blue from effluents.

13.3.2 Xanthan Gum

Xanthan gum is a natural polymer produced as an extracellular polysaccharide of high molecular weight, and the branched trisaccharide chains of pyruvic acid and glucuronic acid confer them an anionic charge. A plant pathogen, *Xanthomonas campestris*, is the main producer of this kind of biopolymer under aerobic fermentation conditions. Maintaining an appropriate carbon to nitrogen ratio by providing required carbon and nitrogen sources can enhance its production. Using inexpensive substrates like kitchen wastes (Li et al. 2016) and chicken feathers (Ozdal and Kurbanoglu 2018) can reduce the production cost. Xanthan gum polymers have higher swelling and water retention capacities than other synthetic polymers and may remain stable for a longer period under various incubation conditions. A novel superabsorbent material made of composite XG-g-PAA/loess had a high water retention capacity up to 5 days and still maintained a 32% higher potential for adsorption of multivalent metal ions even after 5 days, indicating its potential for successful use in the treatment of micropollutants in wastewater (Enke Feng, Guofu Ma, Yajuan Wu, Haiping Wang 1989). These polymers may be effective in removing persistent organic pollutants because they are required at a minimum dose for both water and leachate treatment, and they produce less sludge (Pariatamby and Kee 2016). In addition, the nanocomposite developed using xanthan gum/silica to remove cationic dyes from aqueous solutions has demonstrated excellent stability and high adsorption capacity (Thakur, Pandey, and Arotiba 2017).

13.4   MARINE-BASED POLYMERS

Most of these biopolymers are produced by marine crustaceans and other microorganisms and are used as alternatives to synthetic polymers for industrial applications. Some marine-based polymers are discussed here.

13.4.1 Chitosan

It is the second most abundant linear polysaccharide composed of randomly distributed β-(1→4)-linked D-glucosamine (deacetylated unit) and N-acetyl-D-glucosamine (acetylated unit), commonly found in the exoskeleton of arthropods (shrimp, lobsters, crabs) and fungi. The amino group in chitosan makes it an effective biosorbent to remove multiple pollutants from wastewater. Chitosan can be easily modified by grafting, cross-linking, and blending and can be prepared as hydrogels or biofilms for use in different fields. Cross-linked chitosans are more efficient than raw ones due to the presence of an interconnected porous network structure. Crustaceans like crab and shrimp have more chitin in their exoskeletons and can be used as a commercial source for chitosan production (Dima, Sequeiros, and Zaritzky 2017). David reported that 20% of the gross weight of shrimp is being discarded as industrial waste by shrimp processing industries worldwide and that such waste can be used as an economical raw material for chitosan production since it contains approximately 8–10% chitin (Kun and Pukánszky 2017). Chitosan polymers chelate heavy metal ions to the surface by releasing hydrogen ions and therefore this sorption process depends primarily on pH. Hematite/chitosan nano-composite produced to remove heavy metals had high removal efficiencies of 94, 76, and 83% for Pb, Cu, and Cd, respectively, and can be used for environmental remediation (Saad et al. 2020). Another composite $Fe_3O_4 - CS$ was employed for the treatment of mixed textile dyes and had excellent adsorption capacity and stability for both cationic and anionic dyes (Chen et al. 2020). Bor Shuang proposed an innovative strategy for the removal of heavy metal ions from wastewater using a composite of hydroxyapatite/chitosan scaffold (Liaw et al. 2020). (Tummino et al. 2020) Maria et al. prepared a hybrid gel using chitosan and alginate for use in wastewater remediation and found it to be an effective adsorbent for the removal of different dyes (crystal violet, rhodamine B, and orange II) (Rocha et al. 2019).

13.4.2 ALGINATES

Alginates are linear unbranched polysaccharides of (1→4′)-linked β-D-mannuronic acid and α-L-guluronic acid, produced extracellularly by brown algae and certain bacteria such as *Azotobacter vinelandii* (Reyes, Peña, and Galindo 2003) and *Pseudomonas fluroscens* (Maleki et al. 2017). These alginate polymers are highly stable and can be successfully used to remove micropollutants from wastewater. Matthew used alginate produced from immobilized *Chlorella vulgaris* for nutrient removal (Kube et al. 2019). Alginate-based nanocomposites showed a high sorption capacity of about 137 mg/g for cadmium removal with a stable and maximum breakthrough time of 48 hours (Aziz et al. 2019). N-doped carbon dots embedded on alginate beads were excellent for removing rare earth elements and were reusable for five successive regeneration cycles (Guo et al. 2020). Laccase immobilized on modified copper alginate beads had high stability and recyclability showing removal efficiencies of about 89.6% for triclosan and 75.8% for Remazol Brilliant Blue R after 8 hours and 4 hours of treatment (Le et al. 2016).

13.4.3 AGAR

Agar is a mixture of two polysaccharides: agarose (70%) and agaropectin. Agarose is a linear polymer of repeating units of agarose, a disaccharide made up of D-galactose and 3,6-anhydro-L-galactopyranose. Agaropectin is a heterogeneous mixture of alternating units of D-galactose and L-galactose, modified with sulfate and pyruvate (Lee et al. 2017). All the agar-producing seaweeds in different marine ecosystems and conditions are known as agrophytes. They are commonly produced by red algae like *Gelidales, Ahnfeltia plicata* (Ismail and Labib 2013) and *Gracilariaceae* (Rocha et al. 2019) and recently some other genus of *Bangiophyceae* like *Pyropia yezoensis* (Sasuga et al. 2018). The high gelling capacity of agar enhances the efficiency of hydrogels prepared from various sources. Hydrogel prepared by immobilizing *Chlorella saccrarophila* within an agar matrix was used for nitrogen and phosphorous removal from wastewater and was recyclable up to eight cycles with removal efficiencies of about 94% and 66% for nitrogen and phosphorous (Hu et al. 2020). Graphene oxides loaded agar beads as hydrogels had high adsorption towards organic compounds and could remove 321.7 mg/g of rhodamine and 196.4 mg/g of aspirin from wastewater (Cheng et al. 2018).

13.5 PLANT-BASED POLYMERS

These polymers are generally obtained from plant-based materials and can be easily modified by chemical or other method to obtain polymers of desired properties. These natural polymers are sustainable and can be produced at a lower cost in larger quantities.

13.5.1 STARCH

Plant starch, one of the most abundant polymeric carbohydrates in nature, is made up of glucose units joined by glycosidic bonds. It consists of linear and helical amylose and the branched amylopectin. Some of the common plant sources of starch are potato, wheat, maize, sorghum, and tapioca. Microalgal cultures such as *Chlamydomonas applanata, Chlamydomonas oblanga, Chlamydomonas moewusii, Chlorella sorokiniana* (Gifuni et al. 2017; Gifuni et al. 2018) and macro algae like *Ulva ohnai* (Prabhu et al. 2019) also act as potential sources of starch. A starch-based temperature resistant organic/inorganic composite coagulant demonstrated excellent textile dye removal capability from textile effluent and required up to 50% less dosage than conventional coagulation materials. Magnetic nanocomposite using starch and polyacrylamide-co-sodium xanthate significantly decreased turbidity and demonstrated a 78.3 and 63% absorption of Pb^{2+} and Cu^{2+} from aqueous solution (Wang, Zhang, and Chang 2017). Likewise, low-cost flocculant developed by grafting polyacrylamide onto leguminous starch has shown rapid sedimentation of flocculants while treating wastewater from calamine (Li et al. 2017). Since starch-based composites have good adsorptive properties and could reduce 90 and 95% of total phosphorous and turbidity, respectively, they can be used to develop a combined water treatment system to remove turbidity and phosphorus (Ren et al. 2020).

13.5.2 CELLULOSE

Cellulose is a linear natural exopolysaccharide having anhydro-β-D-glucopyranose units linked by a (1→4) glycosidic bond, found extensively in plants in combination with other polymers. Although it can mostly be produced from plant woods, this source is not widely selected for industrial production because it can cause many environmental problems. Other raw materials such as bagasse, wheat straw, and rice straw are used as alternative substrates for cellulose production on a large scale. Many bacterial strains such as *Agrobacteria, Sarcina, Acetobacter, Pseudomonas,* and *Alcaligene*s can act as efficient producers of bacterial celluloses. Nowadays, due to their interesting characteristics, cellulose nanocomposites are gaining more attention and are finding more applications in the field of wastewater treatment. They are used either as adsorbents such as nano cellulose-based adsorbents, carboxylated nanocellulose materials, succinylated nano cellulose, amino-modified nano celluloses or in composite forms such as magnetic cellulose nanocomposites, cellulose/silsesquiox-ane nanocomposite, cellulose/clay nanocomposite, and cellulose/polymer nanocom-posites (Basavarajaiah 2016). A stable and reusable (up to five generations) cellulose/polyaniline derivative nanocomposite was synthesized to remove anionic dye from simulated industrial effluents, which showed maximum adsorption capacities of 117 and $56\,mg.g^{-1}$ for acid red 4 and direct red 23, respectively (Abbasian, Niroomand, and Jaymand 2017). Cellulose-based polymers can also be used as efficient mem-branes for the filtration-based removal of metal ion pollutants (Karim et al. 2016).

13.5.3 PECTIN

Pectins are anionic polysaccharides with α-1,4 linked D-galacturonic acids and are of two types: high methoxy pectins (HMP) with their degree of methylation greater than 50% and low methoxy pectins (LMP) with a degree of methylation lesser than 50%. When the hydrophobic groups and hydrogen bonds present in the HMPs, they form a polymeric structure, whereas, in the case of LMP, the polymeric gel is formed by the ionic linkages (Ansarifar et al. 2017). A pectin-based quaternary amino anion exchanger developed for treating aqueous solutions showed excellent toxic phosphate ion removal capacity by monolayer adsorption mechanisms (Naushad et al. 2018). Ethylenediamine modified pectins were used for treating Pb^{2+} contaminated waters with 94% maximum removal efficiency (Liang et al. 2020) and composite beads of pectin-cellulose microfibers were used for multi-metal ions in wastewater (Lessa et al. 2020). A pectin combination may also be used to treat pollutant mixtures such as metals, dyes, and pesticides. Biocomposite of magnetic pectin from *Chlorella vulgaris* was used for the adsorptive removal of nine dyes and zinc in batch and column studies (Khorasani and Shojaosadati 2019).

13.6 PROTEIN-BASED POLYMERS

Protein polymers consist of polypeptide blocks and are produced either by biotech-nological synthesis or from natural sources such as yeast, marine algae, cyanobac-teria, and plants. Plants are one of the major sources of proteins as they contain a diverse group of proteins within each of their structures. Domeradzka et al. (2016) synthesized a protein-based polymer from *Pichia pastoris* and purified protein of about 2 g/L. A protein-based super adsorbent hydrogel was also synthesized and characterized by canola, a genetically modified plant, and it was reported that the synthesized polymer had high swelling properties of 448 g/g in distilled water (Shi, Dumont, and Ly 2014). These protein-based polymers are used widely in the nano-coupled process: functionalized aluminum nanosheet as packing material for a fixed bed reactor to treat textile effluent containing crystal violet and congo red dyes (Ealias and Saravanakumar 2020). A stable nano-hybrid flower made of sericin-based protein-inorganic material removes heavy metal ions from wastewater uti-lizing needle-shaped nanostructures on their surface. Further, β-lactoglobulin nanofibril from whey protein exhibited selective adsorption towards Pb(II) with an effective treatment capacity of ~22 m^3 contaminated water/kg adsorbent (Zhang et al. 2020). A soy protein isolate/polyethyleneimine composite hydrogel facili-tates the removal of heavy metals such as Cu, Zn, Cd, and Pb (Liu et al. 2017).

13.7 USE OF BIOPOLYMERS AGAINST WATER TREATMENT

Two types of sorption mechanisms, namely adsorption and absorption, are used by the biopolymers to remove contaminants or micropollutants from wastewater.

13.7.1 ADSORPTION

It is an important surface phenomenon that removes contaminants by two types of interactions. While physisorption attracts and retains the contaminants through van der Waals forces, hydrogen bonds, electrostatic interactions, and π-π interaction, chemical bond formation between the functional groups of biopolymers and the water contaminants promotes chemisorption (Han et al. 2016). At first, the adsorbate molecules are transferred from the bulk solution of the wastewater onto the surface of the adsorbent. Then they travel by film diffusion mechanisms and undergo intra- particle diffusion within the pores and finally get adsorbed onto the active sites of the adsorbent (Badawi et al. 2017). Their adsorptive capacity is calculated by the following equation:

$$\text{Sorption capacity}: q_t = \frac{C_0 - C_t}{m} V \tag{13.1}$$

where q_t = amount sorbed at a time (mg/g), C_0 and C_t are the concentrations present initially and at a time t (mg/L), V is the volume (L) and m is the mass of adsorbent (g) (Paulino et al. 2006).

13.7.2 ABSORPTION

Biopolymers are also undergoing absorption phenomena to remove contaminants. In this case, the functional groups present in the polymeric structures absorb water and provide excellent swelling properties, and the swollen structure of the polymer is then used to remove contaminants from the wastewater.

13.7.3 ISOTHERMS AND KINETICS OF ADSORPTION

The efficacy or extent to which active adsorption is observed on or within a biopolymer can be well studied by the use of isotherms. It provides much useful information when the optimization of parameters needs to be carried out in a study. The rate of the mechanism and its rate of determination of the steps during the sorption process of biopolymers can be well understood by kinetic models. Table 13.1 shows the main type of isotherms that are exhibited by different biopolymers when used to remove contaminants from wastewater.

13.8 BIOPOLYMERS: SUSTAINABLE TECHNOLOGY OFFERS CLEAN WATER

Sustainable technology refers to the quality eco-friendly technology that serves many people at a low cost. Though many synthetic polymers used in water treatment are aggressive and cost-effective, they have a residual impact on the ecosystem

TABLE 13.1

Different isotherms and their equations (Ergene et al. 2009) (Savran et al. 2017)

Isotherms	Equation
Langmuir	$\dfrac{C_e}{q_e} = \dfrac{1}{q_{max}K_L} + \dfrac{C_e}{q_{max}}$
Freundlich	$q_e = K_F C_e^{\,n}$
Temkin	$q_e = B\ln A + B\ln C_e$, where $B=RT/b$
Dubinin Radushkevich	$\ln q_e = \ln q_m - K_{DR}\varepsilon^2$
Flory-Huggins	$\ln (\theta/C_o) = \ln K_{KF} + n \ln (1 - \theta)$
Brunauer Emmett Teller	$\dfrac{C_e}{q_e(C_s - C_e)} = \dfrac{1}{q_m C_{BET}} + \dfrac{(C_{BET} - 1)}{q_m C_{BET}}\dfrac{C_e}{C_s}$
Harkins Jura	$\dfrac{1}{q_e^{\,2}} = \dfrac{B_{HJ}}{A_{HJ}} - \dfrac{\log C_e}{A_{HJ}}$
Fowler-Guggenheim	$K_{FG}C_e = \dfrac{\theta_{FG}}{1-\theta_{FG}}\exp\left(\dfrac{2\theta_{FG}W}{RT}\right)$
Hill de Boer	$\ln\left(\dfrac{C_e(1-\theta)}{\theta}\right) - \dfrac{\theta}{1-\theta} = -\ln K_{HB} - \dfrac{K_1\theta}{RT}$
Frumkin	$\ln\left[\left(\dfrac{\theta}{1-\theta}\right)\dfrac{1}{C_e}\right] = \ln K_{Fr} + 2a\theta$
Hasley	$\log q_e = \dfrac{1}{n_H}\log K_H - \dfrac{1}{n_H}\log C_e$
Henderson	$\ln[-\ln(1 - C_e)] = \ln K_{Hn} + n_{Hn}\ln q_e$
Smith	$q_e = W_b - W\ln(1 - C_e)$
Jovanovic	$\ln q_e = -K_J C_e + \ln q_m$
Scatchard	$\dfrac{q_e}{C_e} = -K_{SC}q_e + K_{SC}q_m$
Kinetic equations and their significance	
Pseudo first order model	$\log (q_e - q_t) = \log q_e - \dfrac{k_1}{2.303t}$
Pseudo second order model	$\dfrac{t}{q_t} = \dfrac{1}{k_2 q_e^{\,2}} + \dfrac{1}{q_e}t$
Intra particle diffusion model	$q_e = k_i t^{1/2}$
Elovich	$q_t = \left(\dfrac{1}{\beta}\right)\ln(\alpha\beta) + \left(\dfrac{1}{\beta}\right)\ln t$

that paves the way for the use of biopolymers as an alternative sustainable technology. These biopolymers are biodegradable at a much shorter time than synthetic polymers. Biopolymer enhanced sand filtration is a technology working towards a decentralized treatment method that facilitates the clean water supply to the common people in economically weaker countries. The chelating ability of the biopolymers is a unique aspect to remove the dissolved metals from water. In closed systems of plant operation, where there is no exchange of water or release of effluents, biopolymer-based filtration systems provide an opportunity for the recirculation of water towards sustainability. The easy availability of biopolymers, biodegrading nature, and effectiveness help to develop indigenous sustainable technologies towards clean water.

13.9   FUTURE PERSPECTIVES

With a growing demand for biopolymers, synthesizing those using renewable sources can make them cost-effective and contribute to sustainable development. Opting for nonthermal extraction technologies such as pulsed electric field assisted extraction can help in reducing the extraction time of biopolymers at large-scale levels and the application of a high electric field in this technique may help in easy disruption of cell membranes. Research should also focus on manufacturing novel polymers with interesting properties such as increased durability, recyclability, flexibility, specificity, and stability. Such properties can be integrated into biopolymers by casting them in the form of composite polymers, by blending with other effective cross-linking agents, or by modifying them with specific functional groups. Applying additive-based chemistry will help in bringing such modifications to the biopolymers and in improving the performance and properties of them. Developing nano-based reinforcements or combining nanofillers with these biopolymers, may produce polymers with enhanced properties and produce new functional biopolymers. A wide range of industries can benefit from the use of emerging and innovative technologies to produce novel advanced biopolymers. 3-D printing technology is an upcoming technology that can be employed to manufacture biopolymers with user-defined and unique complex structures based on our interests in environmental applications. The casting of these biopolymers in the form of ultra-filtration or nano-filtration membranes with improved antifouling properties will also be beneficial when applied to water polishing and water treatment operations. Stable filter membranes that are resistant to biofilm formation, improve the removal of toxic metals from wastewater, and are most suitable for industrial wastewater treatment are prepared by immobilizing nanoparticles in biopolymers. Because of their water-retaining capacities, these biopolymers can be employed for sludge management systems; that they can help in reducing and dewatering the bulk volumes of sludge, by which their handling would become easier.

13.10   CONCLUSION

Biopolymers are identified as a potential alternative to synthetic polymers and chemical coagulants in water treatment-related applications that help push towards a circular economy. A brief note on the various biopolymers used and the mechanisms by which the micropollutants are trapped and removed from both the effluents and the wastewater was also discussed. The chapter also outlined the use of biopolymers in ancient water treatment plants, and how the properties have been modified through recent research for a wide range of applications. Biopolymers can therefore serve as eco-friendly alternatives to help develop research-based products to maintain sanitary water quality for public health.

REFERENCES

Abbasian, Mojtaba, Pouneh Niroomand, and Mehdi Jaymand. 2017. "Cellulose/Polyaniline Derivatives Nanocomposites: Synthesis and Their Performance in Removal of Anionic Dyes from Simulated Industrial Effluents." *Journal of Applied Polymer Science* 134 (39). doi:10.1002/app.45352.

Ansarifar, Elham, Mohebbat Mohebbi, Fakhri Shahidi, Arash Koocheki, and Navid Ramezanian. 2017. "Novel Multilayer Microcapsules Based on Soy Protein Isolate Fibrils and High Methoxyl Pectin: Production, Characterization and Release Modeling." *International Journal of Biological Macromolecules* 97: 761–69. doi:10.1016/j. ijbiomac.2017.01.056.

Aziz, Faissal, Mounir El Achaby, Amina Lissaneddine, Khalid Aziz, Naaila Ouazzani, Rachid Mamouni, and Laila Mandi. 2019. "Composites with Alginate Beads: A Novel Design of Nano-Adsorbents Impregnation for Large-Scale Continuous Flow Wastewater Treatment Pilots." *Saudi Journal of Biological Sciences* 27 (10): 2499–2508. doi:10.1016/j.sjbs.2019.11.019.

Badawi, M. A., N. A. Negm, M. T. H. Abou Kana, H. H. Hefni, and M. M. Abdel Moneem. 2017. "Adsorption of Aluminum and Lead from Wastewater by Chitosan-Tannic Acid Modified Biopolymers: Isotherms, Kinetics, Thermodynamics and Process Mechanism." *International Journal of Biological Macromolecules* 99: 465–76. doi:10. 1016/j.ijbiomac.2017.03.003.

Olivera, Sharon, Handanahally Basavarajaiah Muralidhara, Krishna Venkatesh, Vijay Kumar Guna, Keshavanarayana Gopalakrishna, and Yogesh Kumar. 2016. "Potential applications of cellulose and chitosan nanoparticles/composites in wastewater treatment: a review." *Carbohydrate polymers* 153 (2016): 600–618. doi: 10.1016/j. carbpol.2016.08.017.

Chen, Bo, Fengxia Long, Sijiang Chen, Yangrui Cao, and Xuejun Pan. 2020. "Magnetic Chitosan Biopolymer as a Versatile Adsorbent for Simultaneous and Synergistic Removal of Different Sorts of Dyestuffs from Simulated Wastewater." *Chemical Engineering Journal* 385. doi:10.1016/j.cej.2019.123926.

Cheng, Chongling, Yongqing Cai, Guijian Guan, Leslie Yeo, and Dayang Wang. 2018. "Hydrophobic-Force-Driven Removal of Organic Compounds from Water by Reduced Graphene Oxides Generated in Agarose Hydrogels." *Angewandte Chemie - International Edition* 57 (35): 11177–81. doi:10.1002/anie.201803834.

Dima, Jimena Bernadette, Cynthia Sequeiros, and Noemi Zaritzky. 2017. "Chitosan from Marine Crustaceans: Production, Characterization and Applications." In *Biological Activities and Application of Marine Polysaccharides,* pp. 39-56. InTech, Croatia doi:10.5772/65258.

Domeradzka, Natalia E., Marc W. T. Werten, Renko de Vries, and Frits A. de Wolf. 2016. "Production in Pichia Pastoris of Complementary Protein-Based Polymers with Heterodimer-Forming WW and PPxY Domains." *Microbial cell factories*, 15(1), 1–11 doi:10.1007/978-1-4939-0563-8_5.

Ealias, Anu Mary, and Manickam Puratchiveeran Saravanakumar. 2020. "Application of Protein-Functionalised Aluminium Nanosheets Synthesised from Sewage Sludge for Dye Removal in a Fixed-Bed Column: Investigation on Design Parameters and Kinetic Models." *Environmental Science and Pollution Research* 27 (3): 2955–76. doi:10.1007/ s11356-019-07139-x.

Feng Enke, Guofu Ma, Yajuan Wu, Haiping Wang, Ziquang Lei. 2014. "Preparation and Properties of Organic-Inorganic Composite Superabsorbent Based on Xanthum Gum and Loess." *Carbohydrate Polymers 111: 463–468.* doi: 10.1016/j. carbpol.2014.04.031

Ergene, Aysun, Kezban Ada, Sema Tan, and Hikmet Katircioğlu. 2009. "Removal of Remazol Brilliant Blue R Dye from Aqueous Solutions by Adsorption onto Immobilized Scenedesmus Quadricauda: Equilibrium and Kinetic Modeling Studies." *Desalination* 249 (3): 1308–14. doi:10.1016/j.desal.2009.06.027.

Gifuni, Imma, Giuseppe Olivieri, Antonino Pollio, Telma Teixeira Franco, and Antonio Marzocchella. 2017. "Autotrophic Starch Production by Chlamydomonas Species." *Journal of Applied Phycology* 29(1): 105–114. doi:10.1007/s10811-016-0932-2.

Gifuni, Imma, Giuseppe Olivieri, Antonino Pollio, and Antonio Marzocchella. 2018. "Identification of an Industrial Microalgal Strain for Starch Production in Biorefinery Context: The Effect of Nitrogen and Carbon Concentration on Starch Accumulation." *New Biotechnology* 41: 46–54. doi:10.1016/j.nbt.2017.12.003.

Guo, Zhiwei, Quan Li, Zhiyue Li, Cong Liu, Xuerui Liu, Yuanyuan Liu, Genlai Dong, Tao Lan, and Yun Wei. 2020. "Fabrication of Efficient Alginate Composite Beads Embedded with N-Doped Carbon Dots and Their Application for Enhanced Rare Earth Elements Adsorption from Aqueous Solutions." *Journal of Colloid and Interface Science* 562: 224–34. doi:10.1016/j.jcis.2019.12.030.

Han, Hekun, Wei Wei, Zhifeng Jiang, Junwei Lu, Jianjun Zhu, and Jimin Xie. 2016. "Removal of Cationic Dyes from Aqueous Solution by Adsorption onto Hydrophobic/Hydrophilic Silica Aerogel." *Colloids and Surfaces A: Physicochemical and Engineering Aspects* 509: 539–549. doi:10.1016/j.colsurfa.2016.09.056.

Hu, Jun, Hao Liu, Pratyoosh Shukla, Weitie Lin, and Jianfei Luo. 2020. "Nitrogen and Phosphorus Removals by the Agar-Immobilized Chlorella Sacchrarophila with Long-Term Preservation at Room Temperature." *Chemosphere* 251:126406. doi:10.1016/j.chemosphere.2020.126406.

Jahn, Samia Al Azharia. 1999. "From Clarifying Pearls and Gems to Water Coagulation with Alum. History, Surviving Practices, and Technical Assessment." *Anthropos* 419-430

Kang, Jianxiong, Bo Li, Jing Song, Daosheng Li, Jing Yang, Wei Zhan, and Dongqi Liu. 2011. "Defluoridation of Water Using Calcined Magnesia/Pullulan Composite." *Chemical Engineering Journal* 166(2): 765–771. doi:10.1016/j.cej.2010.11.031.

Karim, Zoheb, Aji P. Mathew, Vanja Kokol, Jiang Wei, and Mattias Grahn. 2016. "High-Flux Affinity Membranes Based on Cellulose Nanocomposites for Removal of Heavy Metal Ions from Industrial Effluents." *RSC Advances* 6 (25): 20644–53. doi:10.1039/c5ra27059f.

Khorasani, Alireza Chackoshian, and Seyed Abbas Shojaosadati. 2019. "Magnetic Pectin-Chlorella Vulgaris Biosorbent for the Adsorption of Dyes." *Journal of Environmental Chemical Engineering* 7(3); 103062. doi:10.1016/j.jece.2019.103062.

Kube, Matthew, Arash Mohseni, Linhua Fan, and Felicity Roddick. 2019. "Impact of Alginate Selection for Wastewater Treatment by Immobilised Chlorella Vulgaris." *Chemical Engineering Journal* 358:1601–1609. doi:10.1016/j.cej.2018.10.065.

Kun, Dávid, and Béla Pukánszky. 2017. "Polymer/Lignin Blends: Interactions, Properties, Applications." *European Polymer Journal* 93:618–641. doi:10.1016/j.eurpolymj.2017.04.035.

Le, Thao Thanh, Kumarasamy Murugesan, Chung Seop Lee, Chi Huong Vu, Yoon Seok Chang, and Jong Rok Jeon. 2016. "Degradation of Synthetic Pollutants in Real Wastewater Using Laccase Encapsulated in Core-Shell Magnetic Copper Alginate Beads." *Bioresource Technology* 216: 203–10. doi:10.1016/j.biortech.2016.05.077.

Lee, Wei Kang, Yi Yi Lim, Adam Thean Chor Leow, Parameswari Namasivayam, Janna Ong Abdullah, and Chai Ling Ho. 2017. "Biosynthesis of Agar in Red Seaweeds: A Review." *Carbohydrate Polymers* 164: 23–30. doi:10.1016/j.carbpol.2017.01.078.

Lessa, Emanuele F., Aline L. Medina, Anderson S. Ribeiro, and André R. Fajardo. 2020. "Removal of Multi-Metals from Water Using Reusable Pectin/Cellulose Microfibers Composite Beads." *Arabian Journal of Chemistry* 13 (1): 709–20. doi:10.1016/j.arabjc.2017.07.011.

Li, Panyu, Ting Li, Yu Zeng, Xiang Li, Xiaolong Jiang, Yabo Wang, Tonghui Xie, and Yongkui Zhang. 2016. "Biosynthesis of Xanthan Gum by Xanthomonas Campestris LRELP-1 Using Kitchen Waste as the Sole Substrate." *Carbohydrate Polymers* 151: 684–91. doi:10.1016/j.carbpol.2016.06.017.

Li, Shanshan, Lan Zheng, Yuqi Wang, Xiaolong Han, Wen Sun, Yijun Yue, Danning Li, Jinting Yang, and Yongqiang Zou. 2017. "Polyacrylamide-Grafted Legume Starch for Wastewater Treatment: Synthesis and Performance Comparison." *Polymer Bulletin* 74 (11): 4371–92. doi:10.1007/s00289-017-1959-5.

Liang, Rui Hong, Ya Li, Li Huang, Xue Dong Wang, Xiao Xue Hu, Cheng Mei Liu, Ming Shun Chen, and Jun Chen. 2020. "Pb2+ Adsorption by Ethylenediamine-Modified Pectins and Their Adsorption Mechanisms." *Carbohydrate Polymers* 234: 115911. doi:10.1016/j.carbpol.2020.115911.

Liaw, Bor Shuang, Ting Ting Chang, Haw Kai Chang, Wen Kuang Liu, and Po Yu Chen. 2020. "Fish Scale-Extracted Hydroxyapatite/Chitosan Composite Scaffolds Fabricated by Freeze Casting—An Innovative Strategy for Water Treatment." *Journal of Hazardous Materials* 382: 121082. doi:10.1016/j.jhazmat.2019.121082.

Liu, Jie, Dihan Su, Jinrong Yao, Yufang Huang, Zhengzhong Shao, and Xin Chen. 2017. "Soy Protein-Based Polyethylenimine Hydrogel and Its High Selectivity for Copper Ion Removal in Wastewater Treatment." *Journal of Materials Chemistry A* 5 (8): 4163–71. doi:10.1039/c6ta10814h.

Maleki, Susan, Mali Mærk, Radka Hrudikova, Svein Valla, and Helga Ertesvåg. 2017. "New Insights into Pseudomonas Fluorescens Alginate Biosynthesis Relevant for the Establishment of an Efficient Production Process for Microbial Alginates." *New Biotechnology* 37: 2–8. doi:10.1016/j.nbt.2016.08.005.

Naushad, Mu, Gaurav Sharma, Amit Kumar, Shweta Sharma, Ayman A. Ghfar, Amit Bhatnagar, Florian J. Stadler, and Mohammad R. Khan. 2018. "Efficient Removal of Toxic Phosphate Anions from Aqueous Environment Using Pectin Based Quaternary Amino Anion Exchanger." *International Journal of Biological Macromolecules* 106: 1–10.

Ozdal, Murat, and Esabi Basaran Kurbanoglu. 2018. "Valorisation of Chicken Feathers for Xanthan Gum Production Using Xanthomonas Campestris MO-03." *Journal of Genetic Engineering and Biotechnology* 16(2): 259–263. doi:10.1016/j.jgeb.2018.07.005.

Pariatamby, Agamuthu, and Yang Ling Kee. 2016. "Persistent Organic Pollutants Management and Remediation." *Procedia Environmental Sciences* 31: 842–848. doi:10.1016/j.proenv.2016.02.093.

Paulino, Alexandre T., Marcos R. Guilherme, Adriano V. Reis, Gilsinei M. Campese, Edvani C. Muniz, and Jorge Nozaki. 2006. "Removal of Methylene Blue Dye from an Aqueous Media Using Superabsorbent Hydrogel Supported on Modified Polysaccharide." *Journal of Colloid and Interface Science* 301(1): 55–62. doi:10.1016/j.jcis.2006.04.036.

Pernicova, Iva, Dan Kucera, Ivana Novackova, Juraj Vodicka, Adriana Kovalcik, and Stanislav Obruca. 2019. "Extremophiles – Platform Strains for Sustainable Production of Polyhydroxyalkanoates." *Materials Science Forum* 955: 74–79. doi:10.4028/www.scientific.net/MSF.955.74.

Prabhu, Meghanath, Alexander Chemodanov, Ruth Gottlieb, Meital Kazir, Omri Nahor, Michael Gozin, Alvaro Israel, Yoav D. Livney, and Alexander Golberg. 2019. "Starch from the Sea: The Green Macroalga Ulva Ohnoi as a Potential Source for Sustainable Starch Production in the Marine Biorefinery." *Algal Research* 37: 215–227. doi:10.1016/j.algal.2018.11.007.

Pritchard, M., T. Mkandawire, A. Edmondson, J. G. O'Neill, and G. Kululanga. 2009. "Potential of Using Plant Extracts for Purification of Shallow Well Water in Malawi." *Physics and Chemistry of the Earth* 34(13–16):799–805. doi:10.1016/j.pce.2009.07.001.

Ren, Jie, Na Li, Hua Wei, Aimin Li, and Hu Yang. 2020. "Efficient Removal of Phosphorus from Turbid Water Using Chemical Sedimentation by FeCl3 in Conjunction with a Starch-Based Flocculant." *Water Research* 170: 115361. doi:10.1016/j.watres.2019.115361.

Reyes, César, Carlos Peña, and Enrique Galindo. 2003. "Reproducing Shake Flasks Performance in Stirred Fermentors: Production of Alginates by Azotobacter Vinelandii." *Journal of Biotechnology* 105 (1–2): 189–198. doi:10.1016/S0168-1656(03)00186-X.

Rishi, Valbha, Armaan Kaur Sandhu, Arashdeep Kaur, Jaspreet Kaur, Sanjay Sharma, and Sanjeev Kumar Soni. 2020. "Utilization of Kitchen Waste for Production of Pullulan to Develop Biodegradable Plastic." *Applied Microbiology and Biotechnology.* 104(3), 1307–1317 doi:10.1007/s00253-019-10167-9.

Rocha, Cristina M. R., Ana M. M. Sousa, Jang K. Kim, Júlia M. C. S. Magalhães, Charles Yarish, and Maria do Pilar Gonçalves. 2019. "Characterization of Agar from Gracilaria Tikvahiae Cultivated for Nutrient Bioextraction in Open Water Farms." *Food Hydrocolloids* 89: 260–271. doi:10.1016/j.foodhyd.2018.10.048.

Zhang, Yu, Xiaoting Fu, Delin Duan, Jiachao Xu, and Xin Gao. 2013. "Preparation and Characterization of Agar, Agarose, and Agaropectin from the Red Alga Ahnfeltia Plicata." *Journal of Oceanology and Limnology* 37(3): 815–824.

Jahn, S. A. A. 1988. "Using 'Moringa' Seeds as Coagulants in Developing Countries." *Journal - American Water Works Association* 80(6): 43–50.

Saad, Abdel Halim A., Ahmed M. Azzam, Shaimaa T. El-Wakeel, Bayaumy B. Mostafa, and Mona B. Abd El-Latif. 2020. "Industrial Wastewater Remediation Using Hematite@chitosan Nanocomposite." *Egyptian Journal of Aquatic Biology and Fisheries* 24(1): 13–29. doi:10.21608/ejabf.2020.67580.

Sasuga, Keiji, Tomoya Yamanashi, Shigeru Nakayama, Syuetsu Ono, and Koji Mikami. 2018. "Discolored Red Seaweed Pyropia Yezoensis with Low Commercial Value Is a Novel Resource for Production of Agar Polysaccharides." *Marine Biotechnology* 20 (4): 520–30. doi:10.1007/s10126-018-9823-7.

Savran, Ali, Nilüfer Çiriğ Sclçuk, Şenol Kubilay, and Ali Rıza Kul. 2017. "Adsorption Isotherm Models for Dye Removal by Paliurus Spinachristi Mill. Frutis and Seeds in a Single Component System." *IOSR Journal of Environmental Science, Toxicology and Food Technology* 11 (4): 18–30. doi:10.9790/2402-1104021830.

Shi, Weida, Marie Josée Dumont, and Elhadji Babacar Ly. 2014. "Synthesis and Properties of Canola Protein-Based Superabsorbent Hydrogels." *European Polymer Journal* 54(1): 172–180. doi:10.1016/j.eurpolymj.2014.03.007.

Su, Ting, Lipeng Wu, Xihao Pan, Cheng Zhang, Mingyang Shi, Ruru Gao, Xiaoliang Qi, and Wei Dong. 2019. "Pullulan-Derived Nanocomposite Hydrogels for Wastewater Remediation: Synthesis and Characterization." *Journal of Colloid and Interface Science* 542: 253–62. doi:10.1016/j.jcis.2019.02.025.

Terán Hilares, R., J. Resende, C. A. Orsi, M. A. Ahmed, T. M. Lacerda, S. S. da Silva, and J. C. Santos. 2019. "Exopolysaccharide (Pullulan) Production from Sugarcane Bagasse Hydrolysate Aiming to Favor the Development of Biorefineries." *International Journal of Biological Macromolecules* 127: 169–77. doi:10.1016/j.ijbiomac.2019.01.038.

Thakur, Sourbh, Sadanand Pandey, and Omotayo A. Arotiba. 2017. "Sol-Gel Derived Xanthan Gum/Silica Nanocomposite—A Highly Efficient Cationic Dyes Adsorbent in Aqueous System." *International Journal of Biological Macromolecules* 103: 596–604. doi:10.1016/j.ijbiomac.2017.05.087.

Wang, Shening, Chun Zhang, and Qing Chang. 2017. "Synthesis of Magnetic Crosslinked Starch-Graft-Poly(Acrylamide)-Co-Sodium Xanthate and Its Application in Removing Heavy Metal Ions." *Journal of Experimental Nanoscience,* 12(1), 270-284. doi:10.1080/17458080.2017.1321793.

Zeng, Qiankun, Xiaoliang Qi, Mengying Zhang, Xianqin Tong, Ning Jiang, Wenhao Pan, Wei Xiong, et al. 2020. "Efficient Decontamination of Heavy Metals from Aqueous Solution Using Pullulan/Polydopamine Hydrogels." *International Journal of Biological Macromolecules* 145: 1049–1058. doi:10.1016/j.ijbiomac.2019.09.197.

Zhang, Qingrui, Shuaiqi Zhang, Zhixue Zhao, Meng Liu, Xiaofeng Yin, Yanping Zhou, Yun Wu, and Qiuming Peng. 2020. "Highly Effective Lead (II) Removal by Sustainable Alkaline Activated β-Lactoglobulin Nanofibrils from Whey Protein." *Journal of Cleaner Production* 255: 120297. doi:10.1016/j.jclepro.2020.120297.

14 Bioreactors for Degradation of Toxic Pollutant in Wastewater System

Monalisa Satapathy, Aparna Yadu and J. Anandkumar*
Department of Chemical Engineering,
National Institute of Technology Raipur, India

Biju Prava Sahariah
University Teaching Department, Chhattisgarh
Swami Vivekanand Technical University, India

CONTENTS

14.1 INTRODUCTION

Rapid industrialization and indiscriminate utilization of natural resources such as air, water, and soil have created a highly stressed environment. Water is an essential resource for all living creatures. However, due to the increased market demand for various products, industrial activities have increased across the globe, which has resulted in the contamination of groundwater, surface water, soil, and sediments with toxic and hazardous pollutants. In general, pharmaceutical products, personal care products, endocrine disrupting compounds, phenolic compounds, polycyclic aromatic hydrocarbons (PAHs), surfactants, inorganic chemicals, flame retardants, and some trace metal elements fall into the category of toxic and hazardous pollutants (Kanaujiya et al., 2019). These toxic pollutants are present in various aquatic sources such as drinking, surface, and groundwater with concentrations ranging from a few ppb to several ppm, which impairs the characteristics of water. The characteristics of wastewater generated from various types of industries in terms of organic and inorganic pollutants are described in Table 14.1.

Despite the regulatory guidelines issued by several governments and nongovernment organizations such as the United States' Environmental Protection Agency (EPA), the International Program on Chemical Safety (IPCS), the European Union (EU) and the World Health Organization (WHO) for pollution control procedures (Ahmed et al., 2017), wastewater containing toxic pollutants is directly discharged into water bodies without adequate treatment, which poses a serious threat to the

TABLE 14.1

Toxic pollutants (organic and inorganic) present in various kind of industrial wastewater

Type of wastewater	Organic pollutants	Inorganic pollutants	References
Petrochemical	PAHs, phenols, chlorinated nitrobenzene, anilines, indole, dimethylpyrimidine	Hydrogen sulfide, ammonia, heavy metals, mercaptans, cyanides	*Varjani et al., 2020*
Cokeoven	Phenols, benzene, toluene, ethylbenzene (BTEX), xylene, PAHs	Sulfates, ammonia, cyanides	*Gui et al., 2019*
Refinery	BTEX, PAHs, methanol, ether, surfactants, phenols	Ammonia, cyanides, radioactive elements, sulfides, halides	*Barthe et al., 2015*
Oilfield-produced water	BTEX, biocides, PAHs, phenols, surfactants	Mineral salts, heavy metals, radioactive elements	*Bakke et al., 2013*
Metallurgy	Phenols, BTEX, PAHs, quinolones, hydrazine, thiophenes	Heavy metals, ammonia, cyanide	*Wu et al., 2017*

environment. At present, it is not possible to stop the discharge of wastewater into the ecosystem, although the harmful effects associated with these toxic pollutants can be minimized by the involvement of suitable treatment techniques. Earlier chapters discussed how various conventional processes including precipitation, oxidation, adsorption, filtration, and extraction were successfully applied to treat some toxic pollutants. However, due to several limitations of traditionally used conventional processes, there is an urgency to develop suitable treatment techniques for the complete removal of toxic pollutants from the environment.

As a feasible alternative, biological treatment processes have gained much attention due to their distinguished characteristics: cost-effectiveness, lower energy consumption, and eco-friendly nature for the treatment of any kind of industrial wastewater (Singh and Singh, 2019). The biological treatment process is considered to be a vital and essential part of any wastewater treatment plant. The biological treatment process is governed by utilizing microorganisms such as bacteria, algae, fungi, yeast, and plants to transform or degrade toxic pollutants into less toxic or simpler compounds, thus helping to remediate or eliminate toxic pollutants from the environment. The rate and extent of the biological treatment processes to efficiently degrade pollutants in wastewater largely depends on the source and type of wastewater. Microorganisms possess the capability to break down or degrade the toxic pollutants for their growth or energy requirement. Additionally, it is desirable to combine the treatment of wastewater simultaneously with waste utilization. In this kind of situation, it is imperative to develop and propose efficient treatment processes for the improvement of the total economy by minimizing the energy consumption. Also, a single unit would not be able to fulfill all the requisite criteria for the effective treatment of wastewater. Hence, there is a great demand for the development and exploitation of integrated approaches to meet stringent water quality standards. To comply with all the above-mentioned conditions, the extensive use of an engineered bioreactor is preferred. The use of a well-designed bioreactor offers optimum conditions for the growth of microorganisms during the biodegradation process, which help to accomplish the desired remediation goals. Bioreactors can be operated in various modes such as batch, fed-batch, and continuous, and they are specifically designed for the optimization of microbial processes regarding the types of media and characteristics of pollutants (Jeon and Madsen, 2013; Quintero et al., 2007). This chapter discusses the different kinds of bioreactors used for the treatment of toxic pollutants in wastewater and innovations in the field of bioreactors for effective wastewater treatment processes.

14.2 BIOREACTORS

14.2.1 Different Types of Bioreactor

Many bioreactors are available in order to comply with environmental engineering practices. In general, particular bioreactors are designed to emphasize suspended or attached growth systems. The bioreactors based on the suspended growth processes are recognized as slurry-phase reactors, dispersed growth, or suspended-floc bioreactors, although the reactors that use the technique of attached growth processes or make use of biofilm are known as attached growth reactors, immobilized cell reactors, or fixed-film reactors. In the design and operation of bioreactors, a series of reactors can be used in combination with suspended as well as attached growth processes. To optimize the bioreactor performance of any wastewater treatment system, one should understand the kinetic of substrate removal with the use of different types of microorganisms by considering the basic properties of the bioreactor system. Several factors that may influence the choice of bioreactors in between the attached or suspended growth reactors are the concentration of toxic pollutants, availability of oxygen, physical and chemical characteristics of wastewater, the efficacy of treatment required, and climate conditions suitable to the bioreactor system. In the following section, the different types suspended and attached growth bioreactor systems are briefly discussed.

14.2.1.1   Bioreactors Assisted with Suspended Growth Process

i. Continuous Stirred-tank reactors:

Continuous stirred-tank bioreactors generally comprise a cylindrical vessel with a motor-shaft that supports one or more impellers (see Figure 14.1). These bioreactors are largely employed for the submerged cultures. The advantages of stirred-tank bioreactors include vigorous mixing of the substances, effective and flexible operating conditions. The requirement of a high amount of energy during mechanical agitation can create a shear strain on microorganisms, which is the only constraint of a stirred-tank bioreactor. It is used in the activated sludge process for the treatment of industrial wastewater in aerobic conditions. However, in many cases, it was applied for the anaerobic treatment of municipal and domestic waste because of its large holdup capacity and ease of operation and installation (Narayanan and Narayan, 2019). A stirred-tank bioreactor was successfully implemented for the treatment of hydrocarbon-rich industrial wastewater by using an acclimatized microbial consortium (Gargouri et al., 2011).

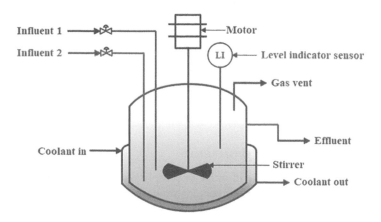

FIGURE 14.1   Schematic diagram of a continuous stirred-tank bioreactor.

ii. Sequencing batch reactor (SBR) and fed batch reactor

SBR possesses certain merits over conventional activated sludge processes (ASP) for the treatment of any kind of wastewater. SBR has extensively used as a very effective, well-designed, and outstanding approach for the treatment of industrial and domestic wastewater owing to its cost-efficiency and ease of operation (Ghazani and Taghdisian, 2019). SBR can be operated in many ways such as physicochemical, biological, or physicochemical-biological. The basic steps usually carried out in the SBR process are filling/equalization, mixing, reaction, clarification, and decanting (Mallick and Chakraborty, 2019). The development of SBR technology is based on the assumption that intermittent exposure of microbial culture to the defined process conditions can be proficiently accomplished in fed-batch operations at a particular retention time, the rate of exposure and concentration of pollutants can be independent of inlet conditions. Studies show that enhanced phosphate removal was achieved in SBRs because of the accumulation of phosphate in the anaerobic condition that was subsequently used further in the aerobic condition in sequence (Zhao et al., 2016). In addition to that, SBR technology has been fruitfully used for the treatment of dye wastewater, landfill leachates, and complex chemical wastewater (Mohan et al., 2005). The schematic diagram of a SBR is shown in Figure 14.2.

iii. Anaerobic digesters

The technology of anaerobic digesters can be defined as a biological process that produces biogas (a mixture of methane and carbon dioxide) through the action of mixed microbial cultures under strictly anaerobic conditions. This is one of the oldest technologies used for the treatment of high solids-containing municipal wastewater and in sewage sludge stabilization (Vinardell et al., 2020). Generally, anaerobic digesters can be categorized in two ways as a standard rate and high rate-based digesters. To make a difference between these two reactors, the standard digestion process is carried out without the application of heat or stirring. This is because the application of heat or mixing during the digestion process can lead to longer reaction time, therefore, expanding the overall detention time by 30 to 60 days. However,

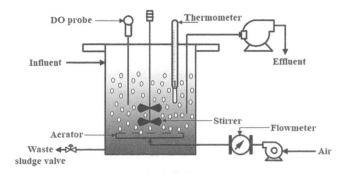

FIGURE 14.2 Schematic diagram of a sequencing batch reactor (SBR).

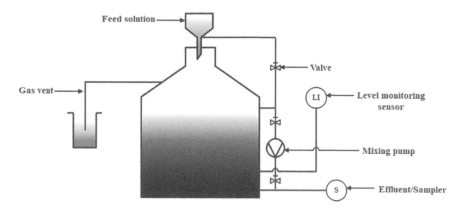

FIGURE 14.3   Schematic representations of anaerobic digesters.

in the case of high rate anaerobic digesters, heat, as well as stirring, is provided for the homogeneous mixing. This allows the reduction of standard detention time to less than 15 days. Sometimes based on the demand or depending upon the application, a two-stage system consisting of both standard and high rate processes is used to take advantage of both the processes with its pros and cons. In that, initially, the sludge is thoroughly mixed with the help of a stirrer, simultaneously, with heating up the contents and further, it was allowed to settle down to get two discriminate layers as a supernatant layer and digested sludge solids (Turovskiy and Mathai, 2006). The schematic representation of anaerobic digesters is represented in Figure 14.3.

14.2.1.2   Bioreactors Assisted with Attached Growth Process

i. Packed bed reactor

A packed bed reactor is the most common biofilm reactor in which microbial cultures are attached in a fixed carrier medium. packed bed reactors are the most desirable systems due to compact nature and more reliable performance. A packed bed reactor offers a certain advantage in context to remediation of dilute aqueous solutions at higher biomass concentration without separation of biomass and treated effluent (Tekere et al., 2001). Unlike suspended growth processes, a packed bed reactor does not require special arrangements like centrifugation and membrane filters to preserve the biomass. This feature of the packed bed reactor makes it useful among all the available bioreactors where large substrate flow is required. The selection of suitable packing material is valuable for the better immobilization of microorganisms during the treatment process. Packing material that has been used in the packed bed bioreactors are porous ceramics, silicon tubing, activated carbon, stainless steel, sintered glass, propylene, agarose, agar gel beads and clay balls (Talha, et al., 2018). Figure 14.4 represents the schematic view of the packed bed reactor.

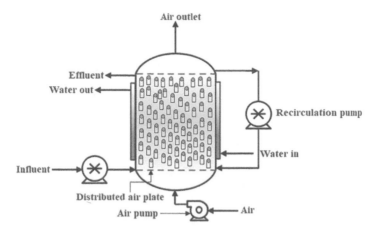

FIGURE 14.4   Schematic view of a packed bed reactor.

ii. Fluidized bed reactor

The fluidized bed reactor employs attachment of the microorganism to the particles maintained in suspension through an upward flow condition of the liquid to be treated. The particles in the suspended condition are known as biofilm carriers. These may be granular activated carbon (GAC), sand grains, or any other solids that show resistance towards abrasion. The upward velocity of the fluid should be sufficient enough to maintain the biofilm carriers in suspension, which tremendously depends upon the density of carrier material, the shape and size of the carrier material, and the quantity of attached biomass (Shieh and Keenan, 1986). Fluidized bed reactors can remain anywhere between a plug-flow and continuously stirred-tank reactors. When the bioreactor system is operated in once-through mode, the fluid regime is almost similar to the plug flow reactor. On the other side, when effluent is recycled to attain sufficient upward velocities for fluidization, the liquid regime is more like a continuous stirred-tank reactor. One of the major drawbacks of the fluidized bed reactor is the maintenance and control of the Fluidization velocity. The upwards flow velocity must be adequate for fluidization, though too high flow rate can cause washout conditions of carriers from the reactor. This reactor is significantly used for the aerobic treatment of low organics-containing wastewater (Narayanan and Narayan, 2019). The pictorial view of the fluidized bed reactor is given in Figure 14.5.

iii. Biofilters

Biofilters consist of a large porous media bed over which pollutants are passed and degraded by microorganisms. The biofilter is known to be the oldest reactor system having a significant role in wastewater treatment. These kinds of reactors are widely used for industrial and municipal effluents containing organic and inorganic contaminants. However, it is primarily used in the control of air pollution chiefly from the gases of sewage treatment plants (Schmidt and Anderson, 2017). The gases are ammonia, mercaptan, disulfides,

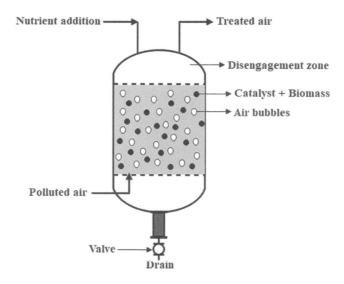

FIGURE 14.5   Schematic view of a fluidized bed reactor.

hydrogen sulfides, and volatile organic compounds such as phenols, propane, styrene, butane, and ethylene chloride. The materials generally used in the biofilter media bed are peat, bark, composted yard waste, coarse soil, and gravel. Figure 14.6 providesa graphical view of a biofilter bioreactor unit.

14.2.2 DIFFERENT OPERATING ENVIRONMENTS OF A BIOREACTOR

The evaluation of the bioreactor performance under different environmental conditions is an essential part of the biodegradation process. The bioreactor can be operated in a different environment such as anaerobic, anoxic, and aerobic. In the current

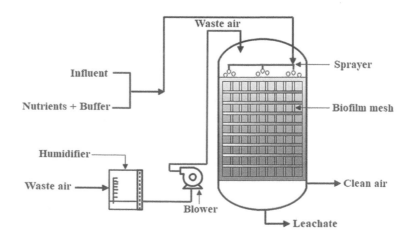

FIGURE 14.6   Schematic representation of a biofilter bioreactor unit.

scenario, the anaerobic reactor has gained considerable attention for its ability to handle high-strength wastewater. In anaerobic conditions, microbes metabolized the organic compounds in the absence of oxygen and produced biogas (CH_4 and CO_2). However, in some of the anaerobic conditions sulfate and iron act as terminal electron acceptors. Anaerobic degradation follows four basic steps: hydrolysis, acidogenesis, acetogenesis, and methanogenesis. Anaerobic degradation has numerous advantages against the aerobic condition such as less sludge production, flexible application, less energy required for operation, and production of valuable by-product. Sharma and Philip (2014) have used the anaerobic batch reactor to treat coke-oven wastewater that contains phenol, cresol, quinoline, and indole.

In anoxic conditions, microbial metabolism occurs in the presence of bound oxygen in which the organic compounds are the electron donor and nitrate acts as the terminal electron acceptor. In anoxic conditions, nitrate and nitrite are converted into nitrogen gas. Biological denitrification is cost-effective and the most commonly used nitrogen removal technique due to the concern for eutrophication (Metcalf and Eddy, 2003). Mallick and Chakraborty (2020) used the anoxic sequential bioreactor to treat oil, phenol, sulfide, and ammonia.

In aerobic conditions, the oxygen acts as a terminal electron acceptor. The aerobic degradation is also known as nitrification. Nitrification is further classified as autotrophic and heterotrophic nitrification. The heterotrophs consume carbon compounds or nitrogenous carbon compounds for their respiration. In autotrophic nitrification, the ammonium oxidizing bacteria convert the ammonia into nitrate. The toxic organics and chlorinated organic compounds are able to degrade in aerobic conditions, promoting the growth of the heterotrophs present in the aerobic reactor (Metcalf and Eddy, 2003). Sharma et al. (2018) have investigated the performance of the continuous aerobic reactor to degrade the phenol quinoline and indole as a single substrate as well as mixed substrate conditions.

14.2.3 Factors Affecting Bioreactor Performance

Several factors play a key role while operating bioreactors in the bioremediation process: concerns about the microbial community, the type of pollutant, and design and optimization of the treatment process. Several factors should be kept in the mind during the design and operation of a bioreactor in order to enhance the bioreactor performance. The following factors have an immense effect on the microbial growth and degradation capabilities in the bioreactors: pH, temperature, moisture content, nutrients, and mixing time.

14.2.3.1 pH

pH is one environmental factor with a direct influence on the performance of the bioreactor, as a little fluctuation in pH can affect the growth and degradation efficiency of the microorganism. The microbial cell's ionic properties directly depend on the pH of the media, which ultimately affects microbial growth. Microorganisms such as acidophiles possess minimum, maximum, and optimum pH values for their growth and survival in the bioreactor system; they grow well at an acidic pH whereas alkaliphiles thrive best at an alkaline pH. Sulfate-reducing bacteria can survive at neutral pH values whereas denitrifying and nitrifying bacteria can be very active at pH 7–7.5 and 6–8.5, respectively. The behavior of the pollutants has a direct correlation with pH, which affects biodegradation efficiency. The optimum pH for benzene, toluene, ethylbenzene, and xylene (BTEX) was found to be in the range of 7.0–8.0 (Lee et al., 2002; Lu et al., 2002).

14.2.3.2 Temperature

Similar to pH, temperature is also a crucial factor for the achievement of an effective bioremediation process in a bioreactor system. Microorganisms require a temperature range for their survival and growth. Also in any bioreactor system, there is always the necessity of an optimum temperature for microorganisms to give better reactor performance (Tekere et al., 2001). The fluctuation in temperature (low or high) is drastically affected by the growth of microorganisms and also enzymatic activity, which eventually affects the metabolic activity inside the bioreactor system. The solubility and rate of diffusion of some chemicals increase with the increase in temperature, such as PAHs and heavy metals (Gargouri et al., 2011; Sharma, 2012). By considering all these conditions, bioreactors are now designed with the ability to control the temperature.

14.2.3.3 Moisture

The performance of the bioreactor system also depends on the moisture content, which is facilitated by the amount of water present in the bioreactor. Water is an essential element needed for the provision of microbial growth and the catalysis process. This is because the cellular level chemical reactions happen in aqueous conditions and water is required to ensure the exact osmotic pressure for the maintenance of microbial growth. In the case of a biofilter, which is a kind of bioreactor, the moisture content of the filter bed is a very important factor to ensure the bioreactor performance, since microorganisms need the water content to accomplish the metabolic activity. Not enough water in the microorganism ratio results in a substantial decline in the rate of biodegradation. Too much water within the filter bed hinders the oxygen transfer and hydrophobic compounds, thus promoting the development of an anaerobic zone inside the filter bed in the biofilter bioreactor. Excess water can also cause a bad odor owing to the lower oxygen level and channeling of gas inside the filter bed (Mudliar et al., 2010).

14.2.3.4 Nutrients

Nutrients are an important element for microbial growth and metabolic activity of microorganisms. In that, carbon is considered the most basic element for living organisms and it is needed in higher quantities compared to other elements. Oxygen, nitrogen, sulfur, iron, calcium, nickel, and magnesium are required in trace amounts to ensure balanced availability of nutrients inside the bioreactor system to achieve maximum biodegradation efficiency (Naik and Duraphe, 2012; Srivastava et al., 2014).

14.2.3.5 Mixing

Mixing is one of the most significant factors that affect both reactor performance and scale-up efficiency in the bioreactor. Mixing is employed to facilitate the fermentative process and it is linked with several fermenter scaling issues (Bonvillani et al., 2006). In general, mixing is used in the bioreactor for eliminating the concentration and temperature gradient, and other factors. Also, due to mixing in the bioreactor, thermal stratification can be minimized, helps to maintain uniform conditions in the bioreactor, and provides good contact between microbial culture and media. However, poor mixing can affect the biodegradation efficiency in the bioreactor.

14.2.3.6 Retention Time

Retention time plays a crucial role in the design of the bioreactor system. Exact or optimum retention time is required in the bioreactor system in order to obtain good removal efficiency. Longer hydraulic retention time (HRT) results in the decrease of the bioreactor performance due to less availability of substrates and depletion of nutrients, which diminishes the microbial population. In contrast to this, lesser retention time may cause washout of a cell and does not allow the microorganism to degrade efficiently (Sheoran et al., 2010). A retention time of three to four days was reported by Neculita et al. (2007) for complete removal of metals and sulfates in an anaerobic reactor.

14.3 RECENT ADVANCEMENTS IN BIOREACTOR SYSTEMS AND THEIR SUSTAINABILITY

14.3.1 Two-Phase Portioning Bioreactor

Biodegradation of highly loaded toxic and hazardous pollutants-containing gas and water are the biggest hurdle for bioremediation. The biological system shows poor performance for hydrophobic pollutants, such as PAHs and VOCs. The accumulation of toxic inhibitory compounds in the system restricts the growth of microorganisms. Therefore, controlled delivery of the substrate to the reactor is required to overcome the above-mentioned problems. Two-phase partitioning bioreactors (TPPB) are designed to separate the pollutants in two phases: organic or non aqueous phase (NAP) and aqueous phase. In TPPB, the aqueous phase contains the biomass and the organic phase is present on the surface of the aqueous phase containing toxic hazardous compounds. The toxic organics are dissolved with some biocompatible solvents. A schematic representation of TPPB is depicted in Figure 14.7. Traditionally, many liquid solvents are used as NAP, while many solid polymers have gained interest for their low cost, better separation, and reusability (Morrish and Daugulis, 2008). The NAP plays a major role in the TPPB process robustness. Therefore, the selection of organic solvent is important since it depends on high affinity towards the pollutant. In TPPB, the cell never experiences high organic loading, even though the high substrate is available in the system. Due to the thermodynamically stable nature of the system, the microbes use an adequate amount of the substrate for their metabolism until the whole pollutant is depleted into the aqueous phase. Moreover, the TPPB shows better performance for VOCs removal than the other available conventional method (Wu et al., 2018; Dorado et al., 2015; Muñoz et al., 2012).

TPPB is used to treat different wastewater in various reactor configurations such as trickling filter, airlift bioreactor, stirred-tank bioreactor, and bubble column bioreactor (Wu et al., 2018; San-Valero et al., 2017; Munoz et al., 2012). The presence of NAP reduces the surface tension between the two phases. Therefore, the rates of mass transfer and oxygen transfer substantially increase, and the system also

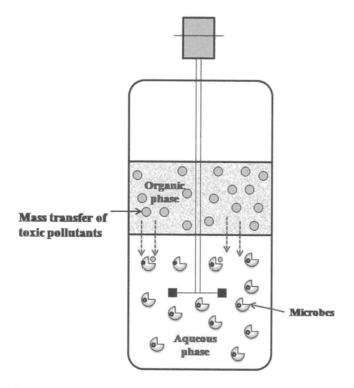

FIGURE 14.7 Two-phase portioning bioreactor systems.

responds well in shock-loading conditions. The use of TPPB for the treatment of various wastewaters is summarized in Table 14.2. There are three mechanisms by which the microbes use the substrate in TPPB: (1) uptake of the available dissolved substrate from the aqueous phase, (2) uptake of substrate available in the interfacial area, and (3) uptake of the substrate directly from the organic phase (Mahanty et al., 2008). The rate agitation and phase volume ratio are the rate-limiting parameters for degradation in TPPB (Baskaran et al., 2020).

TABLE 14.2
Treatment of different types of pollutants by TPPB

S. no.	Pollutants	Microbial source	Organic solvent/Solid polymers	Reference
1	Trichloroethylene (1000 mg/L)	*Oleaginous Rhodococcus opacus*	Silicone oil	Baskaran et al., 2020
2	4-chlorophenol (1000–2500 mg/L), Cr(VI)(100 mg/L)	-	Hytrel 8206	Angelucci et al., 2017
3	Pyrene	*Mycobacterium frederiksbergense*	Silicone oil	Mahanty et al., 2008

14.3.2 MEMBRANE BIOREACTOR

Membrane bioreactor (MBR) is a combination of the physical-biological process where the membrane acts as a physical separating barrier to remove the solids from the liquid. The MBR comes into the picture after the successful utilization of other conventional membrane techniques. MBRs can degrade many toxic emerging pollutants that are difficult to degrade in the activated sludge process. The high solid retention time promotes the growth of slow-growing microbes to degrade the pollutant. In a membrane bioreactor, the microbes consume the pollutant, and the solid in the system increases, which is further filtered by the membrane. Based on the location of the membrane module membrane are of two types: submerged membrane bioreactor and external membrane bioreactor. MBR is a well-known biological system used to treat many industrial wastewaters due to its simple, efficient, reliable, and cost-effective nature. However, many drawbacks like membrane fouling reduces the permeate flux in the utility of membrane bioreactor on a large scale. Hence, recent researches on membrane bioreactor are in the direction of the low-cost membrane module, easy fouling control, process cost reduction, and lower energy requirements. MBR can operate in aerobic or anaerobic environmental conditions and can be used in different reactor configurations. MBR performance is affected by membrane fouling, presence of recalcitrant compounds, the concentration of mixed liquor suspended solids, HRT, and SRT. Aerobic MBR is a well-established technique to treat many toxic pollutants because of its reduced startup time, high biomass concentration, and higher effluent quality than the anaerobic system. The anaerobic MBRs open a new gateway in wastewater treatment for their lower energy requirement, ability to bear high organic loading, better energy recovery, and reduced sensitivity to the temperature (Abuabdou et al., 2020). Many successful research works have been carried out on the use of MBR to treat high-strength industrial and toxic wastewater such as textile, landfill leachate, pharmaceutical, petrochemical, and tannery (Abuabdou et al., 2020; Berkessa et al., 2020).

Many researchers analyze the life cycle assessment of MBR technology to know about the process sustainability. The cost analysis is one of the important parameters to determine the adaptability potential of any new techniques. The high-energy demand for aerobic MBR is a major drawback that draws a negative impact on the sustainability viewpoint (Song et al., 2020). Even though the MBR has some drawbacks, the high market demand, high effluent quality, and reduction in manpower requirement makes MBR application as a more reliable technique for wastewater treatment. Submerged membranes can be replaced by the side-stream MBR and have immense potential applications. MBR technology will be one of the best biological treatment techniques to treat high-strength wastewater.

14.3.3 MOVING BED BIOFILM REACTOR

Moving bed biofilm reactor (MBBR) is an advanced bioreactor system in which microbes is being attached with the porous carrier media that are in irregular chaotic motion inside the reactor to enable the microbes to access the substrate present in the surrounding liquid. The use of the conventional biological system to treat wastewater is difficult because of its drawbacks. Therefore, MBBR techniques are widely accepted to treat complex wastewater. Moving bed biofilm reactors have some additional advantages such as less operational and maintenance costs, less land area requirement, high sludge retention time, lower sensitivity to shock loading, no sludge bulking, and appropriateness for slow-growing microbes, which makes its utility more immense. In MBBR, the microbes are being attached with the inert carrier media and form a biofilm layer that protects the microbe from the inhibitory effect of toxic compounds. The inert carrier material provides a high surface area for the growth of the microbes, which keep moving inside the reactor. The mixing is done by aeration in an aerobic reactor and by external mechanical mixing in anaerobic and anoxic reactors. The mixing enhances the rate of substrate transfer from wastewater to the microbes through biofilm. The carrier filling ratio is one of the key parameters in the MBBR operation. The schematic representation of the MBBR is represented in Figure 14.8. The carrier filling must be below 70% for better performance (Levya-Diaz et al., 2017). MBBR technology rehabilitates the existing conventional activated sludge process and is also used for the effective treatment of some industrial wastewater, which is summarized in Table 14.3. The research developments on different types of cheaper carrier media to attach more biomass and increase self-life have been continued in the last few years. Sonwani et al. (2019) investigated different novel carrier media (polypropylene, low-density polyethylene-polypropylene, poly, polyurethane foam-polypropylene) to understand the efficiency of MBBR to treat naphthalene-containing wastewater.

Shokoohi et al. (2017) investigated the effect of filling percentage (30%, 50%, and 70%) of carrier media and comparative evaluation of different carrier media (Kalden K1 packing material and lightweight expanded clay) on the performance of MBBR. Yadu et al. (2020) analyzed the use of mixed microbial culture from local indigenous sources for efficient removal of naphthalene, which enhances the sustainability of the MBBR technique. The reduction of head loss in MBBR, less sludge bulking problem, use of indigenous bacterial culture, and no sludge recirculation reduces the

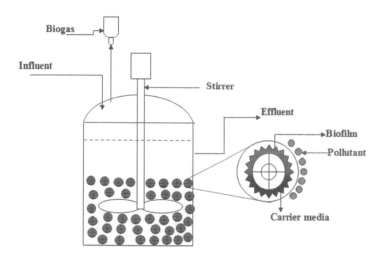

FIGURE 14.8   Schematic representation of an anaerobic moving bed biofilm bioreactor.

TABLE 14.3

Biodegradation of different pollutants using a moving bed biofilm reactor system

Industrial source	Targeted Pollutants	Configuration of bioreactor	Environmental condition	References
Petroleum refinery wastewater	Phenol, sulfide, crudeoil, nitrate, ammonia	Sequential series operated MBBR	Anoxic-Aerobic	Mallick and Chakraborty, 2020
Textile wastewater	Di-phthalate pentachlorophenol	Sequencing batch biofilm reactor	Anaerobic/ Aerobic	Yakamercan and Aygun, 2020
Coke-oven wastewater	Thiocyanate, cresol, pyridine, phenol	Sequential continuous moving bed bioreactor	Anaerobic- Anoxic- Aerobic	Sahariah et al., 2016
Hospital wastewater	BOD and COD	Continuous MBBR	Aerobic	Shokoohi et al., 2017
Petrochemicals	Napthalene	MBBR	Aerobic	Sonwani et al., 2019

operational cost. The biogas produced during the anaerobic process can be further used as an alternative energy source.

14.3.4 MICROBIAL FUEL CELL

A microbial fuel cell is an emerging and environmentally sustainable treatment technique in which the microbes convert the chemical energy produced by the oxidation of organic/inorganic compounds present in the wastewater into electrical energy. MFC consists of two chambers, anodic and cathodic, which are separated by a proton exchange membrane. The microbes present in the anodic chamber metabolize the substrate present in the wastewater and produce an electron and proton. The electron transfers through the circuit and the proton is transfer to the cathode via a semipermeable membrane. The electron and proton are consumed by the electron acceptor for the reduction reaction and produce clean water. The working principle of an MFC is depicted in Figure 14.9. The MFC has been used to treat many toxic pollutants from wastewater. Since it has the potential for the production of sustainable energy, many recent advancement have been carried out to scale up its application in the field. Traditional MFC has some drawbacks, such as use of different carbonaceous material for the anode having less surface area, higher cost of modified composite anodic material, production of less current density with two-dimensional materials, and salt precipitation in the cathode hindering the real scale application of MFC (Gajda et al., 2018; Sonwane et al., 2017). Application of integrated or stacked MFC systems for large-scale wastewater treatment to achieve sufficient electrical output is one option to scale up the MFC. Das et al. (2020) investigated about six MFCs stacked together to treat 720 L of sanitary wastewater; they achieved upto 87±7% of chemical oxygen demand (COD) reduction within 36 hours of HRT and produced 61 mW of power.

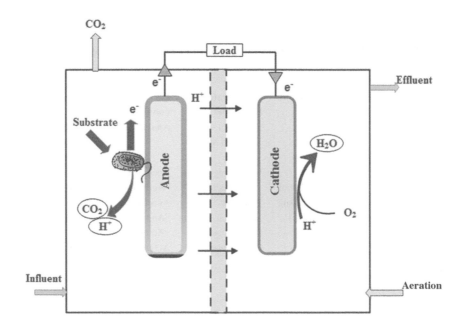

FIGURE 14.9   Schematic representation of a microbial fuel cell.

Many research works are based on the application of a graphene-based electrode to produce high current density (Sonwane et al., 2017). Wu et al. (2019) used graphene oxide-modified carbon cloth to check the removal of Cu (II).

Life cycle assessment is done to know the impact of MFC on the environment during its whole life cycle. Many researchers have also assessed MFC as per the life cycle assessment framework given by ISO 14040: 2006 (Corbella et al., 2017; Foley et al., 2010). Corbella et al. (2017) discussed the life cycle assessment's impact of constructed wetland coupled with the MFCs by using different anodic material, such as gravel and graphite, which is further compared with the conventionally constructed wetland. It was found that the constructed wetland integrated with the graphite-based anode MFC is the most environmentally friendly approach, which can replace the constructed wetland, but it is economically not sound. Similar finding on the life cycle assessment of MFCs is also observed by Foley et al. (2010). They explained the benefits of MFCs—less material for construction and replacement of fossil fuel by generating electric energy—along with some drawbacks, such as the selection of cost-effective material for the anode, which is a topic for future research in this field.

14.3.5 Hybrid Treatment System

A hybrid system is the combination of two or more treatment systems joined together to achieve a high quality of effluent by saving energy. The hybrid treatment system can be of different types, such as a combination of different biological processes or a combination of physical operations and biological processes. The main objective behind the application of a hybrid treatment system is to reuse the reclaimed waste-water and produce bioenergy to replace the use of fossil fuel. Many researchers have combined the number of bioreactor units (such as anoxic-aerobic, anaerobic-anoxic-aerobic, and anaerobic-aerobic) to treat different industrial wastewaters that contain several pollutants (Mallick and Chakraborty, 2020; Sahariah et al., 2016). Each reactor used in the hybrid treatment process, operated in different environments, responded well to treat the particular pollutant present in the wastewater. Therefore, the whole hybrid bioreactor system shows better efficiency as compared to the stand-alone processes. Sharma and Phillip (2016) conducted a comparable study on hybrid treatment processes (i.e., coagulation-biological treatment-photo degradation) along with biological treatment process to treat the real coke-oven wastewater and found that the combination of physicochemical, biological, and photodegradation processes were able to remove 92% of COD from real coke-oven wastewater. The treatment of COD and color from the tannery effluent by the hybrid membrane bioreactor, which is comprised of electrocoagulation-activated sludge process and membrane bioreactor, is investigated by Suganthi et al. (2013) to avoid the fouling characteristic of the membrane. The performance of the hybrid MBR is compared with the MBR and membrane filtration and is achieved by the hybrid membrane treatment process (Suganthi et al., 2013).

14.4 CONCLUSION AND FUTURE SCOPE

Many challenges still remain unresolved in using the advanced bioreactor in real field applications to increase bioproduct output. The use of a single reactor to treat multiple components is quite challenging task. The process parameters, such as F/M ratio, HRT, organic loading rate, temperature, and pH, should be analyzed properly to achieve better performance for any bioreactor system. The conventional bioreactor system needs to be upgraded by the advanced biological technique to achieve better effluent quality. The future research work for wastewater treatment should be in the direction of resource recovery and must have some realistic approach. The life cycle assessment gives an idea about the direction for the development of any bioreactor system, which can further scale up to enhance the application perspective. Future research needs to be projected on the following points by considering the cost of the operation: suitability of the right and low-cost solvent for two-phase portioning reactors, development of 3D and low-cost anodic material, and use of hybrid MFCs for scaling up the MFCs. More research work should be carried out on the use of potential microbes collected from indigenous sources. Modifications on membrane bioreactor design or the use of an energy sustainable hybrid system to avoid biofouling in the MBR system will provide value addition towards its application on a large scale.

ACKNOWLEDGMENTS

The authors would like to acknowledge SERB (DST), Government of India, for permitting the authors to use the information on project SB/EMEQ-107/2014 to make this chapter.

REFERENCES

Ahmed, M. B., Zhou, J. L., Ngo, H. H., Guo, W., Thomaidis, N. S., Xu, J. (2017). Progress in the biological and chemical treatment technologies for emerging contaminant removal from wastewater: a critical review. *Journal of Hazardous Material*, 323, 74–298.

Angelucci, D. M., Stazi, V., Daugulis, A. J., Tomei, M. C. (2017). Treatment of synthetic tannery wastewater in a continuous two-phase partitioning bioreactor: Biodegradation of the organic fraction and chromium separation. *Journal of Cleaner Production*, 152, 321–329.

Bakke, T., Klungsøyr, J., Sanni, S. (2013). Environmental impacts of produced water and drilling waste discharges from the Norwegian offshore petroleum industry. *Marine Environmental Research*, 92, 154–169.

Barthe, P., Chaugny, M., Roudier, S., Sancho, L. D. (2015). *EU Best Available Techniques (BAT): Reference Document forthe Refining of Mineral Oil and Gas*. Publications Office of the European Union, Luxembourg.

Baskaran, D., Paul, T., Pakshirajan, K., Krithivasan, M., Devanesan, M. G., Rajamanickam, R. (2020). Batch degradation of trichloroethylene using oleaginous Rhodococcus opacus in a two-phase partitioning bioreactor and kinetic study. *Bioresource Technology Reports*, 11, 100437.

Berkessa, Y. W., Yan, B., Li, T., Jegatheesan, V., Zhang, Y. (2020). Treatment of anthraquinone dye textile wastewater using anaerobic dynamic membrane bioreactor: Performance and microbial dynamics. *Chemosphere*, 238, 124539.

Abuabdou, S. M., Ahmad, W., Aun, N. C., Bashir, M. J. (2020). A review of anaerobic membrane bioreactors (AnMBR) for the treatment of highly contaminated landfill leachate and biogas production: effectiveness, limitations and future perspectives. *Journal of Cleaner Production*, 255, 120215.

Zhao, W., Zhang, Y., Lv, D., Wang, M., Peng, Y., Li, B. (2016). Advanced nitrogen and phosphorus removal in the pre-denitrification anaerobic/anoxic/aerobic nitrification sequence batch reactor (pre-A2NSBR) treating low carbon/nitrogen (C/N) wastewater. *Chemical Engineering Journal*, 302, 296–304.

Dorado, A. D., Dumont, E., Muñoz, R., Quijano, G. (2015). A novel mathematical approach for the understanding and optimization of two-phase partitioning bioreactors devoted to air pollution control. *Chemical Engineering Journal*, 263, 239–248.

Gargouri, B., Karray, F., Mhiri, N., Aloui, F., Sayadi, S. (2011). Application of a continuously stirred tank bioreactor(CSTR) for bioremediation of hydrocarbon-rich industrial wastewater effluents. *Journal of Hazardous Materials*, 189(1–2), 427–434.

Ghazani, M. T., Taghdisian, A. (2019). Performance evaluation of a hybrid sequencing batch reactor under saline and hyper saline conditions. *Journal of Biological Engineering*, 13, 64.

Gui, X., Xu, W., Cao, H., Ning, P., Zhang, Y., Li, Y., Sheng, Y. (2019). A novel phenol and ammonia recovery process for coal gasification wastewater altering the bacterial community and increasing pollutants removal in anaerobic/anoxic/aerobic system. *Science of the Total Environment*, 661, 203–211.

Jeon, C. O., Madsen, E. L. (2013). In situ microbial metabolism of aromatic hydrocarbon environmental pollutants. *Current Opinion in Biotechnology*, 24(3), 474–481.

Kanaujiya, D. K., Paul, T., Sinharoy, A., Pakshirajan, K. (2019). Biological treatment processes for the removal of organic micropollutants from wastewaters: a review. *Current Pollution Reports*, 5, 112–128.

Lee, E. Y., Jun, Y. S., Cho, K. S., Ryu, H. W. (2002). Degradation characteristics of toluene, benzene, ethylbenzene, and xylene by *Stenotrophomonasmaltophilia* T3-C. *Journal of Air Waste Management Association*, 52, 400–406.

Lu, C., Lin, M.-R., Chu, C. (2002). Effects of pH, moisture, and flow pattern on trickle bed air biofilter performance for BTEX removal. *Advance Environmental Research*, 6, 99–106.

Mahanty, B., Pakshirajan, K., Dasu, V. V. (2008). Biodegradation of pyrene by Mycobacterium frederiksbergense in a two-phase partitioning bioreactor system. *Bioresource Technology*, 99(7), 2694–2698.

Mohan, S. V., Rao, N. C., Prasad, K. K., Madhavi, B. T. V., Sharma, P. N. (2005). Treatment of complex chemical wastewater in a sequencing batch reactor (SBR) with an aerobic suspended growth configuration. *Process Biochemistry*, 40, 1501–1508.

Morrish, J. L. E., Daugulis, A. J. (2008). Improved reactor performance and operability in the biotransformation of carveol to carvone using a solid–liquid two-phase partitioning bioreactor. *Biotechnol Bioenginering*, 101, 946–956.

Mudliar, S., Giri, B., Padoley, K., Satpute, D., Dixit, R., Bhatt, P., Pandey, R., Juwarkar, A., Vaidya, A. (2010). Bioreactors for treatment of VOCs and odours: A review. *Journal of Environmental Management*, 91, 1039–1054.

Muñoz, R., Daugulis, A. J., Hernández, M., Quijano, G. (2012). Recent advances in two-phase partitioning bioreactors for the treatment of volatile organic compounds. *Biotechnology Advances*, 30(6), 1707–1720.

Naik, M. G., Duraphe, M. D. (2012). Review paper on-parameters affecting bioremediation. *International Journal of Life Science and Pharma Research*, 2(3), L77-L80.

Narayanan, C. M., Narayan, V. (2019). Biological wastewater treatment and bioreactor design: a review. *Sustainable Environment Research*, 29, 33.

Bonvillani, P., Ferrari, M. P., Ducros, E. M., Orejas, J. A. (2006). Theoretical and experimental study of the effects of scale-up on mixing time for a stirred-tank bioreactor. *Brazilian Journal of Chemical Engineering*, 23(1), 1–7.

Quintero, J. C., Lu-Chau, T. A., Moreira, M. T., Feijoo, G., Lema, J. M. (2007). Bioremediation of HCH present in soil by the white rot fungus *Bjerkanderaadusta* in a slurry batch bioreactor, *International Biodeterioration& Biodegradation*, 60(4), 319–326.

San-Valero, P., Gabaldón, C., Penya-Roja, J. M., Quijano, G. (2017). Enhanced styrene removal in a two-phase partitioning bioreactor operated as a biotrickling filter: Towards full-scale applications. *Chemical Engineering Journal*, 309, 588–595.

Schmidt, T., Anderson, W. A. (2017). Biotrickling filtration of air contaminated with 1-butanol. *Environments*, 4(3), 57.

Sharma, S. (2012). Bioremediation: Features, strategies and applications. *Asian Journal of Pharmacy and Life Science*, 2231, 4423.

Sheoran, A. S., Sheoran, V., Choudhary, R. P. (2010). Bioremediation of acid-rock drainage by sulfate-reducing prokaryotes: a review. *Minerals Engineering*, 23, 1073–1100.

Shieh, W. K., Keenan, J. D. (1986). Fluidized bed biofilm reactor for wastewater treatment. Bioproducts. *Advances in Biochemical Engineering/Biotechnology*, 33, 131–69.

Singh, R. L., Singh, R. P. (2019). Advances in biological treatment of industrial wastewater and their recycling for a sustainable future. *Applied Environmental Science and Engineering for a Sustainable Future*, Springer Nature, Singapore Pvt Ltd., Singapore.

Srivastava, J., Naraian, R., Kalra, S. J., Chandra, H. (2014). Advances in microbial bioremediation and the factors influencing the process. *International Journal of Environmental Science and Technology*, 11(6), 1787–1800.

Talha, M. A., Goswami, M., Giri, B. S., Sharma, A., Rai, B. N., Singh, R. S. (2018). Bioremediation of Congo red dye in immobilized batch and continuous packed bed bioreactor by *Brevibacillus parabrevis* using coconut shell bio-char. *Bioresource Technology*, 252, 37–43.

Tekere, M., Mswaka, A. Y., Zvauya, R., Read, J. S. (2001). Growth, dye degradation and ligninolytic activity studies on Zimbabwean white rot fungi. *Enzyme and Microbial Technology*, 28(4–5), 420–426.

Turovskiy, I. S., Mathai, P. K. (2006). *Wastewater Sludge Processing*. Wiley, New York.

Vinardell, S., Astals, S., Peces, M., Cardete, M. A., Fernández, I., Mata-Alvarez, J., Dosta, J. (2020). Advances in anaerobic membrane bioreactor technology for municipal wastewater treatment: A 2020 updated review. *Renewable and Sustainable Energy Reviews*, 130, 109936.

Wu, P., Jiang, L. Y., He, Z., Song, Y. (2017). Treatment of metallurgical industry wastewater for organic contaminant removal in China: Status, challenges, and perspectives. *Environmental Science: Water Research & Technology*, 3, 1015–1031.

Mallick, S. K., Chakraborty, S. (2019). Bioremediation of wastewater from automobile service station in anoxic-aerobic sequential reactors and microbial analysis. *Chemical Engineering Journal*, 361, 982–989.

Sharma, N. K., Philip, L. (2014). Effect of cyanide on phenolics and aromatic hydrocarbons biodegradation under anaerobic and anoxic conditions. *Chemical Engineering Journal*, 256, 255–267.

Sharma, N. K., Philip, L., Murty, B. S. (2018). Aerobic degradation of complex organic compounds and cyanides in coke oven wastewater in presence of glucose. In *Urban Ecology, Water Quality and Climate Change* (pp. 293–304). Springer, Cham.

Mallick, S. K., Chakraborty, S. (2020). Bioremediation of hydrocarbon containing wastewater in anoxic-aerobic sequential reactors. *Environmental Technology*, 41(22), 2884–2897.

Bonvillani, P., Ferrari, M. P., Ducrós, E. M., Orejas, J. A. (2006). Theoretical and experimental study of the effects of scale-up on mixing time for a stirred-tank bioreactor. *Brazilian Journal of Chemical Engineering*, 23(1), 1–7.

Neculita, C. M., Zagury, G. J., Bussière, B. (2007). Passive treatment of acid mine drainage in bioreactors using sulfate-reducing bacteria: Critical review and research needs. *Journal of Environmental Quality*, 36(1), 1–16.

Wu, C., Li, W., Xu, P., Li, S., Wang, X. (2018). Hydrophobic mixed culture for 1, 2-dichloroethane biodegradation: Batch-mode biodegradability and application performance in two-phase partitioning airlift bioreactors. *Process Safety and Environmental Protection*, 116, 405–412.

Song, W., Xu, D., Bi, X., Ng, H. Y., Shi, X. (2020). Intertidal wetland sediment as a novel inoculation source for developing aerobic granular sludge in membrane bioreactor treating high-salinity antibiotic manufacturing wastewater. *Bioresource Technology*, 314, 123715.

Leyva-Díaz, J. C., Monteoliva-García, A., Martín-Pascual, J., Munio, M. M., García-Mesa, J. J., Poyatos, J. M. (2020). Moving bed biofilm reactor as an alternative wastewater treatment process for nutrient removal and recovery in the circular economy model. *Bioresource Technology*, 299, 122631.

Sonwani, R. K., Swain, G., Giri, B. S., Singh, R. S., Rai, B. N. (2019). A novel comparative study of modified carriers in moving bed biofilm reactor for the treatment of wastewater: Process optimization and kinetic study. *Bioresource Technology*, 281, 335–342.

Shokoohi, R., Asgari, G., Leili, M., Khiadani, M., Foroughi, M., & Hemmat, M. S. (2017). Modelling of moving bed biofilm reactor (MBBR) efficiency on hospital wastewater (HW) treatment: A comprehensive analysis on BOD and COD removal. *International Journal of Environmental Science and Technology*, 14(4), 841–852.

Yadu, A., Sahariah, B. P., Anandkumar, J. (2020). Process optimization and comparative study on naphthalene biodegradation in anaerobic, anoxic, and aerobic moving bed bioreactors. *Engineering Reports*, 2(3), e12127.

Gajda, I., Greenman, J., Ieropoulos, I. A. (2018). Recent advancements in real-world microbial fuel cell applications. *Current Opinion in Electrochemistry*, 11, 78–83.

Sonawane, J. M., Yadav, A., Ghosh, P. C., Adeloju, S. B. (2017). Recent advances in the development and utilization of modern anode materials for high performance microbial fuel cells. *Biosensors and Bioelectronics*, 90, 558–576.

Das, I., Ghangrekar, M. M., Satyakam, R., Srivastava, P., Khan, S., Pandey, H. N. (2020). On-site sanitary wastewater treatment system using 720-L stacked microbial fuel cell: Case study. *Journal of Hazardous, Toxic, and Radioactive Waste*, 24(3), 04020025.

Wu, Y., Wang, L., Jin, M., Kong, F., Qi, H., Nan, J. (2019). Reduced graphene oxide and biofilms as cathode catalysts to enhance energy and metal recovery in microbial fuel cell. *Bioresource Technology*, 283, 129–137.

Corbella, C., Puigagut, J., Garfí, M. (2017). Life cycle assessment of constructed wetland systems for wastewater treatment coupled with microbial fuel cells. *Science of the Total Environment*, 584, 355–362.

Foley, J. M., Rozendal, R. A., Hertle, C. K., Lant, P. A., Rabaey, K. (2010). Life cycle assessment of high-rate anaerobic treatment, microbial fuel cells, and microbial electrolysis cells. *Environmental Science & Technology*, 44(9), 3629–3637.

Sahariah, B. P., Anandkumar, J., Chakraborty, S. (2016). Treatment of coke oven wastewater in an anaerobic–anoxic–aerobic moving bed bioreactor system. *Desalination and Water Treatment*, 57(31), 14396–14402.

Sharma, N. K., Philip, L. (2016). Combined biological and photocatalytic treatment of real coke oven wastewater. *Chemical Engineering Journal*, 295, 20–28.

Suganthi, V., Mahalakshmi, M., Balasubramanian, B. (2013). Development of hybrid membrane bioreactor for tannery effluent treatment. *Desalination*, 309, 231–236.

Varjani, S., Joshi, R., Srivastava, V. K., Ngo, H. H., Guo, W. (2020). Treatment of wastewater from petroleum industry: current practices and perspectives. *Environmental Science and Pollution Research*, 27(22), 27172–27180.

Metcalf & Eddy 2003 *Wastewater Engineering: Collection, Treatment, Disposal*. McGraw-Hill Series in Water Resources and Environmental Engineering, McGraw-Hill, New York, USA.

Yakamercan, E., Aygün, A. (2020). Anaerobic/aerobic cycle effect on di (2-ethylhexyl) phthalate and pentachlorophenol removal from real textile wastewater in sequencing batch biofilm reactor. *Journal of Cleaner Production*, 273, 122975.

Part VI

Integrated Technologies
for Water Treatment

15 Sustainable Technologies for Recycling Greywater: A Shift Towards Decentralized Treatment

Anupam Mukherjee and Anirban Roy*
Water-Energy Nexus Lab, Department of Chemical
Engineering, BITS Pilani Goa Campus, Goa, India

Aditi Mullick and Siddhartha Moulik
Cavitation and Dynamics Lab, Department of Process
Engineering & Technology Transfer, CSIR-Indian
Institute of Chemical Technology, Hyderabad, India

CONTENTS

15.1 INTRODUCTION

The emerging issue of water scarcity has motivated researchers worldwide to develop sustainable management of water resources. Greywater (GW) is recognized as a promising stream produced to a large extent (if the cumulative number is considered) from household activities; a graphical representation of the same is shown in Figure 15.1. GW constitutes about 60–70% of total household water consumption and is considered to be low-grade wastewater in terms of basic water quality parameters (Al-Jayyousi 2003; Edwin, Gopalsamy, and Muthu 2014; Eriksson et al. 2002).

FIGURE 15.1　Venn diagram of greywater system.

The characteristic features of GW are discussed in a later section. Recycling and reuse of such streams are techno-economically feasible and can contribute to progressive, sustainable water management. Reuse of GW has two significant benefits: (i) it will reduce the need for freshwater demand, which will significantly reduce household water bills, and (ii) it will reduce the amount of wastewater entering on-site treatment, thus reducing the load on centralized treatment plants (Boano et al. 2020). A recent development on the treatment of GW and reuse is on a small scale within the household or community in a decentralized fashion because of lower organic load and microbial content (Leong et al. 2017). A decentralized approach can also resolve the problem of a relatively "clean" stream of greywater getting mixed with blackwater and taken to municipal treatment plants. Since this practice is relatively new, only a few off-shelf technologies are available to be tested in a pilot study. The world map in Figure 15.2 shows the evolution of GW treatment and reuse practices.

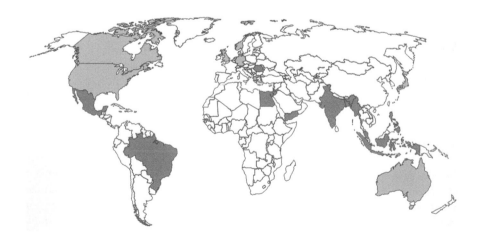

FIGURE 15.2　World map showing the prominent areas of greywater treatment—Green: developed countries; Red: developing countries.

The objective of this work is to highlight for the readers the potential impact of recycling GW along with the reuse standards and conventional treatment technologies. This work also highlights the shift towards decentralized treatment schemes with respect to several factors like environmental sustainability, operating costs, and so on, focusing on field case studies of two major developing countries. An overview of sustainable technologies focusing on the treatment and recycling of GW streams is also highlighted.

15.2 TRANSITION FROM A CENTRALIZED TO DECENTRALIZED WASTEWATER MANAGEMENT SYSTEM

The growing popularity of decentralized treatment schemes in developing countries and areas with low population density or dispersed households is well reported in the literature. A wastewater management system is illustrated in Figure 15.3. This section scientifically enlightens readers about the benefits of a transition from centralized to decentralization water systems: ecological benefits, techno-economic impacts, societal impacts, and increased sustainability.

   i. *Ecological benefits:* It may reduce surface water pollution by minimizing its rapid utilization (Libralato, Volpi Ghirardini, and Avezzù 2012). By implementing a decentralized system, an extensive infrastructure can be divided into several small segments, which may have a much higher potential to avoid catastrophic effects in a large area during any kind of system failure. Furthermore, the decentralized approach may reduce the enormous impact on the aquatic system by decreasing the effluent quantity for the application of segmented treatment (Fane and Fane 2005).

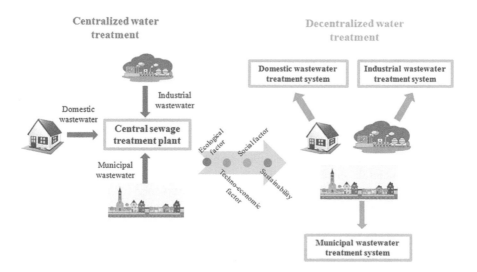

FIGURE 15.3 Graphical representation of centralized and decentralized treatment scheme.

ii. *Techno-economic impacts:* It is a well-known fact that centralized treatment systems mainly include construction and maintenance costs, whereas decentralized systems include most financial activities for treatment unit implementation. Moreover, this type of distributed small-scale system can be constructed gradually as per the requirements to meet the demand, which must reduce the higher cost of the centralized system with equivalent treatment capacity. Fane and Fane (2005) have suggested that most economical investments should be considered when organizing wastewater treatment plants in a cluster approach.

iii. *Societal impacts:* This decentralization method has more potential than a centralized one to get involved in the overall communal system as localized urban systems due to the distribution pattern of wastewater management (Fane and Fane 2005). However, the centralized water management scheme may satisfy the water supply pattern in a highly populated region. Still, it cannot meet the hope and potential of a water recycling and reuse concept (Libralato, Volpi Ghirardini, and Avezzù 2012). Hence, at this point, the decentralization concept is strongly favored, which not only meets the demand for water management shortcomings but is also useful for watering/dewatering and fertilizing/defertilizing of the green lands.

iv. *Increased sustainability:* The decentralized wastewater management structure is undoubtedly a sustainable approach for wastewater treatment. A decentralized system can efficiently reduce water and energy consumption by optimizing the water-energy nexus (Fane and Fane 2005). Likewise, it can explore the recyclability and reusability of the effluent components from household or industrial streams. Thus, the literature reveals that a decentralized wastewater management system can be turned into a necessity for the establishment of sustainable water resource management (Libralato, Volpi Ghirardini, and Avezzù 2012).

The case study by Chelleri et al. (2015) and Suriyachan et al. (2012) on the development of urban resilience and building of sustainability in two different cities—Mexico and Bangkok—are live examples of transitioning to a decentralized water management system. The exemplification mentioned above scientifically demonstrated the undeniable importance and promising technological advancement of decentralized wastewater management in the current scenario, especially in developing countries. In this regard, India is also considered a developing nation with a very high population and poor sanitation where proper freshwater supply and management system is necessary to implement every aspect (cost, technological, societal, sustainability) with an appropriate framework model policy development. In Agra, India, a decentralized wastewater treatment system was installed in association with the Centre for Urban Regional Excellence and maintained BOD levels below 50 ppm. By 2014, in Bengaluru, India, approximately 2000 apartments have installed a decentralized sewage treatment plant with an estimated capacity of 110 MLD. In this regard, treatment and recycling of GW in a decentralized fashion will be a turning point for sustainable water management.

Hence, the subsequent section will deal with the fundamentals of greywater and its treatment scheme concerning the reuse standards and, finally, the state-of-the-art treatment technologies.

15.3 GREYWATER: CHARACTERISTIC FEATURES

15.3.1 QUANTITATIVE CHARACTERISTICS

The volume of GW produced is directly proportional to the total water consumption, and the contaminants present in it vary from source to source, local sanitation, the lifestyle of the occupants, age and gender of the occupants, regional customs and practices, amount of chemical products used, geographical locations, and supply of water (Eriksson et al. 2002). Based on the water consumption per capita, the published literature is divided into two categories: high-income countries (HICs) and low-income countries (LICs) (Ghaitidak and Yadav 2013). The literature data reveals that the GW produced in HICs per capita is lower than the total water consumption of LICs. Data published by Li et al. (2009) suggests that the average volume of GW varies from 90–120 L/p/d in HICs and 20–30 L/p/d in LICs. Shaikh et al. (2019) reported 50 studies and concluded that the mean GW production is less than 100 lL/p/d in most LICs and greater than 100 L/p/d in HICs. It was interesting to note that people from HICs consume more water for bath and showers while those from LICs use it for other domestic activities in the kitchen. A report published from NEERI suggests that from the mean total water consumption of 50 L/p/d, GW of 29 L/p/d is produced (National Environmental Engineering Research Institute and WHO/UNICEF 2007). Abdul and Sharma (2007) surveyed seven Indian cities (Delhi, Mumbai, Kolkata, Hyderabad, Kanpur, Ahmadabad, and Madurai) and learned that the mean total water consumption is 91.56 L/p/d and the mean GW production is 57.77. Edwin et al. (2014) studied 32 samples of various sources of GW from 8 different families comprising people of different ages. The study revealed that the mean total water consumption is 114 L/p/d, and the mean GW production is 71. The survey from Boano et al. (2020) also showed that different sources of household activities contribute in their own way to GW production. Three primary locations for GW production are bathrooms, kitchens, and laundry, each contributing around 49%, 27%, and 24%, respectively. The volume of GW generated in the summer is higher than in the winter; thus, GW production also varies seasonally. Therefore, it is clear that proper studies need to be conducted on a particular region to understand GW's characteristics and its production to design a sustainable technique for recycling GW (Eriksson et al. 2002; Boano et al. 2020; Edwin, Gopalsamy, and Muthu 2014).

15.3.2 QUALITATIVE CHARACTERISTICS

As discussed above, the quantity of GW produced depends on multiple factors; the qualitative characteristics of GW are also highly variable. Though there is considerable variation in the quality of GW from different sources, it is clear that the kitchen and laundry have higher organic loading than the bathroom and showers (Edwin, Gopalsamy, and Muthu 2014; Boano et al. 2020; Eriksson et al. 2002). Greywater stored for more than 48 hours will result in exponential bacterial growth, depletion of dissolved oxygen, foul smell, and color, ultimately turning to blackwater. The physical characteristics of GW include temperature, suspended solids, color, and turbidity. The temperature of GW is slightly higher than the usual water and usually ranges from 18 to 30°C. A higher temperature range favors bacterial growth and causes precipitation (for example, $CaCO_3$) in transportation and storage systems. The reliable content of GW generally includes food particles and oils from kitchen sinks, hairs, and fibers from the laundry, and they are significant causes of turbidity (Eriksson et al. 2002). These particles and colloids result in the clogging of pumps and other systems installed for the treatment process. The concentration of suspended solids in mixed GW is expected to be lower than the usual streams from the kitchen and laundry and usually ranges from 30 to 300 mg/L. GW stored even for a short stint will be cloudy in nature (Pidou et al. 2007; Mukherjee, Mullick, Teja, et al. 2020). The chemical characteristics of GW include pH, alkalinity, electrical conductivity, hardness, biological and chemical oxygen demand (BOD, COD), nutrient content and heavy metals, and xenobiotic organic compounds (XOCs). Generally, the pH of GW ranges from 6.5 to 8.4, whereas streams originating from laundry have a pH range of 9.3–10. GW's alkalinity is indicated by the concentration of $CaCO_3$, which generally lies in the range of 20–340 mg/L. Additionally, the measurement of hardness and alkalinity is analogous to the measurement of turbidity and solid content and is a measure of the risk of clogging. The content of BOD and COD will measure the oxygen depletion rate due to the degradation of organics during transportation and storage (Boano et al. 2020). As reported by Eriksson et al. (2002) the heavy metal content and XOCs are of particular importance during the storage of GW. The qualitative characteristics of GW produced from various sources of domestic activities and the physicochemical properties of GW produced from different countries are reported in Tables 15.1 and 15.2.

15.4   GREYWATER TREATMENT TECHNOLOGIES

The extent of treatment and recycling of greywater is based on the "fit-for-purpose" mode, which depends on the quality of treated effluents for reuse and the treatment techniques available to achieve them (Pidou et al. 2007). The quality requirement for GW reuse depends on multiple factors ranging from the GW origin to possible human contact with the recycled water. The recycled GW should fulfill four essential criteria for reuse: (a) safety and hygiene, (b) aesthetics, (c) should be environment friendly, and (d) should be feasible economically.

The lack of proper guidelines for treated GW effluent quality has hampered the appropriate use of GW reuse as different reuse applications meet different water quality specifications. There are no unique regulations published globally to maintain the standards for GW reuse (Pidou et al. 2007; Shaikh, Mansoor Ahammed, and Sukanya Krishnan 2019), though most countries have their guidelines based on geographical locations and the required criteria for reuse are based on the microbial

TABLE 15.1

Qualitative characteristics of greywater

Parameter	Unit	Observed Values (6 Weeks of Data)				Literature (Li, Wichmann, and Otterpohl 2009)			Literature (Edwin, Gopalsamy, and Muthu 2014)		
		BITS Pilani Canteen	BITS Pilani Laundry	BITS Pilani Institute Cafe	Canteen Hyderabad	Bathroom	Laundry	Kitchen	Bathroom	Laundry	Kitchen
Turbidity	NTU	30.51–33.71	35.093–38.787	23.02–25.44	51.43–56.83	44–375	50–444	298	122.67	108.6	347.2
Chemical Oxygen Demand (COD)	mg/L	770–820	905–910	440–460	628.2–751.9	100–633	231–2,950	26–2,050	357.9	1,545.8	1,122.8
Biological Oxygen Demand (BOD)	mg/L	312–423	480–510	235–255	425–437	50–300	48–472	536–1,460	135	186.5	932.4
Total Organic Carbon (TOC)	mg/L	544.4–601.7	391.59–432.81	201.78–223.02	521.74–576.66	–	–	–	65	189.2	542
Total Suspended Solids (TDS)	mg/L	684–756	1178–1302	118.75–131.25	1,662.5–1,837.5	7–505	68–465	134–1,300	122.7	141.2	398.7
E. coli	CFU/100 mL	ND	ND	ND	ND	–	–	–	2.98–3.06	ND	ND
Total Coliform	CFU/100 mL	$3.14{\pm}0.15{*}10^4$	$3.7{\pm}0.18{*}10^4$	$3.4{\pm}0.17{*}10^4$	$4.2{\pm}0.21{*}10^4$	$10–2.4{*}10^7$	$200.5–7{*}10^5$	$>2.4{*}10^8$	3.95–6.28	3.04–5.6	3.38–5.11

TABLE 15.2
Physicochemical features of domestic greywater (data reproduced with permission from Boano et al. (2020)

Country	pH Min-Max (Avg)	TSS (mg/L) Min-Max (Avg)	BOD$_5$ (mg/L) Min-Max (Avg)	COD (mg/L) Min-Max (Avg)	TN (mg/L) Min-Max (Avg)	TP (mg/L) Min-Max (Avg)	TC (MPN/100 mL) Min-Max (Avg)	FC (MPN/100 mL) Min-Max (Avg)	E. coli (MPN/100 mL) Min-Max (Avg)
Australia	—	(74)	(104)	—	(5.3)	(3)	—	—	—
Canada	6.7–7.6	—	—	278–435	—	0.24–1.02	—	4.7E+04–8.3E+05	—
Egypt	6.05–7.96 (7)	70–202 (116)	220–375 (298.6)	301–557 (388)	—	8.4–12.1 (10.54)	—	—	—
France	6.46–7.48 (7.28)	23–80 (59)	85–155 (110)	176–323 (253)	—	—	1.7E+08–1.4E+09 (4.9E+08)	4.0E+03–5.7E+06 (1.3E+06)	—
Germany	(7.6)	—	(59)	(109)	(15.2)	(1.6)	—	(1.4E+05)	—
Ghana	5.00–9.00 (6.89)	192–414 (296.8)	87–301 (204.1)	207–1299 (643.8)	—	1–3 (2.3)	2.5E+06–4.9E+06 (3.7E+06)	0–6.9E+06(1.80E+06)	—
India	5.90–8.34 (7.4)	53.80–788.00 (337.2)	17.10–290.00 (244.2)	43.90–733.00 (705.4)	17.00–28.82 (17.8)	0.01–3.84	—	5.0E+01–1.2E+02	—
Indonesia	(6.85)	(18.00)	(8.50)	(15.00)	—	—	—	—	—
Israel	6.3–8.2	30–298	74–890	840–1,340	10–34.3	1.9–48	—	3.5E+04–4.0E+06	(5.0E+04)
Japan	—	—	—	(675)	(25.6)	(1.1)	—	—	8.5E3–1.2E4
Jordan	6.4–9	23–845	36–1,240	58–2,263	6.44–61	0.69–51.58	250–1.0E+07	1.3E+01–3.0E+05	(2.0E+05)
Malaysia	6.5–7.2 (6.85)	19–175 (114.4)	1.1–309 (188.85)	16–1,103 (328.9)	—	(4.5)	—	0–1.9E+06(2.9E+05)	0–6.7E+03 (1.1E+03)

Niger	(6.9)	–	(106)	–	–	–	–	–	–
Norway	(7.1)	(39)	(129)	(241)	(10.61)	(1.03)	(6.8E+06)	–	(4.9E+06)
Oman	6.7–8.5 (7.5)	11–505	25–562	58–486	–	–	2.0E+02–3.5E+03	(2.0E+02)	(2.0E+02)
Pakistan	(6.2)	(155)	(56)	(146)	–	–	–	–	–
Palestine	5.8–8.26 (7.8)	304–4,952 (1290)	407–512 (470.6)	863–1,240 (995)	111–322 (199)	5.8–15.16 (10.45)	–	–	–
Republic of Korea	(7.4)	(2180)	(255)	–	–	–	–	–	–
Spain	(7.39)	(336.09)	(130.32)	151–177 (405.11)	10.00–11.00 (16.17)	–	–	(1.0E+03)	(1.0E+03)
Sweden	(7.8)	–	(425)	(890)	(75)	(4.2)	–	(1.7E+05)	–
Taiwan	6.5–7.5	(29)	(23)	(55)	–	–	(5.1E+03)	–	–
Tunisia	(7.5)	(33)	(97)	(102)	(8.1)	–	–	–	–
Turkey	–	(54)	(91)	190–350	(7.6)	(7.2)	–	(1.1E+04)	–
United Kingdom	6.6–7.8	37–153	8.7–155	33–587	4.6–10.4	0.4–0.9	1.8E+03–2.2E+07	1.0E+01–2.2E+05	1.0E+01–3.9E+05
United Arab Emirates	–	–	–	1,020	–	–	(1.0E+07)	–	–
United States	(6.4)	(17)	(86)	–	(13.5)	(4)	–	–	–
Western Europe	6.1–9.6 (7.5)	20–361 (89)	20–756 (221)	25–1,583 (362)	3–75 (14)	0–11 (4)	–	–	(5.4E+02)
Yemen	(6)	(511)	(518)	(2000)	–	–	–	(1.9E+07)	–

TABLE 15.3

GW reuse standards from different countries across the globe (values are reported depending on available data)

Countries	Parameters for GW Reuse							
	pH	EC (µS/cm)	BODs (mg/L)	COD (mg/L)	TSS (mg/L)	Turbidity (NTU)	Anionic Surfactants (mg/L)	FIB (MPN/100 mL)
Africa	7–8	-	37–69	101–143	-	-	-	1.25–3.7
Australia	9	-	< 20	-	10	-	-	< 4
Canada	7–9	-	200	280	< 100	< 2	-	2–200 (fecal)
China	6–9	-	10–20	< 15	10–50	< 10	0.5–1	-
Costa Rica	6–9	~ 400	167	-	-	-	-	$1.5–4.6 \times 10^8$
France	6–9	-	70	180	60	Near clear	-	Almost none
Germany	6–9	-	20	-	Near free	Near clear	-	< 100 (fecal)
Great Britain	5–9.5	-	-	-	-	< 10	-	25 (E. coli)
India	6–9	10,000	< 30	< 250	< 200	–	< 10	-
Israel	-	-	10	-	10	-	-	< 1
Italy	6–9.5	3000	20	100	10	-	0.5 (total)	100 (E. coli)
Japan	6–9	-	< 20	-	-	Clear	30	< 1.105
Jordan	6–9	-	30–300	100–500	50–150	2–10	30–100	100 (E. coli)
Malaysia	6–9	-	129	212	76	-	-	-
Nepal	6–9	-	200	411	98	-	-	-
Niger	6.9	-	106	-	-	85	-	-
Pakistan	6.2	-	56	146	155	-	-	-
Palestine	6.7–8.35	1,585	590	1,270	1,396	-	-	3.1×10^4
Slovenia	7–9	-	-	200	80	-	1	-
Spain	-	-	10	-	3	2	-	-
United Kingdom	6.6–7.6	32.7	39–155	96–587	37–153	26.5–164	-	-
United States	6–9	-	< 10	-	-	< 5	-	< 0
Yemen	6	-	518	2,000	511	619	-	-

contents, organics, and suspended solids (Pidou et al. 2007). These regulations focus mainly on the potential human health risk associated. Table 15.3 portrays the standards for GW reuse from various countries across the globe. These standards are based on the associated human risks and the need to protect public health. These standards are laid down by some well-known organizational policies based on the type of reuse. The United States Environmental Protection Agency (EPA) has laid down a different type of reuse, including urban reuse, restricted area irrigation, agricultural reuse, construction, and indirect potable reuse (EPA 2004). Policies laid down by WHO indicate the decision-making criteria for GW reuse based particularly on public health, environmental protection, and food security (WHO 2006). Australia's national guidelines specify the policies for managing health risks, mainly chemical and microbial, in recycled GW (National Water Quality Management Strategy Australian 2006). The European guideline laid down by the EPA focuses on GW reuse in agriculture and integrating water reuse into water planning and management (European Commission 2001). BS8525 guidelines by the United Kingdom describe the policies for reusing GW for garden and lawn irrigation and toilet flushing (Great Britain The Water Supply (Water Fittings) Regulations 2022). The Central Pollution Control Board for India highlights two different standards for discharged GW into inland surface water and for irrigation (NOTIFICATION 1990). Evaluation of these standards for reuse, however, depends upon the type of treatment applied. The exact choice of the most appropriate technology depends on many factors ranging from scalability, end-use of water, economics, regional customs, and practices, etc. However, the potential of treating and reusing GW for sustainable applications dates back to the 1970s. The chronological development of technologies is shown in Figure 15.4. The literature reveals that for physcial treatment processes, membranes were excellent removers of suspended and dissolved solids. Still, the performance was limited in the removal of organic contents. The membrane's pore size played a crucial impact on treatment efficiency, and fouling became the main issue. However, the emerging process of a two-stage system followed by a disinfection step played a critical role in pathogenic removal. Though there was significant growth in the biological process, it was expensive, and thus, a cheaper alternative was looked for, and this is when extensive technologies comprising constructed wetlands like reed beds and ponds emerged. These systems were advantageous in terms of long-term storage and volume reduction of biosolids and minimized unwanted and harmful solids so that the recycled clean water could be utilized in secondary household water, recreation, gardening, and many other ways. MBRs were designed with microfiltration or a nanofiltration membrane combined with a suspended growth bioreactor to produce low sludge with a high-quality treated effluent. Thus, it is inferred that GW needs to be treated well so that the water can be reused without any health risk and adverse effects on aesthetics and the environment. The main objective of GW reclamation and reuse is to diminish the organics, solid fractions, and the microorganisms content present. The long-standing physical treatment process is limited in

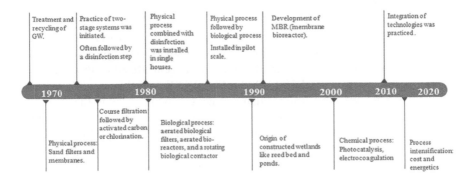

FIGURE 15.4 Chronological development of greywater treatment process.

its performance efficiency, and solo physical methods of treatment are not recommended for recycling greywater. It is seen that in comparison to the physical form of treatment, chemical processes can diminish the organic load and turbidity to a higher degree for low- to medium-strength greywater. However, for greywater, having a higher load of organic content, the chemical method is not feasible as it is both expensive and does not meet the standard for GW reuse. It can be used as a polishing step with existing physical treatment methods and is an attractive option for single households. The biological process for GW treatment proved to be a boon as aerobic biological processes achieved excellent performance in the reduction of organic loads and turbidity for medium- to high-strength greywater.

Nevertheless, it has demonstrated an inadequate removal of microorganisms and suspended solids demands a post-treatment step. Hydraulic retention times (HRTs) for the biological process were very high, ranging from one hour to three days with an average of 19 hours for treating greywater of high strengths with BOD concentrations ranging from 300–1200 mg.L^{-1} (Pidou et al. 2007). Though constructed wetlands are regarded as the most eco-friendly and cost-effective technology, it has a potential limitation in removing pathogens, requires a disinfection step, and demands ample space, making it not feasible to use in urban regions. MBRs achieve excellent performance in terms of removal efficiency without a post-treatment step and perfectly meet the GW reuse standards. However, MBRs are relatively expensive to install and operate. In particular, the main drawback lies in membrane fouling, high energy consumption, greenhouse gas production, and the bulk amount of sludge generation.

Therefore, the best performance for recycling greywater can be achieved by combining different types of treatment steps to ensure proper treatment of all fractions. The selection of appropriate treatment technology is crucial for diminishing human health risks, aesthetic issues, and hygienic for further uses (toilet flushing and irrigation). Compared worldwide, GW recycling practices have been deployed in both centralized and decentralized fashions in developed countries like the United States, Australia, the United Kingdom, and Japan (Libralato, Volpi Ghirardini, and Avezzù 2012; Pidou et al. 2007). India is still in its infancy in recycling GW. To date, there are only a limited number of companies (Greywater Solutions Pvt. Ltd., K. Pack System Pvt. Ltd.) in India that work on treating and recycling GW in a centralized fashion based on a combined treatment scheme with either a physical process (membranes, carbon filters), a chemical process (ozonation, advanced oxidation process) or with the biological process (constructed wetland). The GW technology providers in India are limited, and their solutions are precise. The existing GW treatment units in India are designed in a decentralized method capable of treating a capacity of around 500–1000 m^3/hr of GW, and the price of such units is about 2.5–6.5 lakh. This implies that if water can be recycled at the source and put to reuse, the system will be twice as effective in meeting the increasing demand for freshwater. Keeping in mind the gaining popularity of decentralized treatment methods in India, there is a clear need to overcome the challenges of existing treatment methods and develop a sustainable process for greywater treatment from Indian households.

15.5   EXPLORATION OF TECHNOLOGIES: SUSTAINABILITY, ENVIRONMENTAL ASSESSMENT, AND TECHNO-ECONOMIC FACTORS

The sustainability of technology depends on three primary aspects: technical, economical, and environmental. This section will describe in brief four technologies, which, if arranged in a particular sequence and combination, can be a promising solution to treat and recycle GW sustainably.

   i. *Ozonation:* Ozonation, a very well-known advanced oxidation process, mainly refers to the infusion of ozone into the water. The effectiveness of ozonation over chlorination was described by the EPA in 1999 as a sustainable treatment method, having almost zero impact on the environment. The cost of an ozone disinfection system depends on the manufacturer, the site, the capacity of the plant, and the characteristics of the wastewater to be disinfected. In our previous studies, the treatment of real-life GW from kitchen sinks was studied with the combination of ozonation coupled with hydrodynamic cavitation, and it was found that the operational cost and energy consumption for 75% reduction of COD was around $2.04/m^3 and 29.17 kWh/m^3, respectively (Mukherjee, Mullick, Teja, et al. 2020). Margot et al. (2013) investigated the combination of ozonation with sand filtration for the removal of micropollutants in wastewater, and the operational cost was estimated to be around 3–4 Swiss cents/m^3. Paulikienė et al. (2020) studied the influence of ozone technology on carrot loss reduction with outstanding efficiency; they discovered that the food sector does not have any vulnerability. Altmann et al. (2012) investigated the impact of ozonation on the endocrine activity of *Oryzias latipe,* in which they proved the technology as a suitable way of treatment without showing adverse effects on eco-toxicological screening.

   ii. *Hydrodynamic Cavitation (HC):* HC is one of the emerging advanced oxidation processes and is the most eco-friendly technology to reduce the pollution load in wastewater without the use of chemicals and the generation of sludge (Mukherjee, Mullick, Vadthya, et al. 2020). As discussed in the case of ozonation, studies were conducted to check the solo performance of HC, too, and it was found that the operational cost and energy consumption for 75% reduction of COD was estimated to be around $2.24/m^3 and 32.05 kWh/m^3, respectively (Mukherjee, Mullick, Teja, et al. 2020). Patil et al. (2020) studied the treatment of laundry wastewater using both hydrodynamic and ultrasonic cavitation and found HC to be better economically in reducing COD with an operational cost of $3.33/m^3. A COD reduction of 24.69 and 38.27% was achieved using only HC and HC+O$_3$. Braeutigam et al. (2012) studied the degradation of antibiotics in water using HC and found that there are no toxicological effects with this technology. Albanese et al. (2019) described the HC technology as a pathway towards sustainability and healthy technology. The technology was applied due to its superior efficiency in resource use, energy consumption, process yield, and exergy balance during the reduction and waste valorization process, having zero impact on human health. Again Maslak et al. (2011) used hydrodynamic cavitation and chlorination techniques synergistically and preferred the HC technique more due to its reduced impact on health, the environment, and fewer costs and less energy consumption during the disinfection of water.

iii. ***Membrane Technology:*** Membrane technology, regarded as one of the most conventional treatment techniques, is a physical form of the GW recycling process. Though it has multiple drawbacks like control of pore size and fouling, if it is used in combination with other treatment techniques, it produces beneficial results (Mukherjee, Lanjewar, et al. 2020). Ahn et al. (1998) studied the performance of both microfiltration and ultrafiltration membranes on the wastewater generated from hotels and concluded that almost 100% removal of turbidity and suspended solids was obtained. It was also concluded that microfiltration membranes provided higher permeate flux and were best suited for treating wastewater for secondary purposes in terms of energy consumption. Similar studies were performed by Ramon et al. (2004) with ultrafiltration and nanofiltration membranes for low strength greywater. Though there was a significant reduction of solid matters and turbidity, the BOD and COD reduction was not that high, and the performance of nanofiltration membranes was better than ultrafiltration membranes for the treatment of GW. Therefore, a sustainable water management membrane coupled with an advanced oxidation process may significantly reduce organic pollutants from GW.

iv. ***Adsorption:*** Adsorption by the packed bed is again recognized as one of the oldest and a familiar processes in the water treatment sector (Das et al. 2020). Adsorption also forms the conventional physical method of GW treatment. Janani et al. (2019) used corn cobs as an adsorbent in a packed bed for the treatment of GW and found that the percentage reduction of BOD, COD, TDS, and surfactant to be 40, 70, 40, and 35%, respectively, in 10 minutes. It was found that using corn cobs as an adsorbent proved beneficial because of their porous structure; they were economically feasible, too. Hernández-Leal et al. (2011) also studied the performance efficiency of packed beds filled with granular activated carbon on the removal efficiency of micropollutants from aerobically treated GW from 32 different sources in the Netherlands. The operation was carried out at various flow rates, and it was found that at a low flow rate of two beds of volume/hr., GAC was efficient in removing micropollutants to the concentration of 10–100 mg/L. It was also concluded that only aerobic treatment is not sufficient to meet the GW reuse standards, and hence advanced treatment with the packed bed as a post-treatment is essential. Apart from these, multiple conventional technologies of physical treatment, as discussed above, males this adsorption process techno-economically feasible as well as sustainable.

15.6   CONCLUSION

From this study, the following conclusions can be made:

1. GW discharged from any sources has a reasonable rate of biodegradability, with the kitchen having the highest sources of organic loads followed by laundry.
2. Decentralized treatment is gaining popularity in India for point of use of GW treatment.
3. Available technologies for GW treatment are limited in their performance efficiency, and a single treatment method does not meet the GW reuse standards.
4. There is a need for state-of-the-art technology comprising different treatment schemes for treating and reusing GW.

ACKNOWLEDGMENTS

Dr. Anirban Roy is thankful for the Additional Competitive Grant (Ref. GOA/ACG/2018–2019/Oct/08 dt. 30.10.2018) by BITS Pilani Goa for carrying out the work. Mr. Anupam

Mukherjee would also like to thank the BRIC Idea Exposition Grant [Ref. BIRAC Innovation

Centre/IKP Knowledge Park/GIM/Goa Cluster dt. 23.02.2020] for the partial support.

REFERENCES

Ahn, Kyu Hong, Ji Hyeon Song, and Ho Young Cha. 1998. "Application of Tubular Ceramic Membranes for Reuse of Wastewater from Buildings." *Water Science and Technology.* 38 (405): 373–382. doi:10.1016/S0273-1223(98)00521-6.

Al-Jayyousi, Odeh R. 2003. "Greywater Reuse: Towards Sustainable Water Management." *Desalination.* 156 (1–3): 181–192. doi:10.1016/S0011-9164(03)00340-0.

Albanese, Lorenzo, and Francesco Meneguzzo. 2019. "Hydrodynamic Cavitation Technologies: A Pathway to More Sustainable, Healthier Beverages, and Food Supply Chains." *Processing and Sustainability of Beverages.* 319–372. doi:10.1016/b978-0-12-815259-1.00010-0.

Altmann, Dominik, Heidemarie Schaar, Cordula Bartel, Dirk Louis P. Schorkopf, Ingrid Miller, Norbert Kreuzinger, Erich Möstl, and Britta Grillitsch. 2012. "Impact of Ozonation on Ecotoxicity and Endocrine Activity of Tertiary Treated Wastewater Effluent." *Water Research.* 46 (11): 3693–3702. doi:10.1016/j.watres.2012.04.017.

Boano, Fulvio, Alice Caruso, Elisa Costamagna, Luca Ridolfi, Silvia Fiore, Francesca Demichelis, Ana Galvão, Joana Pisoeiro, Anacleto Rizzo, and Fabio Masi. 2020. "A Review of Nature-Based Solutions for Greywater Treatment: Applications, Hydraulic Design, and Environmental Benefits." *Science of the Total Environment.* 711: 134731. doi:10.1016/j.scitotenv.2019.134731.

Braeutigam, Patrick, Marcus Franke, Rudolf J. Schneider, Andreas Lehmann, Achim Stolle, and Bernd Ondruschka. 2012. "Degradation of Carbamazepine in Environmentally Relevant Concentrations in Water by Hydrodynamic-Acoustic-Cavitation (HAC)." *Water Research.* 46 (7): 2649–2477. doi:10.1016/j.watres.2012.02.013.

Central Pollution Control Board, The Environment (Protection) Rules, Schedule-VI, Ministry of Environment and Forests, New Delhi, 2012, 1986.

Chelleri, L., T. Schuetze, and L. Salvati. 2015. "Integrating Resilience with Urban Sustainability in Neglected Neighborhoods: Challenges and Opportunities of Transitioning to Decentralized Water Management in Mexico City." *Habitat International.* 48: 122–130. doi:10.1016/j.habitatint.2015.03.016.

Das, Radha, Anupam Mukherjee, Ishita Sinha, Kunal Roy, and Binay K. Dutta. 2020. "Synthesis of Potential Bio-Adsorbent from Indian Neem Leaves (Azadirachta Indica) and Its Optimization for Malachite Green Dye Removal from Industrial Wastes Using Response Surface Methodology: Kinetics, Isotherms and Thermodynamic Studies." *Applied Water Science* 10 (5): 117. doi:10.1007/s13201-020-01184-5.

Edwin, Golda A., Poyyamoli Gopalsamy, and Nandhivarman Muthu. 2014. "Characterization of Domestic Gray Water from Point Source to Determine the Potential for Urban Residential Reuse: A Short Review." *Applied Water Science.* 4 (1): 39–49. doi:10.1007/s13201-013-0128-8.

Eriksson, Eva, Karina Auffarth, Mogens Henze, and Anna Ledin. 2002. "Characteristics of Grey Wastewater." *Urban Water.* 4 (1): 85–104. doi:10.1016/S1462-0758(01)00064-4.

Guidelines on Integrating Water Reuse into Water Planning and Management in the context of the WFD, The European Commission, Amsterdam, June 2016.

Fane, A. G., and Simon A. Fane. 2005. "The Role of Membrane Technology in Sustainable Decentralized Wastewater Systems." *Water Science and Technology.* 51 (10): 317–325. doi:10.2166/wst.2005.0381.

Ghaitidak, Dilip M., and Kunwar D. Yadav. 2013. "Characteristics and Treatment of Greywater: A Review." *Environmental Science and Pollution Research.* 20 (5): 2795–2809. doi:10.1007/s11356-013-1533-0.

Great Britian The Water Supply (Water Fittings) Regulations. 2022. "The Water Supply (Water Fittings) Regulations 1999." *The Water Supply (Water Fittings) Regulations 1999.*

Hernández-Leal, L., G. Zeeman, H. Temmink, and C. J. N. Buisman. 2011. "Grey Water Treatment Concept Integrating Water and Carbon Recovery and Removal of Micropollutants." *Water Practice and Technology.* 6 (2). doi:10.2166/wpt.2011.035.

Janani, T., J. S. Sudarsan, and K. Prasanna. 2019. "Grey Water Recycling with Corn Cob as an Adsorbent." In *AIP Conference Proceedings.* (2112). doi:10.1063/1.5112366.

Leong, Janet Yip Cheng, Kai Siang Oh, Phaik Eong Poh, and Meng Nan Chong. 2017. "Prospects of Hybrid Rainwater-Greywater Decentralised System for Water Recycling and Reuse: A Review." *Journal of Cleaner Production.* 142: 3014–3027. doi:10.1016/j.jclepro.2016.10.167.

Li, Fangyue, Knut Wichmann, and Ralf Otterpohl. 2009. "Review of the Technological Approaches for Grey Water Treatment and Reuses." *Science of the Total Environment.* 407 (11): 3439–3449. doi:10.1016/j.scitotenv.2009.02.004.

Libralato, Giovanni, Annamaria Volpi Ghirardini, and Francesco Avezzù. 2012. "To Centralise or to Decentralise: An Overview of the Most Recent Trends in Wastewater Treatment Management." *Journal of Environmental Management.* 94 (1): 61–68. doi:10.1016/j.jenvman.2011.07.010.

Margot, Jonas, Cornelia Kienle, Anoÿs Magnet, Mirco Weil, Luca Rossi, Luiz Felippe de Alencastro, Christian Abegglen, et al. 2013. "Treatment of Micropollutants in Municipal Wastewater: Ozone or Powdered Activated Carbon?" *Science of the Total Environment.* 461–462: 480–498. doi:10.1016/j.scitotenv.2013.05.034.

Maslak, Dominik, and Dirk Weuster-Botz. 2011. "Combination of Hydrodynamic Cavitation and Chlorine Dioxide for Disinfection of Water." *Engineering in Life Sciences.* 11 (4): 350–358. doi:10.1002/elsc.201000103.

Mukherjee, Anupam, Shubham Lanjewar, Ridhish Kumar, Arijit Chakraborty, Amira Abdelrasoul, and Anirban Roy. 2020. "Role of Thermodynamics and Membrane Separations in Water-Energy Nexus." In *Modeling in Membranes and Membrane-Based Processes*. 145–199. doi:10.1002/9781119536260.ch4.

Mukherjee, Anupam, Aditi Mullick, Ravi Teja, Pavani Vadthya, Anirban Roy, and Siddhartha Moulik. 2020. "Performance and Energetic Analysis of Hydrodynamic Cavitation and Potential Integration with Existing Advanced Oxidation Processes: A Case Study for Real Life Greywater Treatment." *Ultrasonics Sonochemistry*. 66: 105116. doi:10.1016/j.ultsonch.2020.105116.

Mukherjee, Anupam, Aditi Mullick, Pavani Vadthya, Siddhartha Moulik, and Anirban Roy. 2020. "Surfactant Degradation Using Hydrodynamic Cavitation Based Hybrid Advanced Oxidation Technology: A Techno Economic Feasibility Study." *Chemical Engineering Journal*. 398: 125599. doi:10.1016/j.cej.2020.125599.

National Environmental Engineering Research Institute and WHO/UNICEF. 2007. "Greywater Reuse in Rural Schools." *Wise Water Management*. https://www.ircwash. org/sites/default/files/Devotta-2007-Greywater.pdf

National Water Quality Management Strategy Australian (NWQMS). 2006. *National Guidelines for Water Recycling: Managing Health and Environmental Risks. National Water Quality Management Strategy.* **New South Wales, Australia, 2006.**

Patil, Vishal V., Parag R. Gogate, Akash P. Bhat, and Pushpito K. Ghosh. 2020. "Treatment of Laundry Wastewater Containing Residual Surfactants Using Combined Approaches Based on Ozone, Catalyst and Cavitation." *Separation and Purification Technology*. 239: 116594. doi:10.1016/j.seppur.2020.116594.

Paulikienė, Simona, Kęstutis Venslauskas, Algirdas Raila, Renata Žvirdauskienė, and Vilma Naujokienė. 2020. "The Influence of Ozone Technology on Reduction of Carrot Loss and Environmental IMPACT." *Journal of Cleaner Production*. 244: 118734. doi:10.1016/j.jclepro.2019.118734.

Pidou, Marc, Fayyaz Ali Mamon, Tom Stephenson, Bruce Jefferson, and Paul Jeffrey. 2007. "Greywater Recycling: Treatment Options and Applications." *Proceedings of the Institution of Civil Engineers: Engineering Sustainability*. 160 (3): 119–131. doi:10.1680/ensu.2007.160.3.119.

Ramon, Guy, Michal Green, Raphael Semiat, and Carlos Dosoretz. 2004. "Low Strength Graywater Characterization and Treatment by Direct Membrane Filtration." *Desalination*. 170 (3): 241–250. doi:10.1016/j.desal.2004.02.100.

Shaban, Abdul, and R. N. Sharma. 2007. "Water Consumption Patterns in Domestic Households in Major Cities." *Economic and Political Weekly*. 42 (23): 2190–2197.

Shaikh, Irshad N., M. Mansoor Ahammed, and M. P. Sukanya Krishnan. 2019. *Graywater Treatment and Reuse. Sustainable Water and Wastewater Processing*. Elsevier Inc. 19–54. doi:10.1016/B978-0-12-816170-8.00002-8.

Suriyachan, Chamawong, Vilas Nitivattananon, and Nurul T. M. N. Amin. 2012. "Potential of Decentralized Wastewater Management for Urban Development: Case of Bangkok." *Habitat International*. 36 (1): 85–92. doi:10.1016/j.habitatint.2011.06.001.

United States Environmetal Protection Agency. 2004. "Guidelines for Water Reuse. EPA/625/R-04/108." 26 (September): 252.

World Health Organization. 2006. "WHO Guidelines for the Safe Use of Wastewater, Excreta and Greywater: Volume I - Policy and Regulatory Aspects." *World Health*. doi:10.1007/ s13398-014-0173-7.2.

16 Advanced Membrane Bioreactor Hybrid Systems

Nirenkumar Pathak and Hokyong Shon
University of Technology Sydney (UTS), Broadway,
Ultimo, NSW, Australia

Saravanamuthu Vigneswaran*
Faculty of Engineering and IT, University of Technology
Sydney (UTS), Broadway, Ultimo, NSW, Australia

CONTENTS

16.1   INTRODUCTION

In this century, the natural reserves of clean freshwater are depleting at an alarming rate as a consequence of an extremely high increase in demand. The global usage of water is growing at a rate of more than twice the population growth in the last 10 decades. The global population reached almost 7.8 billion in 2021 and is projected to rise to 13.1 billion at the end of this century. In addition to the growing global population, increasing industrialization and high living standards are also contributing to the rise in water demand (Beddington, 2011, Suwaileh et al., 2020). Therefore, a severe water scarcity problem is inevitable. Without any major policy shifts, it is estimated that 2.3 billion additional people will be living in water-stressed regions with a 55% rise in water demand by 2050 (Leflaive et al., 2012). Water stressed regions are those where the annual water supply per person falls below 1,700 m³. When the water supply per person drops further to 1,000 m³, the region experiences water scarcity (Molden, 2013). However, the global water shortage problem causes serious consequences for public health and sanitation. As such 1.2 billion people are deprived of potable water and 2.6 million people suffer from proper sanitation. In addition, globally about 3,900 children die every day due to waterborne diseases (Shannon et al., 2010, Stefan, 2017).

The global freshwater reserve is only 2.5% of total natural water resources, whereas about 96.5% is seawater (Shiklomanov, 1993, Trenberth et al., 2007). Therefore, desalination techniques show promising options in minimizing the water shortage problem. However, at present, the available conventional seawater purification technologies are energy-intensive and the produced water is still beyond the affordability of people with lower incomes (Chekli et al., 2016, Ziolkowska, 2015). Reuse of impaired water can be another potential measure to address the water scarcity issue. However, the conventional treatment of wastewater effluent to produce high-quality water is also a high energy demanding process (Linares et al., 2014). Therefore, an alternative technology is urgently needed to economically recover freshwater from these unconventional sources for the growing global population. As reclaimed water use is increasing, its safety attracts growing attention, particularly with respect to the health risks associated with the wide range of organic micropollutants (OMPs) found in the reclaimed water (Ma et al., 2018). However, the ubiquitous presence of OMPs in reclaimed water and wastewater is often a major obstacle to water reuse (Luo et al., 2017, Zhang et al., 2017).

16.1.1 Occurrence, Fate, and Transport of OMPs in WWTPs and Impact on People and Their Environments

During the last decade, OMPs have been detected in water resources all over the world, and it has become a worldwide issue of great importance for environmental protection strategies (Bodzek and Konieczny, 2018). This is due to their potential to cause undesirable side effects on the ecosystem (Tran and Gin, 2017) and public health authorities, the whole industrial world, and the agricultural sector (Hamza et al., 2016, Priac et al., 2017). OMPs are derived from either anthropogenic activities, such as industrial effluents, discharges of treated effluents from domestic and hospital effluents, agricultural runoff, septic tanks, or natural activities. In addition, other anthropogenic sources include landfills, inappropriately disposed wastes, surface runoff, sewer overflow and sewer leaking (Hamza et al., 2016, Pal et al., 2014, Tran and Gin, 2017). It has been shown that even conventional wastewater treatment plants (WWTPs) are able to remove efficiently some OMPs, although there is still a significant group of compounds with a recalcitrant behaviour (Alvarino et al., 2018). Actually, current WWTPs are not designed to eliminate or degrade OMPs; therefore, many of these OMPs because of their persistence can pass through the treatment system and enter into the natural aquatic system (Asif et al., 2017b). The presence of some OMPs and their metabolites can inhibit the biological activity of microorganisms present in activated sludge and thus produce non-consistent and inadequate removal of OMPs by conventional treatment (Goh et al., 2015, Morrow et al., 2018). OMP removal techniques include adsorption on activated carbon, ultraviolet disinfection, and other advanced oxidation processes such as ozonation and use of hydrogen peroxide. The capital and operating cost and chemical sludge disposal are some of the issues associated with such treatment (Chtourou et al., 2018). Also, membrane-based processes such as MF, UF, NF, RO, and most recently FO and MD are employed for OMP removal (Pathak et al., 2020b). Further, the activated sludge process, when coupled with any of the above-mentioned membranes (most commonly MF/UF), the membrane bioreactor is a promising alternative in OMP removal as it offers higher, consistent, and comparatively cheaper removal (Besha et al., 2017, Calero-Díaz et al., 2017, Morrow et al., 2018, Wei et al., 2018).

Membrane bioreactor (MBR) is an attractive alternative for wastewater reuse applications. Recently, high-retention membrane bioreactor (HRMBR) systems gained more attention in wastewater treatment. This review examines recent developments in forward-osmosis MBR (FO-MBR) and membrane distillation bioreactor (MDBR) for OMPs removal. The MBR, OMBR, and MDBR technologies are compared. Finally, the life cycle assessment of MBR and advanced hybrid MBRs are then discussed.

16.2   MEMBRANE BIOREACTOR OPERATING CONDITIONS

MBR is a promising option in wastewater treatment as it generates pure permeate in terms of suspended solids, lack of microorganisms, nutrients and OMPs, lower space requirements, and a reduced sludge disposal cost as compared to conventional biological treatment. Moreover, it can easily accommodate unstable flow (Bui et al., 2016, Chtourou et al., 2018, Cornelissen et al., 2008, Luo et al., 2014b, Wang et al. 2011). In the beginning, the activated sludge process was coupled with side stream MF/UF membrane and then after membrane was directly submerged into the mixed liquor (Huang and Lee, 2015).

The physicochemical properties of OMPs such as hydrophobicity and biosorption of OMPs, microbial activity and biodegradation, molecular weight and functional groups of OMPs, and other major operating parameters such as biomass concentration and characteristics, hydrodynamic parameters of solids retention time (SRT), and hydraulic retention time (HRT), cometabolism and influence of the redox potential, mixed liquor pH, and mixed liquor temperature all affect OMP removal in MBR (Hai et al., 2018, Zheng et al., 2019, Zolfaghari et al., 2015). The hydrophilicity and hydrophobicity is an important physicochemical property for OMP removal, and hydrophobicity of an organic molecule is defined by the octanol-water partitioning coefficient (Kow) or the solid water partitioning coefficient (Kd) (Hai et al., 2018, Stevens-Garmon et al., 2011). The more hydrophilic OMP is retained in the water phase while more hydrophobic OMP is attached to the sludge surface (Pathak et al., 2020a). Moreover, as compared to negatively charged or neutral OMPs, the positively charged pharmaceutical class OMPs showed more affinity towards sludge adsorption (Hai et al., 2018, Joss et al., 2005). The combined effect between biosorption and biodegradation forms another important OMP removal mechanism realized in the presence of microorganisms and could achieve better OMP removal as higher biosorption provides longer retention time and further opportunities for biodegradation to occur (Stevens-Garmon et al., 2011). Low molecular weight OMPs could not be retained by MF membranes effectively. Actually, higher molecular weight OMPs can be better retained by membrane leads to higher removals by biodegradation in the MBR process. The higher molecular weight OMPs may possess more functional groups, and this provides opportunities for diverse microbial communities to target selective sites to commence biodegradation (Hai et al., 2018, Pathak et al., 2020a).

Higher mixed liquor suspended solids (MLSS) concentration, longer SRT, and smaller floc size of sludge particles are always favorable operating parameters for both sorption and biodegradation processes demonstrated by several studies (Alvarino et al., 2018, Kimura et al., 2007, Verlicchi et al., 2012, Zheng et al., 2019). As compared to the activated sludge process (ASP), the MBR can retain more biomass and allow to proliferate diverse microbial communities for OMPs removal (Verlicchi et al., 2012). Hydraulic retention time (HRT) is another significant operating parameter and it can influence food to microorganism ratio (F/M ratio) and organic loading rate (OLR) in the bioreactor. Too low of HRT negatively affects the biological process performance of the MBR process (Prasertkulsak et al., 2019). Moreover, in anoxic–aerobic MBRs, due to the recirculation of biomass under varying redox conditions of anoxic and aerobic MBRs, an entirely different biological environment has been realized that helps in improved removal of OMPs (Phan et al., 2016). Mixed liquor pH and temperature are other important operational parameters that affect MBR process performance. The average 15–20°C temperature of MBRs is suitable in cold countries and seasonal temperature variation can also affect process performance (Mert et al., 2018). However, the MBR report mentions that a 10–35°C temperature fluctuation did not affect OMP removal in MBRs (Verlicchi et al., 2012).

An aerobic process takes place in presence of oxygen while the anaerobic process does not require oxygen. An anaerobic process is suitable to treat high-strength water and it takes a long time to start up. Anaerobic microorganisms are more sensitive to shock loads (Mutamim et al., 2013). MBR for biogas as an alternative and renewable fuel production is still an emerging concept and limited industrial applications have employed an anaerobic membrane bioreactor (AnMBR) for effluent polishing (Neoh et al., 2016). It has been reported that about 98% of raw sewage chemical oxygen demand (COD) can be efficiently transformed into methane gas using AnMBR. The biogas produced from AnMBRs having a composition of more than 80% methane content means it can be used as a fuel. This methane composition is more favorable than that obtained through conventional anaerobic digesters that produce upto 65% methane, which can be attributed to shorter HRT in AnMBRs (Guo et al., 2016, Skouteris et al., 2012). However, methane production has a linear correlation with the methanogenesis step, which is the slow growth rate process and therefore methanogen can be possibly easy to wash out (Neoh et al., 2016). Monsalvo et al. (2014) kept mesophilic conditions (30°C) in an AnMBR as compared to the higher thermophilic range and very low 0.25 days HRT than the usual HRT of 1–25 days, which are favorable for the anaerobic process. However, methane production details were missing in this report. Song et al. (2016) investigated the effects of increased salt concentration on OMPs removal in an AnMBR. It has been reported that salt accumulation up to 15 g/L (as NaCl) adversely affected its performance in terms of methane production and hydrophilic OMPs. The authors further reported that salt accumulation had no pronounced effect on high removal of hydrophobic OMPs.

16.3   ADVANCED MEMBRANE BIOREACTOR HYBRID SYSTEMS

The hybrid MBR can produce better quality permeate, lessen membrane fouling, and thereby reduce cleaning cycles (Neoh et al., 2016). Nonetheless, the technology has certain drawbacks that need to be resolve for commercialization. For example, in an osmotic membrane bioreactor, draw solution accumulates into the feed tank and due to this reverse diffusion of draw salts salinity builds up in the reactor. It leads to concentration polarization, flux decline, and increases fouling propensity. In order to mitigate such issues, novel membranes with less fouling propensity and the development of bacterial consortia that can withstand hypersaline conditions need to be explored (Luo et al., 2014a). Table 16.1 compares conventional MBRs with two major advanced hybrid MBR systems, namely osmotic membrane bioreactors (OMBR) and membrane distillation bioreactors (MDBR), in their performance in wastewater treatment.

16.3.1   OSMOTIC MEMBRANE BIOREACTOR

The technology of osmotic membrane bioreactors (OMBRs) is employed in wastewater treatment systems and used to reclaim and reuse indirect and direct potable water sources (Achilli et al., 2009, Alturki et al., 2012, Van Huy Tran and Shon, 2020) by integrating the use of semi-permeable forward osmosis membranes (Nguyen et al., 2018, Li et al., 2016). OMBR achieves better permeate quality with less dissolved organic matters, less fouling tendency and higher reversibility of membrane fouling, high removal and rejection of organics by enhanced biodegradation, nitrogen, phosphorus and improved organic micropollutants, and low electrical utilities (Alturki et al., 2013, Jin et al., 2012, Li et al., 2016, Luo et al., 2018). Nevertheless, reverse salt flux (RSF) of draw solutes (DS) leads to salinity rise in the bioreactor, adversely affecting microorganisms exacerbated by deterioration in the OMBR's performance in terms of water flux and removal of organics, nutrients, and OMPs (Cicek et al., 1999). In forward osmosis and OMBR applications cellulose triacetate (CTA FO) (Achilli et al., 2009, Cornelissen et al., 2008, Sun et al., 2016) and TFC FO (thin film composite) (Luo et al., 2017, Morrow et al., 2018, Zhang et al., 2017) are the most commonly used FO membranes. In comparison to TFC FO (2–12 pH), the CTA FO membranes are more pH sensitive, meaning they can be operated in a narrow range of pH variations (4–10) as well as being more prone to bacterial attack in mixed liquor environments (Yip et al., 2010). In OMBR, the application membrane can be oriented either as active layer facing feed side (AL-FS) or active layer facing draw solution side (AL-DS) (Pathak et al., 2020b). AL-FS orientation is more preferred in OMBR applications based on concentration polarization and fouling propensity aspects. Osmotic process is driven by osmotic pressure difference and to achieve this inorganic (sodium chloride, magnesium chloride) and organic draw solutes (sodium acetate) are commonly used (Bowden et al., 2012). Draw solute screening and selection based on higher flux, lower RSF, and fouling properties are another major area of ongoing research in FO applications (Ansari et al., 2015, Bowden et al., 2012). Organic DS could be a more attractive option with regards to OMBRs due to less RSF and non-toxicity to microbes in the mixed liquor. However, less flux compared to inorganic DS and higher fouling

TABLE 16.1

Comparison for MBR, OMBR and MDBR in wastewater treatment (Barbosa et al., 2016, Mert et al., 2018, Neoh et al., 2016, Goh et al., 2015, Bharwada, 2011)

Parameters/ description	MBR	OMBR	MDBR
Membrane type	Low-pressure MF/UF membranes are employed. (MF MWCO = 1000 kDa) (UF MWCO = 10 - 100 kDa) Liquid (permeate water) inside the lumen.	Forward osmosis (FO) semi-permeable membranes are used. FO (MWCO = 0,1 - 2 kDa) Liquid (permeate water) inside the lumen.	Hydrophobic membrane distillation membrane (MD) is used. MD (MWCO < 150 Dalton Vapor phase (permeate) inside the lumen.
Removal mechanism	Size exclusion is principal removal mechanism.	Steric hindrance and electrostatic repulsion are principal removal mechanisms.	Steric hindrance is the principal removal mechanism.
Pressure	Hydraulic pressure (50–70 bar) is a driving force.	Natural osmotic pressure (27 bar) is a driving force.	Vapor pressure gradient (heat transfer) is driving force.
Temperature	Normally operates at ambient conditions.	Normally operates at ambient conditions.	Normally operates at high temperature (30–80°C).
Effect of operating condition on process performance	Normal operating conditions does not have a significant effect on the DO level in a bioreactor. Hydrophilic MF or UF membranes are preferably employed. Comparatively less rejection is obtained than FO.	DO level reduces and adversely affects the microbial community due to salinity build-up in bioreactor with time. Hydrophilic FO membrane can achieve similar rejection as RO.	DO level reduces and adversely affects the microbial community by thermophilic conditions. Hydrophobic MD membrane can achieve similar rejection as RO.
Flux (LMH-Liters per square meter per hour)	MF/UF MBR operates at 10–25 LMH flux.	OMBR operates at 2–10 LMH flux. Lab-scale hollow fiber FO module can achieve up to 30 LMH initial flux during preliminary testing.	MDBR operates at 2–15 LMH flux. Wetting of membrane is major concern.
Process performance on TOC removal	In MBR 30%–75% TOC removal efficiencies can be achieved.	In the OMBR process the FO membrane with a 98% TOC removal efficiency allows the downstream RO to operate in longer cycles.	In MDBR process, the MD membrane with a 98% TOC removal efficiency.
Process performance on P removal	In MBR removal of P can be achieved by the addition of flocculants followed by larger particle flocs filtration or rejection through the membrane.	OMBR system rejects P more cost-effectively because the removal mechanism is size exclusion without flocculation.	MDBR system rejects P more cost-effectively because the removal mechanism is size exclusion without flocculation.

TABLE 16.1

Comparison for MBR, OMBR and MDBR in wastewater treatment (Barbosa et al., 2016, Mert et al., 2018, Neoh et al., 2016, Goh et al., 2015, Bharwada, 2011) (Cont.)

Parameters/ description	MBR	OMBR	MDBR
Concentration and temperature polarization and fouling	In MBR fouling is major issue and cleaning cost is high.	OMBR offers low fouling characteristics and less cleaning cost due to lack of hydraulic pressure across the membrane. Fouling is largely reversible. However, the salinity build-up is one of the major issues. CECP also adversely affects OMPs removal.	The OMBR offers ultra-low fouling characteristics and lower cleaning cost due to lack of hydraulic pressure across the membrane. However, temperature polarization is one of the issues.
Membrane process influence on economy	In MBR, the design has evolved to continually improve the economy of energy required for scouring, backwashing and aeration.	In OMBR, fine bubble diffusion for oxygen transfer and a longer interval between backwashing and cleaning should require less energy.	In MDBR, the waste heat source can be utilized, thus saving energy and minimizing GHG emission. MD utilizes waste heat directly with a heat exchanger.
Energy consumption (kWh/m³)	In MBRs total energy estimate is 4.2 kWh/m³ water treated.	In OMBR total energy estimate is 2.8 kWh/m³ water treated.	Electrical energy requirement for RO would increase as feed solution salinity increases whereas MD is only minimally affected by feed solution salinity.
OMP removal	MF/UF membrane of the MBR process is commercialized. Hydrophilic OMP removal is too low.	Membrane stability is major concern. CTA membrane can operate in narrow pH range and stability of membrane due to biodegradation being a concern.	Complete rejection of inorganic salts and OMPs. Ammonia and CO_2 can seep through MD membrane.
Sludge/ concentrate production	Less sludge yield as compared to CAS.	Concentrate disposal and relevant cost are disadvantages.	Concentrate disposal and relevant cost are disadvantages.

propensity on the membrane surface as it serves as food to bacteria are downsides for organic DS (Huang et al., 2018). FO membrane fouling can be alleviated by air scouring, physical cleaning, and osmotic backwashing techniques. As compared to pressurized MF/UF membranes, application FO membranes have less fouling propensity in absence of hydraulic pressure and even loose biofilm or inorganic scalant formation takes place (Pathak et al., 2018a).

Table 16.2 summarizes some recently published OMBR research studies for OMPs removal.

UF-OMBR performance was evaluated for the long term (505 days) for treating oil refinery effluent and two different draw solutes: sodium chloride (NaCl) and sodium acetate (CH_3COONa). UF membrane helped in salinity buildup mitigation with sodium acetate (5 times) and sodium chloride (10 times) in comparison to OMBR. However, process efficiency declined due to slow degradable or recalcitrant compounds. The raw refinery effluent indicated the presence of highly toxic and recalcitrant compounds, such as polyaromatic hydrocarbons (1413 cm^{-1}), nitrates (3277–3300 cm^{-1}), amide (1653 cm^{-1}), phenol (1248 cm^{-1}), and sulfur-containing groups (609 cm^{-1}). Acetate DS is more favored as compared to NaCl in refractory compound removal. NaCl DS achieved higher flux than acetate DS due to higher biofouling occurrence on the membrane in presence of organic DS (Moser et al., 2020). Yao et al. (2020) recently examined carbamazepine (CBZ) degradation in submerged OMBR. Authors reported very high COD and ammonia removal and 88.20%–94.45% removal of carbamazepine (CBZ). Further, it was reported that higher carbamazepine concentration was favorable in high COD and ammonia removal. The oxidation, hydroxylation, and decarboxylation were found dominant in CBZ degradation mechanisms in presence of Delftia as a predominant degradation species. Raghavan et al. (2018) investigated the removal of 12 antibiotics (500 ng/L) in OMBR in a 40-day period. They achieved very high removal of TOC (>98%) and ammonium (>97%). The antibiotic removal was observed from 77.7 to 99.8% as the FO membrane achieved >90% rejection. The biodegradation (16.6 to 94.4%) was the dominant removal mechanism followed by biosorption (2 to 30%). Certain antibiotics showed poor biodegradation such as ofloxacin, ciprofloxacin, and roxithromycin. Sulfathiazole, enrofloxacin, and chlortetracycline showed the highest removal via biodegradation at 94.4%, 90.2%, and 78.9% respectively.

Li et al. (2018) studied the performance of an anaerobic OMBR for the biodegradation and decolorization performance of a refractory acid dye, Lanaset red G.GR. Authors reported COD, color, and aniline rejection. COD removal by biodegradation decreased from 73 to 65% in 60 days. Similarly, color removal by biodegradation reduced from 41 to 30%. However overall 99.4 ± 0.1% COD removal and color removal (100%) in OMBR was achieved. The reduced biodegradation efficiency could be attributed to the increased soluble microbial products (SMP) and extracellular polymeric substances (EPS) concentration and salt accumulation adversely affected the bacterial community. However, at a later stage of operation, in FO permeate, dye intermediate molecules were observed including chromophoric groups like aniline. However, the authors noticed that the toxic and oxidative intermediate Aniline rejection by CTA FO membrane was only 50% and aniline like compounds concentration increased by 24% within 60 days. Kim et al. (2017) assessed the removal of three organic micropollutants employing side stream AnMBR combined with side stream CTA-FO membrane, in both cases the active layer facing feed solution (AL-FS) and active layer facing draw solution (AL-DS) mode. In AL-DS orientation severe flux decline was observed and this can be attributed to struvite thin layer formation on the membrane surface when DAP was used as a DS. In both AL-FS and AL-DS mode, DAP as a fertilizer DS outperformed other two fertilizer draw solutions in

TABLE 16.2

Summary of recently published OMBR studies

FO membrane	Draw solution	HRT (h)	SRT (d)	MLSS (g/l)	Water flux (LMH)	Bioreactor conductivity	OMP	Removal (%)	Reference
Plate-and-frame FO membrane Hydration Technology Innovations (HTI, USA) made of cellulose triacetate (CTA)	1 M NaCl	25.25	30	--	11.88	--	CBZ 100 (µg/L)	93.27 ±3.77%	(Yao et al., 2020)
Plate-and-frame FO membrane Hydration Technology Innovations (HTI, USA) made of cellulose triacetate (CTA)	1 M NaCl	25.25	30	--	11.88		CBZ 200 (µg/L)	88.20 ±3.27%	(Yao et al., 2020)
HTI-CTA FO membrane	0.75 M NaCl	30	70	3.5	7–5.5	2.5 (g/L)	Caffeine	94	(Pathak et al., 2018b)
							Atrazine	51	
							Atenolol	100	
Cellulose triacetate (CTA) FO membrane (obtained from Hydration Technologies Inc., Oregon, USA)	49.0 g/L NaCl	27.15	50	5.0	5.15	27.89 mS/cm	500 ng/L		(Raghavan et al., 2018)
							Sulfathiazole	98	
							Sulfamethazine	25	
							Trimethoprim	80	
							Norfloxacin	62	
							Ciprofloxacin	10	
							Lomefloxacin	70	
							Enrofloxacin	100	
							Ofloxacin	35	
							Tetracycline	40	
							Oxytetracycline	30	
							Chlortetracycline	80	
							Roxithromycin	30	

(Continued)

TABLE 16.2
Summary of recently published OMBR studies (Cont.)

FO membrane	Draw solution	HRT (h)	SRT (d)	MLSS (g/l)	Water flux (LMH)	Bioreactor conductivity	OMP	Removal (%)	Reference
Cellulose triacetate (CTA) forward osmosis (FO) membranes dye	1 M MgCl$_2$	48	--	4.8–10.3	9.63–3.65	6800 ± 105 µS cm−1	Lanaset red G. GR refractory acid dye 100 mg/L	100	(Li et al., 2018)*
Cellulose triacetate (CTA) FO Hydration Technology Innovations, HTI (Albany, OR, USA) Side stream AnMBR+FO (AL-FS)	1 M MAP	24	--	--	7.58	--	Caffeine Atenolol, Atrazine	95.3 99.5 96.6	(Kim et al., 2017)
Cellulose triacetate (CTA) FO Hydration Technology Innovations, HTI (Albany, OR, USA) Side stream AnMBR+FO (AL-FS)	1 M DAP	24	--	--	7.35	--	Caffeine Atenolol, Atrazine	95.9 99.3 96.4	(Kim et al., 2017)
Cellulose triacetate (CTA) FO Hydration Technology Innovations, HTI (Albany, OR, USA) Side stream AnMBR+FO (AL-FS)	1M KCl	24	--	--	11.20	--	Caffeine Atenolol, Atrazine	94.1 99.5 95.1	(Kim et al., 2017)
Cellulose triacetate (CTA) FO Hydration Technology Innovations, HTI (Albany, OR, USA) Side stream AnMBR+FO (AL-DS)	1M MAP	24	--	--	7.58	--	Caffeine Atenolol, Atrazine	90.6 99.7 92.2	(Kim et al., 2017)

Cellulose triacetate (CTA) FO Hydration Technology Innovations, HTI (Albany, OR, USA) Side stream AnMBR+FO (AL-DS)	1M DAP	24	--	7.58	--	Caffeine Atenolol, Atrazine	97.5 99.2 98.0	(Kim et al., 2017)
Cellulose triacetate (CTA) FO Hydration Technology Innovations, HTI (Albany, OR, USA) Side stream AnMBR+FO (AL-DS)	1M KCl	24	--	7.58	--	Caffeine Atenolol, Atrazine	91.8 99.1 94.1	(Kim et al., 2017)

(Li et al., 2018)* Synthetic dye as feed solution rest of all synthetic wastewater removal of OMPs. Authors concluded that the trade-off between high dilution of draw solution (i.e., high water flux and low flux decline) and high OMPs rejection (i.e., low OMPs forward flux) should be considered in fertilizer drawn forward osmosis (FDFO) design and optimization.

16.3.2 Membrane Distillation Bioreactor

Membrane distillation incorporates a hydrophobic microporous membrane operating at a low temperature, which involves solely the transfer of water vapor from the feed side to the distillate side through membrane pores. Due to gas-phase mass transfer, only volatiles could pass through and thus MD completely retains non-volatiles in the feed solution (Curcio and Drioli, 2005, Wijekoon et al., 2014). More recently the osmotic membrane bioreactor has been studied for its ability to integrate membrane distillation and conventional biological system in a single reactor. The direct contact membrane module submerges into the activated sludge tank (Figure 16.1b) (Phattaranawik et al., 2008, Yeo et al., 2015). By adjusting the 30–38°C temperature range (40 ± 10°C optimum for thermophiles), the temperature gradient provides a driving force for water vapor to pass through the hydrophobic membrane of the MD process. In the submerged configuration, the membrane is submerged inside the mixed liquor of the feed solution and the outer surface of hollow fiber remains in the contact with feed. Pure permeate withdrawn from the membrane is collected in the product tank (Phattaranawik et al., 2008). In the side stream arrangement (Figure 16.1a), the reactor feed is continuously pumped to the membrane unit and returned to the bioreactor. During this operation at a moderately high temperature, water vapor has been produced from the feed that passes through the MD membrane to the collection tank (Neoh et al., 2016). Stricter statutory requirements in particular for OMPs removal could make MDBR a promising option in wastewater treatment. MD membrane rejects low molecular weight cut-off (MWCO) and refractory hydrophilic nature of OMPs, thereby increasing its organic retention time in a bioreactor (Table 16.3) (Asif et al. 2017 a,b, Song et al., 2018, Wijekoon et al., 2014, Yeo et al., 2015, Asif et al. 2018). In the MDBR process, heat transfers from the feed side through the hydrophobic membrane element and then to the mixed liquor of the biological reactor. Hence, naturally higher mixed liquor temperatures in the reactor reduces water viscosity and thereby increases initial water flux through the MD membrane. This also increases fouling and scouring of the MD membrane, which consequently adversely affects permeate quality and leads to elevated fouling and increases operating cost. Furthermore, nitrifiers are too much affected by temperature variations, and a temperature rise also adversely affects the nitrogen removal process (Morrow et al., 2018).

The enzyme laccase in the presence of oxygen as a co-substrate can catalyze recalcitrant molecules including organic micropollutants. Asif et al. (2017b) evaluated the performance of membrane distillation with an enzymatic bioreactor (MD-EMBR) to examine the removal of phenolic and non-phenolic OMPs. This hybrid reactor achieved 90–99% OMPs removal. The removal was correlated with the electron donating group (EDG) and electron withdrawing group (EWG) of the OMPs. The OMPs having EDG demonstrated more than 90% removal while EWG OMPs achieved 40–75% removal. Further, the addition of redox mediators and OMPs removal was studied. Violuric

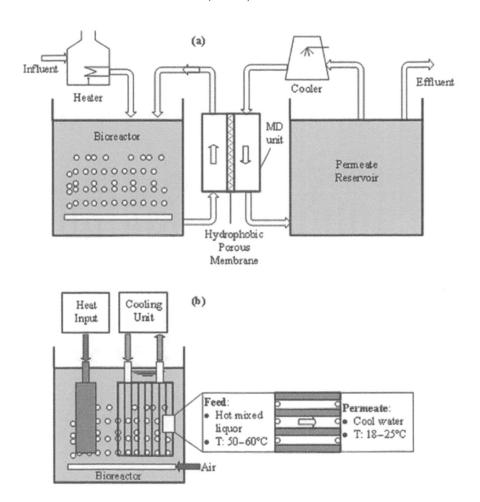

FIGURE 16.1 Schematic diagram of (a) side stream MDBR process (Neoh et al., 2016) and (b) submerged process (Goh et al., 2015).

acid (VA) redox mediator outperformed among syringaldehyde (SA), violuric acid (VA), and 1-hydroxybenzotriazole (HBT) mediators in OMPs removal.

Wijekoon et al. (2014) evaluated the performance of MDBR in OMPs removal and concluded that 95% of OMPs can be removed by this process; biodegradation contributed to 70% of OMPs removal. Actually, high temperature and salinity can adversely affect the performance of MDBR. Triclosan, fenoprop, atrazine, clofibric acid, diclofenac, and carbamazepine compound could be retained by the MD membrane in the range of 42 to 94%. The hydrophilic compounds having EWGs in their structure are more resistant to biodegradation and they were poorly degradable compounds (up to only 53%). Song et al. (2018) examined the performance of AnBR membrane distillation in OMPs removal. This hybrid process accomplished 75 to 100% removal of 26 OMPs studied. Authors notice that recalcitrant compounds such

TABLE 16.3

Summary of recently published MDBR studies

Feed Solution	MD Membrane	Temperature		HRT (h)	SRT (d)	MLSS (g/L)	Water flux (LMH)	DO (mg/L)	OMP	Removal (%)	Reference
		Feed °C	Permeate °C								
Synthetic wastewater	PTFE 0.22 μm	50	20–25	--	--	--			Ibuprofen	99	(Asif et al., 2017a)
									Naproxen	99	
									Ketoprofen	99	
									Diclofenac	97	
									Primidone	99	
									Carbamazepine	98	
									Salicylic acid	98	
									Metronidazole	97	
									Gemfibrozil	99	
									Amitriptyline	98	
									Triclosan	100	
									Benzophenone	97	
									Oxybenzone	97	
									Octocrylene	99	
									Fenoprop	98	
									Pentachloro-phenol	98	
									Atrazine	90	
									Propoxur	99	
									Ametryn	97	
									Clofibric acid	99	
									DEET	100	
									4-tert-butylphenol	90	
									Bisphenol A	98	
									Estrone	98	
									17β-estradiol	99	
									17β-estradiol 17-Acetate	100	
									17α - Ethinylestradiol	100	
									Estriol (E3)	100	
									Enterolactone	99	

Synthetic wastewater	PTFE membrane	45	20	--	10	--	--*		
								Amtriptyline	99
								Atrazine	74
								Bisphenol A	85
								Caffeine	99
								Carazolol	97
								Carbamazepine	90
								Clozapine	99
								Diazinon	99
								Diclofenac	75
								Diuron	99
								Gemfibrozil	99
								Ibuprofen	99
								Ketoprofen	99
								Linuron	93
								Naproxen	97
								Paracetamol	99
								Phenylphenol	80
								Primidone	99
								Propylparaben	91
								Simazine	79
								Sulfamethoxazole	89
								TCEP	92
								Triamterene	98
								Triclocarban	95
								Triclosan	85
								Trimethoprim	99

(Song et al., 2018)

(Continued)

TABLE 16.3
Summary of recently published MDBR studies (Cont.)

Feed Solution	MD Membrane	Temperature		HRT (h)	SRT (d)	MLSS (g/L)	Water flux (LMH)	DO (mg/L)	OMP	Removal (%)	Reference
		Feed °C	Permeate °C								
Synthetic wastewater	PTFE side stream	40	14	9.6 d		5.3	1.2	2.8	17α-Ethinylestradiol	99	(Wijekoon et al., 2014)
									17β-Estradiol	100	
									17β-Estrodiol-17-acetate	100	
									4-Tert-butyphenol	98	
									Ametryn	99	
									Amitriptyline	99	
									Atrazine	96	
									Benzophenone	97	
									Carbamazepine	96	
									Clofibric acid	100	
									Diclofenac	95	
									Estriol	98	
									Estrone	100	
									Fenoprop	97	
									Gemfibrozil	98	
									Ibuprofen	100	
									Ketoprofen	99	
									Naproxen	100	
									Octocrylene	97	
									Oxybenzone	99	
									Pentachlorophenol	97	
									Primidone	100	
									Propoxure	100	
									Salicylic acid	96	
									Triclosan	98	

						Compound	Removal	Reference
Synthetic wastewater	PTFE side stream MD	30	10	3.75	3	Sulfamethoxazole	>99%	(Asif et al., 2017b)
						Carbamazepine	>99%	
						Diclofenac	>99%	
						Oxybenzone	>99%	
						Atrazine	>99%	
Synthetic wastewater		30	10	4 d	3	17α– Ethinylestradiol	98	(Asif et al., 2018)
						17β–Estradiol	98	
						17β-Estradiol-17-acetate	98	
						4-tert-Butylphenol	98	
						Ametrine	94	
						Amitriptyline	99	
						Atrazine	92	
						Benzophenone	99	
						Bisphenol A	96	
						Carbamazepine	99	
						Clofibric acid	99	
						DEET	94	
						Diclofenac	96	
						Enterolactone	96	
						Estriol	97	
						Estrone	99	
						Fenoprop	99	
						Gemfibrozil	98	
						Ibuprofen	98	
						Ketoprofen	98	
						Metronidazole	98	
						Naproxen	99	
						Octocrylene	99	
						Oxybenzone	94	
						Pentachlorophenol	97	
						Primidone	99	
						Propoxur	95	
						Salicylic acid	96	
						Triclosan	97	

as bisphenol A, diclofenac, ibuprofen, and primidone are effectively retained by the MD membrane followed by further degradation in the AnBR process.

16.3.2.1 Life Cycle Assessment of Hybrid MBRs

Life cycle assessment (LCA) is a significant tool to measure the environmental impact of different wastewater treatment schemes to compare their performances in terms of energy, greenhouse emission, and cost components (Krzeminski et al., 2017). However, the LCA of MBRs and advanced hybrid MBRs are limited to a few studies. Ortiz et al. (2007) evaluated the environmental impact analysis of different wastewater treatment schemes designed to accommodate 13,200 population equivalent (PE). In order to find the lowest environmental load of treatment schemes, the authors used Simapro 5.1 software. Authors considered activated sludge process (ASP) standalone, ASP combined with advanced treatment, and membrane bioreactors both side stream and submerged. LCA analysis results suggested that ASP with advanced treatment had the highest environmental loads. It was apparent that side stream MBR had a higher environmental load as compared to submerged MBR due to its high energy consumption in pump operation. In general, the environmental impact resulting from plant operation remains higher than arising from construction, maintenance, and final disposal aspects.

Further, Krzeminski et al. (2017) evaluated performances of four membrane processes: aerobic MBR, anaerobic MBR, biofilm MBR, and OMBR. Aerobic MBR can produce the highest quality of permeate of direct reuse applications and is able to meet stringent prescribed standards with less effect on marine and freshwater eutrophication. Nevertheless, operating costs in terms of frequent membrane cleaning and higher energy for aeration implies larger carbon emission, which is higher for MBRs. In order to reduce environmental loads sludge to biogas production and use of renewable energy, alternatives could be explored by installing anaerobic MBRs that operate in absence of oxygen (no aeration cost). Yet, in sewage treatment plants 30–40% methane seepage was observed that further contributed to GHG emissions. In AnMBR the COD and nutrient removal are low compared to aerobic MBRs, more sensitive to shock loads, and demands more skillful operations. The biofilm MBRs consume more DO to achieve complete nitrification. It is more energy-intensive and a larger carbon footprint. In recent years OMBRs have gained more attention in both academia and industrial application due to very low energy consumption, excellent effluent quality for direct and indirect reuse applications, and efficient organic micropollutant removal. OMBR exploits osmotic pressure of draw solution as a driving force; hence, compared to other hydraulic MF/UF membranes processes, it consumes very little energy. OMBR can effectively remove phosphorous and ammonium but it cannot remove nitrite totally. Table 16.4 below shows various MBR configurations and their anticipated environmental load.

Holloway et al. (2016a) compared the performance of the UF-OMBR hybrid system (Figure 16.2) with advanced wastewater treatment in sewage treatment to obtain pure water for reuse application that employed activated sludge treatment combined with MF/UF membrane, reverse osmosis (RO), and ultraviolet reactor. The UF-OMBR consisted of UF and FO membranes in a bioreactor. UF produced non-potable reuse water while FO produced potable quality water.

TABLE 16.4

Comparison of the performance of MBR, AnMBR, BF-MBR and FO-MBR against energy demand and their impact on climate change, fresh water and marine eutrophication (Krzeminski et al., 2017)

Process type	Energy related emissions	Climate change impact	Fresh water eutrophication	Marine eutrophication
MBR	High	High	Low	Low
AnMBR	Low	High	High	High
BF-MBR	High/medium	High/medium	Medium	Low
FO-MBR	Medium	Medium	Low	Low/medium

The wastewater-energy sustainability tool (WWEST), a LCA program, considered energy use, GHG emissions, and other environmentally relevant emissions to compare process performances. Authors compared both treatment technologies by taking into account construction cost, chemical inventory, and electricity to assess energy demand and environmental impact of both treatment schemes. The hybrid advanced treatment process outperformed UF-OMBR based on this LCA analysis. UF-OMBR exhibited a higher environmental load arising from the larger footprint and lower permeability of FO membranes and higher energy consumption from RO regeneration. However, UF-OMBR optimization was based on 40 g/L NaCl DS concentrations among (20, 30, 40, and 50 g/L NaCl) and demonstrated that both processes had very similar environmental loads when higher permeability FO membranes and RO energy recovery system were implemented. This outcome further reinforces the scope of OMBR in water reuse applications (Holloway et al., 2016b). Further, the authors reported that the UF-OMBR process could have accomplished a much lower environmental

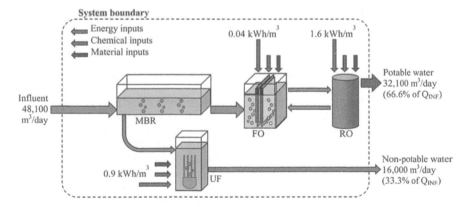

FIGURE 16.2 Schematic drawing of system boundary, flows, unit processes, and energy, materials, and chemical inputs used for the LCA of a UFO-MBR treatment plant. The illustrated RO power is for an RO system without energy recovery, producing an RO brine of 40 g/L NaCl. The energy usage for the MBR component (activated sludge and UF/FO membrane aeration) of the UFO-MBR is included in the UF energy (Holloway et al., 2016b).

impact if nitrogen/phosphorous recovery had been realized with scale-up of the process. However, UF-OMBR has the potential to become a fourth-generation advanced wastewater reclamation alternative provided FO membrane development and OMBR process optimization are accomplished (Holloway et al., 2016b).

16.4 CONCLUSION

This chapter on advanced membrane bioreactor hybrid systems explored two recently examined advanced hybrid systems of osmotic membrane bioreactor and membrane distillation bioreactor. The chapter focused on water scarcity issues and wastewater reuse alternatives followed by the fate and transport of organic micropollutants in wastewater. MBR, MDBR, and OMBR were compared. Recent reports on OMBR and MDBR in OMPs removal were tabulated. The life cycle assessment for MBR and advance hybrid technologies were discussed. The high permeability FO/MD membrane, scale-up module, and optimization of reactor design to produce reclaimed water with nutrient recovery applications now need to be explored to reduce the environmental load (Blandin et al., 2018).

REFERENCES

ACHILLI, A., CATH, T. Y., MARCHAND, E. A. & CHILDRESS, A. E. 2009. The forward osmosis membrane bioreactor: a low fouling alternative to MBR processes. *Desalination*, 239, 10–21.

ALTURKI, A., McDONALD, J., KHAN, S. J., HAI, F. I., PRICE, W. E. & NGHIEM, L. D. 2012. Performance of a novel osmotic membrane bioreactor (OMBR) system: flux stability and removal of trace organics. *Bioresource Technology*, 113, 201–206.

ALTURKI, A. A., McDONALD, J. A., KHAN, S. J., PRICE, W. E., NGHIEM, L. D. & ELIMELECH, M. 2013. Removal of trace organic contaminants by the forward osmosis process. *Separation and Purification Technology*, 103, 258–266.

ALVARINO, T., SUAREZ, S., LEMA, J. & OMIL, F. 2018. Understanding the sorption and biotransformation of organic micropollutants in innovative biological wastewater treatment technologies. *Science of the Total Environment*, 615, 297–306.

ANSARI, A. J., HAI, F. I., GUO, W., NGO, H. H., PRICE, W. E. & NGHIEM, L. D. 2015. Selection of forward osmosis draw solutes for subsequent integration with anaerobic treatment to facilitate resource recovery from wastewater. *Bioresource Technology*, 191, 30–36.

ASIF, M. B., HAI, F. I., KANG, J., VAN DE MERWE, J. P., LEUSCH, F. D., PRICE, W. E. & NGHIEM, L. D. 2018. Biocatalytic degradation of pharmaceuticals, personal care products, industrial chemicals, steroid hormones and pesticides in a membrane distillation-enzymatic bioreactor. *Bioresource Technology*, 247, 528–536.

ASIF, M. B., HAI, F. I., KANG, J., VAN DE MERWE, J. P., LEUSCH, F. D., YAMAMOTO, K., PRICE, W. E. & NGHIEM, L. D. 2017a. Degradation of trace organic contaminants by a membrane distillation—enzymatic bioreactor. *Applied Sciences*, 7, 879.

ASIF, M. B., NGUYEN, L. N., HAI, F. I., PRICE, W. E. & NGHIEM, L. D. 2017b. Integration of an enzymatic bioreactor with membrane distillation for enhanced biodegradation of trace organic contaminants. *International Biodeterioration & Biodegradation*, 124, 73–81.

BARBOSA, M. O., MOREIRA, N. F., RIBEIRO, A. R., PEREIRA, M. F. & SILVA, A. M. 2016. Occurrence and removal of organic micropollutants: an overview of the watch list of EU Decision 2015/495. *Water Research*, 94, 257–279.

BEDDINGTON, S. J. 2011. The future of food and farming. *International Journal of Agricultural Management*, 1, 2–6.

BESHA, A. T., GEBREYOHANNES, A. Y., TUFA, R. A., BEKELE, D. N., CURCIO, E. & GIORNO, L. 2017. Removal of emerging micropollutants by activated sludge process and membrane bioreactors and the effects of micropollutants on membrane fouling: A review. *Journal of Environmental Chemical Engineering*, 5, 2395–2414.

BHARWADA, U. 2011. HTI'S forward osmosis membrane bioreactor process (OsMBR) – A rugged, versatile and ecobalanced process for industrial wastewater plus reuse: Truly sustainable wastewater treatment design for a changing world Scottsdale. *Hydration Technology Innovations. LLC 9311E*. Via de Ventura, Scottsdale, Arizona, 1–7.

BLANDIN, G., LE-CLECH, P., CORNELISSEN, E., VERLIEFDE, A. R., COMAS, J. & RODRIGUEZ-RODA, I. 2018. Can osmotic membrane bioreactor be a realistic solution for water reuse? *NPJ Clean Water*, 1, 1–6.

BODZEK, M. & KONIECZNY, K. 2018. Membranes in organic micropollutants removal. *Current Organic Chemistry*, 22, 1070–1102.

BOWDEN, K. S., ACHILLI, A. & CHILDRESS, A. E. 2012. Organic ionic salt draw solutions for osmotic membrane bioreactors. *Bioresource Technology*, 122, 207–216.

BUI, X., VO, T., NGO, H., GUO, W. & NGUYEN, T. 2016. Multicriteria assessment of advanced treatment technologies for micropollutants removal at large-scale applications. *Science of the Total Environment*, 563, 1050–1067.

CALERO-DÍAZ, G., MONTEOLIVA-GARCÍA, A., LEYVA-DÍAZ, J. C., LÓPEZ-LÓPEZ, C., MARTÍN-PASCUAL, J., TORRES, J. C. & POYATOS, J. M. 2017. Impact of ciprofloxacin, carbamazepine and ibuprofen on a membrane bioreactor system: Kinetic study and biodegradation capacity. *Journal of Chemical Technology & Biotechnology*, 92, 2944–2951.

CHEKLI, L., PHUNTSHO, S., KIM, J. E., KIM, J., CHOI, J. Y., CHOI, J.-S., KIM, S., KIM, J. H., HONG, S. & SOHN, J. 2016. A comprehensive review of hybrid forward osmosis systems: Performance, applications and future prospects. *Journal of Membrane Science*, 497, 430–449.

CHTOUROU, M., MALLEK, M., DALMAU, M., MAMO, J., SANTOS-CLOTAS, E., SALAH, A. B., WALHA, K., SALVADÓ, V. & MONCLÚS, H. 2018. Triclosan, carbamazepine and caffeine removal by activated sludge system focusing on membrane bioreactor. *Process Safety and Environmental Protection*, 118, 1–9.

CICEK, N., FRANCO, J. P., SUIDAN, M. T. & URBAIN, V. 1999. Effect of phosphorus on operation and characteristics of MBR. *Journal of Environmental Engineering*, 125, 738–746.

CORNELISSEN, E., HARMSEN, D., DE KORTE, K., RUIKEN, C., QIN, J.-J., OO, H. & WESSELS, L. 2008. Membrane fouling and process performance of forward osmosis membranes on activated sludge. *Journal of Membrane Science*, 319, 158–168.

CURCIO, E. & DRIOLI, E. 2005. Membrane distillation and related operations—a review. *Separation and Purification Reviews*, 34, 35–86.

GOH, S., ZHANG, J., LIU, Y. & FANE, A. G. 2015. Membrane distillation bioreactor (MDBR)–A lower green-house-gas (GHG) option for industrial wastewater reclamation. *Chemosphere*, 140, 129–142.

GUO, W., NGO, H. H., CHEN, C., PANDEY, A., TUNG, K.-L. & LEE, D. J. 2016. Anaerobic membrane bioreactors for future green bioprocesses. *Green technologies for sustainable water management*. Published by the American Society of Civil Engineers. ISBN 978-0-7844-7978-0 (PDF); Reston, Virginia, USA.

HAI, F. I., YAMAMOTO, K. & LEE, C.-H. 2018. *Membrane biological reactors: theory, modeling, design, management and applications to wastewater reuse*. IWA Publishing, London.

HAMZA, R. A., IORHEMEN, O. T. & TAY, J. H. 2016. Occurrence, impacts and removal of emerging substances of concern from wastewater. *Environmental Technology & Innovation*, 5, 161–175.

HOLLOWAY, R., MILLER-ROBBIE, L., PATEL, M., STOKES, J., MALTOS, R., MUNAKATA-MARR, J., DADAKIS, J. & CATH, T. Three Years Piloting of Osmotic MBR for Potable Reuse: Performance and Life-Cycle Assessment. Proceedings of the American Water Works Association/American Membrane Technology Association Membrane Technology Conference & Exposition, 2016a. 1–4.

HOLLOWAY, R. W., MILLER-ROBBIE, L., PATEL, M., STOKES, J. R., MUNAKATA-MARR, J., DADAKIS, J. & CATH, T. Y. 2016b. Life-cycle assessment of two potable water reuse technologies: MF/RO/UV–AOP treatment and hybrid osmotic membrane bioreactors. *Journal of Membrane Science*, 507, 165–178.

HUANG, J., XIONG, S., LONG, Q., SHEN, L. & WANG, Y. 2018. Evaluation of food additive sodium phytate as a novel draw solute for forward osmosis. *Desalination*, 448, 87–92.

HUANG, L. & LEE, D.-J. 2015. Membrane bioreactor: a mini review on recent R&D works. *Bioresource Technology*, 194, 383–388.

JIN, X., SHAN, J., WANG, C., WEI, J. & TANG, C. Y. 2012. Rejection of pharmaceuticals by forward osmosis membranes. *Journal of Hazardous Materials*, 227, 55–61.

JOSS, A., KELLER, E., ALDER, A. C., GÖBEL, A., MCARDELL, C. S., TERNES, T. & SIEGRIST, H. 2005. Removal of pharmaceuticals and fragrances in biological waste-water treatment. *Water Research*, 39, 3139–3152.

KIM, Y., LI, S., CHEKLI, L., WOO, Y. C., WEI, C.-H., PHUNTSHO, S., GHAFFOUR, N., LEIKNES, T. & SHON, H. K. 2017. Assessing the removal of organic micro-pollutants from anaerobic membrane bioreactor effluent by fertilizer-drawn forward osmosis. *Journal of Membrane Science*, 533, 84–95.

KIMURA, K., HARA, H. & WATANABE, Y. 2007. Elimination of selected acidic phar-maceuticals from municipal wastewater by an activated sludge system and membrane bioreactors. *Environmental Science & Technology*, 41, 3708–3714.

KRZEMINSKI, P., LEVERETTE, L., MALAMIS, S. & KATSOU, E. 2017. Membrane bioreactors–a review on recent developments in energy reduction, fouling control, novel configurations, LCA and market prospects. *Journal of Membrane Science*, 527, 207–227.

LEFLAIVE, X., WITMER, M., MARTIN-HURTADO, R., BAKKER, M., KRAM, T., BOUWMAN, L. & KIM, K. 2012. OECD Environmental Outlook to 2050. *Outlook*, 207.

LI, F., CHENG, Q., TIAN, Q., YANG, B. & CHEN, Q. 2016. Biofouling behavior and per-formance of forward osmosis membranes with bioinspired surface modification in osmotic membrane bioreactor. *Bioresource Technology*, 211, 751–758.

LI, F., XIA, Q., GAO, Y., CHENG, Q., DING, L., YANG, B., TIAN, Q., MA, C., SAND, W. & LIU, Y. 2018. Anaerobic biodegradation and decolorization of a refractory acid dye by a forward osmosis membrane bioreactor. *Environmental Science: Water Research & Technology*, 4, 272–280.

LINARES, R. V., LI, Z., SARP, S., BUCS, S. S., AMY, G. & VROUWENVELDER, J. S. 2014. Forward osmosis niches in seawater desalination and wastewater reuse. *Water Research*, 66, 122–139.

LUO, W., ARHATARI, B., GRAY, S. R. & XIE, M. 2018. Seeing is believing: insights from synchrotron infrared mapping for membrane fouling in osmotic membrane bioreactors. *Water Research*, 137, 355–361.

LUO, W., HAI, F. I., PRICE, W. E., GUO, W., NGO, H. H., YAMAMOTO, K. & NGHIEM, L. D. 2014a. High retention membrane bioreactors: challenges and opportunities. *Bioresource Technology*, 167, 539–546.

LUO, W., PHAN, H. V., XIE, M., HAI, F. I., PRICE, W. E., ELIMELECH, M. & NGHIEM, L. D. 2017. Osmotic versus conventional membrane bioreactors integrated with reverse osmosis for water reuse: biological stability, membrane fouling, and contaminant removal. *Water Research*, 109, 122–134.

LUO, Y., GUO, W., NGO, H. H., NGHIEM, L. D., HAI, F. I., ZHANG, J., LIANG, S. & WANG, X. C. 2014b. A review on the occurrence of micropollutants in the aquatic environment and their fate and removal during wastewater treatment. *Science of the Total Environment*, 473, 619–641.

MA, X. Y., LI, Q., WANG, X. C., WANG, Y., WANG, D. & NGO, H. H. 2018. Micropollutants removal and health risk reduction in a water reclamation and ecological reuse system. *Water Research*, 138, 272–281.

MERT, B. K., OZENGIN, N., DOGAN, E. C. & AYDıNER, C. 2018. Efficient Removal Approach of Micropollutants in Wastewater Using Membrane Bioreactor. *Wastewater and Water Quality*, pp 41-70. Published by Intechopen London, UK. ISBN 978-1-78923-621-7.

MOLDEN, D. 2013. *Water for food water for life: A comprehensive assessment of water management in agriculture*. Published by Routledge.London, UK. ISBN-13: 978-1844073962.

MONSALVO, V. M., McDO NALD, J. A., KHAN, S. J. & LE-CLECH, P. 2014. Removal of trace organics by anaerobic membrane bioreactors. *Water Research*, 49, 103–112.

MORROW, C. P., FURTAW, N. M., MURPHY, J. R., ACHILLI, A., MARCHAND, E. A., HIIBEL, S. R. & CHILDRESS, A. E. 2018. Integrating an aerobic/anoxic osmotic membrane bioreactor with membrane distillation for potable reuse. *Desalination*, 432, 46–54.

MOSER, P. B., DOS ANJOS SILVA, G. R., LIMA, L. S. F., MOREIRA, V. R., LEBRON, Y. A. R., DE PAULA, E. C. & AMARAL, M. C. S. 2020. Effect of organic and inorganic draw solution on recalcitrant compounds build up in a hybrid ultrafiltration-osmotic membrane reactor treating refinery effluent. *Chemical Engineering Journal*, 126374.

MUTAMIM, N. S. A., NOOR, Z. Z., HASSAN, M. A. A., YUNIARTO, A. & OLSSON, G. 2013. Membrane bioreactor: Applications and limitations in treating high strength industrial wastewater. *Chemical Engineering Journal*, 225, 109–119.

NEOH, C. H., NOOR, Z. Z., MUTAMIM, N. S. A. & LIM, C. K. 2016. Green technology in wastewater treatment technologies: integration of membrane bioreactor with various wastewater treatment systems. *Chemical Engineering Journal*, 283, 582–594.

NGUYEN, N. C., CHEN, S.-S., NGUYEN, H. T., CHEN, Y.-H., NGO, H. H., GUO, W., RAY, S. S., CHANG, H.-M. & LE, Q. H. 2018. Applicability of an integrated moving sponge biocarrier-osmotic membrane bioreactor MD system for saline wastewater treatment using highly salt-tolerant microorganisms. *Separation and Purification Technology*, 198, 93–99.

ORTIZ, M., RALUY, R. & SERRA, L. 2007. Life cycle assessment of water treatment technologies: wastewater and water-reuse in a small town. *Desalination*, 204, 121–131.

PAL, A., HE, Y., JEKEL, M., REINHARD, M. & GIN, K. Y.-H. 2014. Emerging contaminants of public health significance as water quality indicator compounds in the urban water cycle. *Environment International*, 71, 46–62.

PATHAK, N., FORTUNATO, L., LI, S., CHEKLI, L., PHUNTSHO, S., GHAFFOUR, N., LEIKNES, T. & SHON, H. K. 2018a. Evaluating the effect of different draw solutes in a baffled osmotic membrane bioreactor-microfiltration using optical coherence tomography with real wastewater. *Bioresource Technology*, 263, 306–316.

PATHAK, N., LI, S., KIM, Y., CHEKLI, L., PHUNTSHO, S., JANG, A., GHAFFOUR, N., LEIKNES, T. & SHON, H. K. 2018b. Assessing the removal of organic micropollutants by a novel baffled osmotic membrane bioreactor-microfiltration hybrid system. *Bioresource Technology*, 262, 98–106.

PATHAK, N., PHUNTSHO, S. & SHON, H. Y. 2020a. Membrane bioreactors for the removal of micro-pollutants. *Current Developments in Biotechnology and Bioengineering*. Published by Elseveir. Oxford, United Kingdom. ISBN 978-0-12-819854-4.

PATHAK, N., TRAN, V. H., MERENDA, A., JOHIR, M., PHUNTSHO, S. & SHON, H. 2020b. Removal of Organic Micro-Pollutants by Conventional Membrane Bioreactors and High-Retention Membrane Bioreactors. *Applied Sciences*, 10, 2969.

PHAN, H. V., HAI, F. I., ZHANG, R., KANG, J., PRICE, W. E. & NGHIEM, L. D. 2016. Bacterial community dynamics in an anoxic-aerobic membrane bioreactor–impact on nutrient and trace organic contaminant removal. *International Biodeterioration & Biodegradation*, 109, 61–72.

PHATTARANAWIK, J., FANE, A. G., PASQUIER, A. C. & BING, W. 2008. A novel membrane bioreactor based on membrane distillation. *Desalination*, 223, 386–395.

PRASERTKULSAK, S., CHIEMCHAISRI, C., CHIEMCHAISRI, W. & YAMAMOTO, K. 2019. Removals of pharmaceutical compounds at different sludge particle size fractions in membrane bioreactors operated under different solid retention times. *Journal of Hazardous Materials*, 368, 124–132.

PRIAC, A., MORIN-CRINI, N., DRUART, C., GAVOILLE, S., BRADU, C., LAGARRIGUE, C., TORRI, G., WINTERTON, P. & CRINI, G. 2017. Alkylphenol and alkylphenol polyethoxylates in water and wastewater: A review of options for their elimination. *Arabian Journal of Chemistry*, 10, S3749-S3773.

RAGHAVAN, D. S. S., QIU, G. & TING, Y.-P. 2018. Fate and removal of selected antibiotics in an osmotic membrane bioreactor. *Chemical Engineering Journal*, 334, 198–205.

SHANNON, M. A., BOHN, P. W., ELIMELECH, M., GEORGIADIS, J. G., MARINAS, B. J. & MAYES, A. M. 2010. Science and technology for water purification in the coming decades. *Nanoscience and Technology* pp 337–346. Nature Publishing Group London United Kingdom.ISBN: 978-981-4287-00-5.

SHIKLOMANOV, I. 1993. *World fresh water resources, water in crisis: A guide to the world's fresh water resources*. Oxford University Press, New York.

SKOUTERIS, G., HERMOSILLA, D., LÓPEZ, P., NEGRO, C. & BLANCO, Á. 2012. Anaerobic membrane bioreactors for wastewater treatment: a review. *Chemical Engineering Journal*, 198, 138–148.

SONG, X., LUO, W., McDONALD, J., KHAN, S. J., HAI, F. I., PRICE, W. E. & NGHIEM, L. D. 2018. An anaerobic membrane bioreactor–membrane distillation hybrid system for energy recovery and water reuse: Removal performance of organic carbon, nutrients, and trace organic contaminants. *Science of the Total Environment*, 628, 358–365.

SONG, X., McDONALD, J., PRICE, W. E., KHAN, S. J., HAI, F. I., NGO, H. H., GUO, W. & NGHIEM, L. D. 2016. Effects of salinity build-up on the performance of an anaerobic membrane bioreactor regarding basic water quality parameters and removal of trace organic contaminants. *Bioresource Technology*, 216, 399–405.

STEFAN, M. I. 2017. *Advanced oxidation processes for water treatment: fundamentals and applications*. IWA Publishing, London.

STEVENS-GARMON, J., DREWES, J. E., KHAN, S. J., McDONALD, J. A. & DICKENSON, E. R. 2011. Sorption of emerging trace organic compounds onto wastewater sludge solids. *Water Research*, 45, 3417–3426.

SUN, Y., TIAN, J., ZHAO, Z., SHI, W., LIU, D. & CUI, F. 2016. Membrane fouling of forward osmosis (FO) membrane for municipal wastewater treatment: A comparison between direct FO and OMBR. *Water Research*, 104, 330–339.

SUWAILEH, W., PATHAK, N., SHON, H. & HILAL, N. 2020. Forward osmosis membranes and processes: A comprehensive review of research trends and future outlook. *Desalination*, 485, 114455.

TRAN, N. H. & GIN, K. Y.-H. 2017. Occurrence and removal of pharmaceuticals, hormones, personal care products, and endocrine disrupters in a full-scale water reclamation plant. *Science of the Total Environment*, 599, 1503–1516.

TRENBERTH, K. E., SMITH, L., QIAN, T., DAI, A. & FASULLO, J. 2007. Estimates of the global water budget and its annual cycle using observational and model data. *Journal of Hydrometeorology*, 8, 758–769.

VAN HUY TRAN, N. P. & SHON, H. Y. 2020. Osmotic membrane bioreactor technology—Concept, potential and challenges. Pp 181-207. *Current Developments in Biotechnology and Bioengineering: Advanced Membrane Separation Processes for Sustainable Water and Wastewater Management-Aerobic Membrane Bioreactor Processes and Technologies*, Published by Elsevier, Oxford, United Kingdom. ISBN: 9780128214176.

VERLICCHI, P., AL AUKIDY, M. & ZAMBELLO, E. 2012. Occurrence of pharmaceutical compounds in urban wastewater: removal, mass load and environmental risk after a secondary treatment—a review. *Science of the Total Environment*, 429, 123–155.

WANG, P., WANG, Z., WU, Z. & MAI, S. 2011. Fouling behaviours of two membranes in a submerged membrane bioreactor for municipal wastewater treatment. *Journal of Membrane Science*, 382, 60–69.

WEI, C.-H., WANG, N., HOPPEJONES, C., LEIKNES, T., AMY, G., FANG, Q., HU, X. & RONG, H. 2018. Organic micropollutants removal in sequential batch reactor followed by nanofiltration from municipal wastewater treatment. *Bioresource Technology*, 268, 648–657.

WIJEKOON, K. C., HAI, F. I., KANG, J., PRICE, W. E., GUO, W., NGO, H. H., CATH, T. Y. & NGHIEM, L. D. 2014. A novel membrane distillation–thermophilic bioreactor system: Biological stability and trace organic compound removal. *Bioresource Technology*, 159, 334–341.

YAO, M., DUAN, L., WEI, J., QIAN, F. & HERMANOWICZ, S. W. 2020. Carbamazepine removal from aastewater and the degradation mechanism in a submerged forward osmotic membrane bioreactor. *Bioresource Technology*, 123732.

YEO, B. J., GOH, S., ZHANG, J., LIVINGSTON, A. G. & FANE, A. G. 2015. Novel MBRs for the removal of organic priority pollutants from industrial wastewaters: a review. *Journal of Chemical Technology & Biotechnology*, 90, 1949–1967.

YIP, N. Y., TIRAFERRI, A., PHILLIP, W. A., SCHIFFMAN, J. D. & ELIMELECH, M. 2010. High performance thin-film composite forward osmosis membrane. *Environmental Science & Technology*, 44, 3812–3818.

ZHANG, B., SONG, X., NGHIEM, L. D., LI, G. & LUO, W. 2017. Osmotic membrane bioreactors for wastewater reuse: Performance comparison between cellulose triacetate and polyamide thin film composite membranes. *Journal of Membrane Science*, 539, 383–391.

ZHENG, W., WEN, X., ZHANG, B. & QIU, Y. 2019. Selective effect and elimination of antibiotics in membrane bioreactor of urban wastewater treatment plant. *Science of the Total Environment*, 646, 1293–1303.

ZIOLKOWSKA, J. R. 2015. Is desalination affordable?—regional cost and price analysis. *Water Resources Management*, 29, 1385–1397.

ZOLFAGHARI, M., DROGUI, P., SEYHI, B., BRAR, S. K., BUELNA, G., DUBÉ, R. & KLAI, N. 2015. Investigation on removal pathways of Di 2-ethyl hexyl phthalate from synthetic municipal wastewater using a submerged membrane bioreactor. *Journal of Environmental Sciences*, 37, 37–50.

17 Development of MBR, MABR and AnMBR Systems for Wastewater Treatment

Buddhima Siriweera and
*Chettiyappan Visvanathan**
Environmental Engineering and Management Program,
School of Environment, Resources and Development,
Asian Institute of Technology, Thailand

CONTENTS

17.1 MEMBRANE BIOREACTOR

Industrialization, urbanization, and agricultural activities have led to a large number of pollutants discharged into bodies of water. Therefore, wastewater treatment units are built to meet the effluent standards, and the activated sludge process was one of the major treatment methods. However, due to the limitations associated with the activated sludge process such as high excess sludge production and low quality of effluent, membrane bioreactor (MBR) was developed as a new efficient biological treatment method. The secondary sedimentation tank in the activated sludge process is replaced by a membrane system in the membrane bioreactor for the solid/liquid separation.

The use of MBR technology enabled direct reuse of treated wastewater for secondary purposes such as watering plants, flushing toilets, etc. In 1969, Smith et al. (1969) used ultrafiltration as a replacement for settling tanks in the activated sludge process. In 1970, Hardt et al. (1970) used an ultrafiltration membrane for the treatment of synthetic wastewater using an anaerobic bioreactor, which was later introduced as the MBR. In 1998, the first large-scale internal MBR system was installed in North America for the treatment of wastewater discharged from the food industry (Sutton, 2003).

In MBR technology, a microfiltration or ultrafiltration membrane is combined with a biological process, such as a suspended growth bioreactor. In MBR systems, more importance is given to the biological processes rather than the filtration, where pollutants are converted to the end products prior to the filtration.

17.1.1 CONFIGURATIONS OF MBR

Based on their configuration, MBR systems are generally divided into two types: internal and external MBR systems. The first generation of MBRs consisted of external crossflow-operated membranes, which were installed outside the aeration tank.

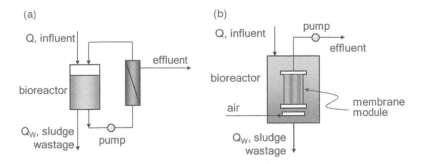

FIGURE 17.1 Different configurations of MBR: (a) external MBR configuration; (b) internal submerged MBR.

A recirculation pump is required for providing the required transmembrane pressure and this method required high energy to maintain both high cross-flow velocity and the required pressure drop for permeation. Also, high pressure and excessive shear generated due to the use of a cross-flow recirculation pump could damage the flocs and stability of the system.

An external submerged MBR system is presented in Figure 17.1(a). Later on, it was decided to submerge the membrane in the aeration tank, and operation was conducted with reduced pressure, which in turn reduced the energy consumption. This type of system was referred to as submerged MBR. In 1990, submerged MBR systems were commercialized, and they were found to have low operational costs compared to the external configuration of MBR. Also, the recirculation pump, which was an essential part of the external MBR, was absent in the submerged MBR configuration. An internal submerged MBR system is presented in Figure 17.1(b).

17.1.2 Types of Wastewater Treated by MBR

Due to the compactness of the system and the reusability of the effluent produced, MBR systems have been used for the treatment of municipal wastewater. Around the 1990s, with the introduction of cost-effective submerged membranes, attention was focused on the utilization of MBR for large-scale municipal wastewater applications. It was also used for the industrial wastewater treatment consisting of high organic loadings as well as for the treatment of landfill leachate that contained high concentrations of organic and inorganic compounds with additional treatment steps such as reverse osmosis for the removal of heavy metals and other inorganics. It has been possible to achieve more than 90% of chemical oxygen demand (COD) removal efficiency in MBR systems for most of the wastewater treatment applications including synthetic domestic wastewater, municipal wastewater, and leachate (Khan et al., 2011; Fudala-Ksiazek et al., 2018; Chae & Shin, 2007).

17.1.3 Nitrogen Removal via MBR

It is possible to achieve good nitrification efficiencies in a submerged MBR due to the possibility of maintaining high biomass concentration, long SRTs employed and high aeration provided for mitigation of membrane fouling. However, denitrification might be limited since the required anoxic conditions might not be easily achieved with the high aeration. To overcome the limitation of denitrification, some modifications have been proposed for MBR such as the installation of an additional anoxic tank with MBR and maintaining interchanging anoxic and aerobic conditions in MBR via intermittent aeration (Sun et al, 2010). Due to the requirement of recycling wastewater between tanks and increased volume requirements for the additional anoxic tank, nitrogen removal via MBR has been observed as an expensive solution. Although the application of membrane provides the advantage of high biomass concentration, the requirement of high energy consumption for the aeration is one of the major drawbacks of MBR systems.

17.2   MEMBRANE AERATED BIOFILM REACTOR

Apart from the removal of organic matter, the removal of nitrogenous compounds from wastewater is of utmost importance for the prevention of severe environmental issues such as eutrophication.

For the oxidation of organic compounds in wastewater, conventional diffusers have been used for a long time in both activated sludge process (ASP) and MBR to dissolve gas in an aqueous medium, which has some disadvantages such as low solubility of gases that requires continuous bubbling.

When introducing the gas at the bottom of the tank, it must be pressurized to overcome the hydrostatic head resulting in high energy consumption. Also, these gas bubbles have a fixed residence time before they reach the surface of the liquid and are lost to the headspace, which results in the waste of the undissolved gases.

In addition to that, in the activated sludge process, nitrification is effectively conducted in the aeration tank. However, when it is intended for total nitrogen (TN) removal, modification with pre-anoxic or post-anoxic tanks is required for the purpose of denitrification. Although suspended growth membrane bioreactors provide a high quality of the affluent, there will be disadvantages such as high energy requirement for aeration and sludge recirculation, very high pre-treatment needs due to frequent membrane fouling as well as the requirement of modification with the separate anoxic area for the denitrification activity similar to the activated sludge process.

Diffusing the gases through the a gas permeable membrane, called a membrane aerated biofilm reactor (MABR), provides an alternative way to supply the gas to the microorganisms. MABR is a recently developed biofilm technology that provides effective heterotrophic oxidation and simultaneous nitrification and denitrification (SND) in a single unit MABR provides the required high hydraulic retention time (HRT) for the slow-growing nitrifiers, through the effective decoupling of sludge retention time (SRT) and HRT (Rouse et al., 2007).

The first reports on the MABR technology concept date back to the 1980s, when the patent for this technology was received by Onishi et al. (1980) for hollow fiber oxygen-based MABR. Due to the air diffusion mechanism employed in MABR, it provides more than 85% of savings of the energy compared to the conventional ASP (Aybar et al., 2014). When the membranes are operated in the dead-end mode, theoretically 100% oxygen transfer efficiency could be achieved (Semmens et al., 2003;

Syron and Casey, 2008). Also, MABR has other benefits such as lower sludge production rates compared to ASP. According to the laboratory scale research work on MABR, it has the ability to achieve high COD and nitrogen removal. However, the full-scale application of MABRs depends on optimizing their cost-effectiveness, oxygen transfer rates, and reactor performance.

Numerous laboratory-scale studies have been conducted for the determination of the potential of MABR as a technology for the biological removal of COD and nitrogen. According to laboratory-scale experiments, MABR can achieve high COD and nitrogen removal. However, this technology has not yet been applied commercially due to the challenges associated with this process.

17.2.1 Unique Microbial Community Structure in MABR

In MABR, the membrane acts as a carrier to immobilize the bacteria as well as the means of oxygen supply to the bacteria in the biofilm formed on the membrane surface. Oxygen and substrates are supplied to the biofilm in counter diffusion mode leading to a dissolved oxygen (DO) concentration gradient along with the thickness of the biofilm, facilitating SND in a single reactor. High DO concentrations exist near the biofilm attachment surface which favors the growth of nitrifying bacteria. High COD and low DO conditions exist in the outermost layers or bulk liquid of MABR, favoring anoxic denitrifying conditions. This spatial stratification of the microbial activity has allowed the application of MABR for simultaneous nitrification/ denitrification processes (Martin & Nerenberg, 2012).

17.2.2 Factors Affecting SND in MABR

There are several factors affecting MABR such as COD/N ratio, HRT, oxygen partial pressure and DO concentration, biofilm thickness, temperature, and pH as well as loading rates.

Controlling HRT is very important in MABR operations because different types of bacteria require different time durations for successful biofilm formation. According to Long (2013), 2–3.3 hours and 12.5 hours of HRT are required for aerobic heterotrophs and aerobic autotrophs, respectively, for successful biofilm formation.

COD/N ratio has an impact on both nitrification and denitrification efficiencies. Extremely high COD/N ratios favor the growth of aerobic heterotrophs, leading to excessive biofilm thickness and slowdown of the activity of nitrifiers. This could result in higher DO consumption by aerobic heterotrophs compared to the nitrifying bacteria.

The DO profile along the thickness of biofilm at high and low COD/N ratios is presented in Figure 17.2. However, very low COD/N ratios reduce the supply of electron donors for the denitrification activity. In the MABR studies conducted by La Para et al. (2006) and Premarathna & Visvanathan (2019), optimum COD/N ratios of 6 and 4 have been recommended.

DO concentration in the bulk liquid is also an important parameter since denitrifiers who live in the outer layer of the biofilm can only exist in anoxic regions with DO less than 0.5 mg/L (Zhu, 2008). Therefore, higher DO concentrations in

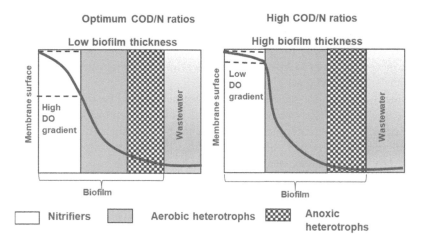

FIGURE 17.2 Variation of DO profile in MABR at different COD/N ratios.

bulk solution should be prevented by controlling the oxygen partial pressure in the membrane. Temperature and pH are responsible for creating an environment for different kinds of bacteria to work optimally. According to Grunditz and Dalhammar (2001), the highest activity of the two kinds of nitrifying bacteria, Nitrosomonas and Nitrobacter, occurs at 35°C and 38°C respectively.

For the optimum growth conditions of both nitrifying bacteria and de-nitrifiers, it is required to maintain pH in the range between 7.0 and 8.5 (Metcalf and Eddy, 2003). In MABR, due to the simultaneous nitrification and denitrification, reduction of the pH caused by the consumption of alkalinity during nitrification could be compensated by the generated alkalinity during the denitrification process (Terada et al., 2006). Natural pH adjustment in the MABR system is represented in Figure 17.3.

COD and NH_4^+-N loading rates are very important parameters that can affect both the composition and growth of the biofilm. According to previous studies, higher COD loading rates will lead to sludgy biofilms with excess bacterial growth. The study conducted by Satoh et al. (2004) concluded that SND was achieved with higher removal efficiencies of 90% and 95% for nitrification and denitrification, respectively, with a low COD loading rate of around 1.0 g-COD/m²/day. Studies carried out by Brindle et al. (1998) and Terada et al. (2006) demonstrated that, at NH_4^+-N loading rates from 0.12 to 6.49 gN/m²/day (DO>0.5), complete nitrification could be achieved.

17.2.3 DIFFICULTIES OF SCALING UP MABR TECHNOLOGY

Numerous laboratory-scale studies have been conducted for the determination of the potential of MABR as a technology for the biological removal of COD and nitrogen. According to laboratory-scale experiments, MABR can achieve high COD and nitrogen removal. However, this technology has not yet been applied commercially due to the challenges associated with this process.

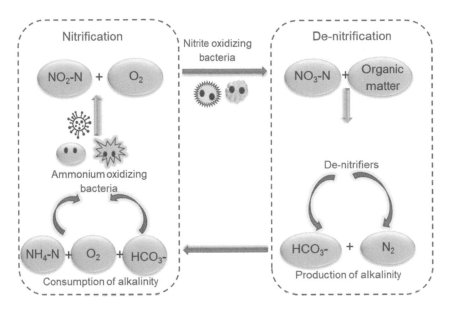

FIGURE 17.3 Natural pH adjustment in SND process in MABR.

One of the major hindrances for upgrading MABR is the low total nitrogen removal efficiencies reported at low HRT values due to low denitrification rates (Premarathna & Visvanathan, 2019). Types of wastewater treated by MABR and the observed removal efficiencies are presented in Table 17.1.

According to these observations, high COD and NH_4^+-N removal efficiencies have been achieved while TN removal efficiencies were around 45–60% at low HRT values in the range between 2.4–7.5 hours (Cote et al., 2016; Li, 2018).

Since high nitrification rates have been reported, low TN removal efficiencies could be due to the low denitrification rates taking place in MABR at low HRTs. Therefore, it is critical to improving the total nitrogen removal efficiency at comparatively low HRT values in MABR through the improvement of denitrification efficiency.

17.2.4 APPLICATION OF POLYVINYL ALCOHOL GEL BIOCARRIERS IN MABR SYSTEMS

Microbial immobilization using bio carriers is gaining interest in the wastewater treatment as a technology with the potential to improve biological process. Among most of the biocarriers, polyvinyl alcohol (PVA) gel has special characteristics such as non-toxicity to the microorganisms and low cost associated with the process. Also, PVA beads have an effective microporous structure in which the microbes could be retained properly (Chaikasem et al., 2015).

A bench-scale study had been conducted to enhance total nitrogen removal from synthetically prepared wastewater using MABR through the immobilization of denitrifiers in PVA gel as biocarriers (Premarathna & Visvanathan, 2019). According

TABLE 17.1

COD and TN removal efficiencies in MABR

Synthetic wastewater	-	Nonporous silicone	2.5	32	-	94.7	Li et al. (2016)
Landfill leachate	0.3–1.5	PDMS HF	60	120–480	-	80–99	Syron et al. (2015)
-	2.4–7.3	-	2.6	12	90	91	Sun et al. (2015)
Oil-field wastewater	15.5	-	-	-	82.3	71.9	Li et al. (2015)
Urban river	-	-	-	15	87	95	Li et al. (2016)
Mixed pharmaceutical wastewater	12.2–43.8	Hydrophobic polypropylene dense membrane	9,216	39–50	90	80	Wei et al. (2012)
Typical municipal RO concentrate	5.8	HF membrane	6	24	92	79.2	Quan et al. (2018)
Synthetic wastewater	4.4	Dense nonporous HF membranes	0.8	6	91.6	46.3	Duvall (2017)
Primary effluent from wastewater treatment plant	17.1	Zeelung module	6,800	7.5	77.5	80.90	Peeters et al. (2016)
Synthetic wastewater	1	Silicone membrane	0.2	14.5	96.9 ± 1	72.0 ± 4.8	Kinh et al. (2017)
Synthetic surface water	8	PVDF HF	470	36	80	60.3	Li and Zhang (2017)
Synthetic domestic wastewater	6	PDMS HF	6	12	>92.2	68.6	Premarathna and Visvanathan (2019)

to the results from this study, the nitrification performance and TN removal efficiencies have been increased by 14% and 13.4%, respectively, after this modification. Also, due to the improved denitrification efficiency, utilization of COD was improved resulting in reduced activity of aerobic heterotrophs, which facilitated enhanced nitrification rates. In addition to that, low biofilm thickness has been observed throughout the experiment due to the faster consumption of COD for the enhanced denitrification process introducing a new biological way of biofilm thickness control in MABR. Therefore, the application of PVA gel in MABR has paved a new way of upscaling the MABR process from laboratory scale to pilot scale.

17.2.5 RECENT DEVELOPMENTS IN MABR TECHNOLOGY

17.2.5.1 Multiple-Stage MABR Systems

A two-stage continuous flow MABR system combined with a pre-hydrolysis acidification process has been used for treating urban river water (Li et al., 2016). According to the results, this combined system has provided 87% and >95% of COD and NH_4-N removal efficiencies, respectively. An anaerobic environment with DO below 0.5 mg/L has been observed in the first MABR while aerobic biofilm was formed better inside the second MABR. Most of the COD entered with the feed had been consumed in the first MABR, and the organic loading of the second MABR has been low. Therefore, in the first MABR, denitrification has occurred, while in the second MABR, nitrification was conducted. From this perspective, this staged MABR system was similar to the conventional activated sludge system in which two tanks were used for nitrification and denitrification. The multiple-stage MABR system is represented in Figure 17.4(**d**).

17.2.5.2 Hybrid MABR Systems

MABR has been integrated with other technologies such as activated sludge process for enhancing nitrogen and organic removal. One of the main benefits of an MABR system coupled with an ASP is the avoidance of excessive biofilm thickness growth.

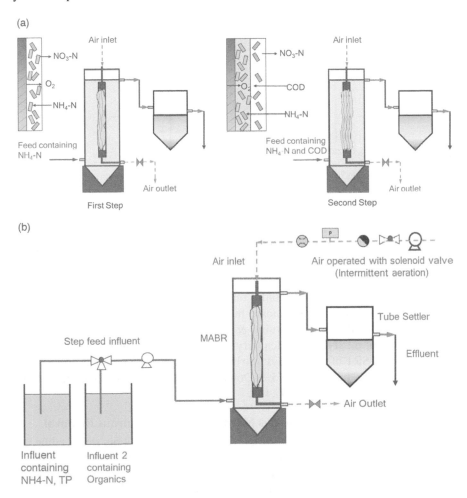

FIGURE 17.4 (a) Two-step startup procedure in MABR; (b) Simultaneous nitrogen and phosphorous removal in MABR with intermittent aeration and step feeding;

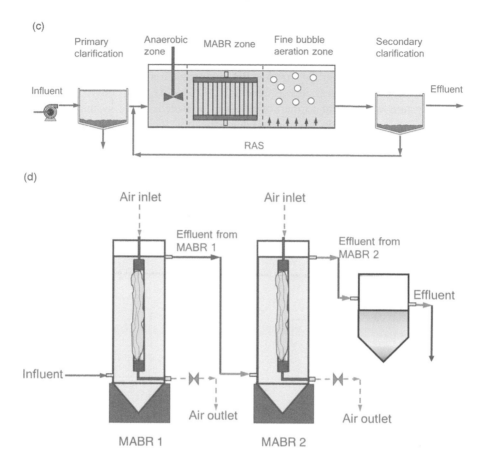

FIGURE 17.4 (Cont.) (c) Hybrid MABR systems; (d) Multiple-staged MABR system.

In the hybrid MABR systems, biofilm on the membrane is devoted to nitrification, and suspended biomass in the anoxic bulk liquid is used for organic compounds removal and denitrification (Downing & Nerenberg, 2007). Hybrid MABR causes the decoupling of nitrifying and denitrifying cultures. In this hybrid MABR system, with bulk liquid SRT of only 5 days, nitrification rate has been around 0.85 gN/m^2.d and the TN removal has reached 75% (Downing & Nerenberg, 2007). Hybrid MABR systems are presented in Figure 17.4(c).

17.2.5.3 MABR for Simultaneous Nitrogen and Phosphorous Removal

A sequencing batch MABR (SBMABR) with intermittent aeration, step-feeding, and periodic backwashing has been used to enhance simultaneous biological phosphorous and nitrogen removal (Sun et al., 2015). The removal efficiencies of COD, NH$_4$-N, TN, and TP in this SBMABR have been maintained above 90%, 96%, 91%, and 85%, respectively. Representation of MABR for simultaneous nitrogen and phosphorous removal is provided in Figure 17.4(b).

17.2.5.4 Shortcut Ammonium Nitrogen Removal over Nitrite (SHARON) in MABR

Several MABR studies have been conducted using the SHARON-ANNAMOX process for nitrogen removal from wastewater (Pellicer-Nacher et al., 2010; Downing & Nerenberg, 2007; Gilmore et al., 2013). Compared to the conventional nitrogen removal pathway, this method has several benefits such as a lower requirement for aeration energy and an external carbon source. Intermittent aeration had been conducted for providing the aerobic condition for aerobic AOB and anaerobic conditions for anaerobic AOB (ANNAMOX bacteria). Autotrophic nitrogen removal up to around 68% and 75% have been reported in intermittently aerated MABR systems, at the O_2/NH_4-N loading ratios of 1.73 and 2.81, respectively (Pellicer-Nacher et al.,2010; Gilmore et al.,2013).

17.2.5.5 Two-Step Startup Procedure in MABR

The traditional procedure of starting up an MABR encourages the growth of both autotrophic nitrifiers and aerobic heterotrophs on the membrane surface in one step. This causes competition between two types of microorganisms for DO consumption. A study has been conducted for developing a two-step startup strategy for the establishment of a layered biofilm in MABR (Wang et al., 2019). Influent with only constant NH_4-N has been provided in the first step and constant NH_4-N and COD have been provided in the second step to the MABR. Nitrifiers have formed the stable biofilm at first and aerobic heterotrophs were grown on top of the nitrifying biofilm in the second step. According to the results, in this two-step startup procedure of MABR, it has been possible to achieve double the specific removal rates of ammonium and COD compared to the reactors that use one-startup step (Wang et al., 2019). The two-step startup procedure is presented in Figure 17.4(a).

17.2.6 Challenges of MABR Systems

17.2.6.1 Comparatively Low TN Removal Efficiency

One of the major hindrances for upgrading MABR is the low total nitrogen removal efficiencies reported at low HRT values due to low denitrification rates as observed in Table 17.1. A bench-scale study had been conducted to enhance total nitrogen removal from synthetically prepared wastewater using MABR through the immobilization of de-nitrifiers in PVA gel as biocarriers (Premarathna and Visvanathan, 2019). A maximum of 14% and 13.4% increase in nitrification and TN removal efficiencies have been reported due to this modification.

17.2.6.2 Biofilm Thickness Controlling in MABR

Mechanical, chemical, and biological methods have all been used for the biofilm thickness controlling in MABR. In most of the MABR studies, intense scouring methods have been applied for the biofilm thickness controlling (Pankhania et al., 1994; Terada and Tsuneda, 2003; Premarathna & Visvanathan, 2019; Semmens et al., 2003). However, biofilm could be damaged due to these methods and it is essential to investigate the microbial activity after biofilm thickness controlling to determine its effect on MABR performance.

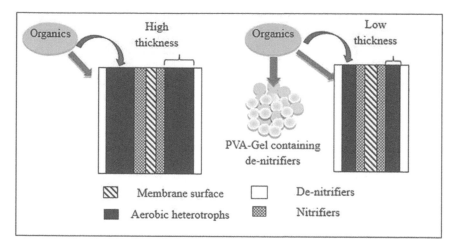

FIGURE 17.5   Hypothesized biofilm thickness reduction with PVA gel biocarriers.

In addition to these methods, using PVA gel biocarriers for enhancing the denitrification rates has been reported to control the biofilm growth on the membrane surface in MABR (Premarathna & Visvanathan, 2019). For effective biofilm thickness controlling, it is required to control the activity of aerobic heterotrophs. When PVA gel containing de-nitrifiers were added, a higher fraction of the supplied COD could be utilized by the de-nitrifiers grown in PVA gel beads due to the enhanced denitrifying activity. This would have caused the reduction of aerobic heterotrophic activity in the biofilm, which led to the reduction of the rate of increasing the biofilm thickness. Hypothesized biofilm thickness controlling in MABR is presented in Figure 17.5.

However, to prove this concept, measuring the biofilm thickness should be conducted before and after adding the PVA gel beads. Also, the relationship between the biofilm thickness and the reactor performance in terms of nitrogen and COD removal could be derived by measuring the biofilm thickness using microsensors together with the removal efficiencies.

17.2.6.3   Improving Oxygen Transfer Efficiency of the MABR System

Conducting an MABR operation with pure oxygen instead of air results in high oxygen transfer efficiencies. Complete oxygen utilization efficiency would be possible due to the complete and bubbleless mass transfer of high purity oxygen through the membrane wall directly to the biofilm. MABR systems supplied with pure oxygen are distinct from conventionally aerated biofilms for two reasons. Firstly, for the same specific surface area, significantly higher biomass concentrations could be achieved in MABR owing to comparatively higher oxygen penetration depths (Syron & Casey, 2013). This provides the opportunity to treat wastewater with comparatively higher strength. In the study conducted by Brindle et al. (1999), it has been possible to achieve good removal efficiencies at high influent concentrations up to 2,400 mg/L of COD. In addition to that, due to the low solubility of oxygen in water, the maximum oxygen diffusion rate in conventionally aerated biofilms is low, causing the limitation of oxygen in many biofilm-based aerobic wastewater processes. When pure oxygen is used, MABR systems have the potential to offer significantly higher oxygen transfer rates (Syron & Casey, 2013).

17.2.6.4 Effect of Membrane Configuration for MABR Performance

A coiled (spiral) membrane provides a dynamic flow path through large bundles of membrane fibers to optimize the fluid contact with the membranes. Therefore, it offers higher mass transfer performance than straight membranes. A secondary flow will appear in a curved membrane due to the frequent change of the main direction of the fluid particles. There will be an adverse pressure gradient generated between the internal and external walls of the membrane from the curvature. This complex flow pattern in a curved pipe/tube is called Dean Vortex, and it improves the mass transfer in a membrane (Moulin et al., 1996).

Due to the coiled/helical shape of the membrane, there will be always a part of the membrane that will be perpendicular to the bulk liquid flow, no matter the direction of the liquid flow. Therefore, the formation of thick laminar boundary layers between the membrane and the bulk liquid is not allowed. This results in the enhanced transfer of oxygen from the membrane lumen to the bulk liquid, which might also improve the transfer rate of NH_4-N ions from the bulk liquid to the inner nitrifying layers, improving the nitrification efficiency. In the MABR study conducted by Castello et al. (2019), a helical membrane has provided considerable enhancement up to 69% of specific volumetric nitrification rates compared to a traditional straight membrane. The advantages of spiral membrane modules for MABR systems have been presented in Figure 17.6.

17.2.6.5 Microbiological Analysis of Biofilm in MABR

Using the Fluorescence In-Situ Hybridization (FISH) analysis and Confocal Laser Scanning Microscopy (CLSM) studies, it is possible to identify the dominant types of bacteria involved in the nitrogen and organic removal at different operating conditions. Variation of their activity in the presence and the absence of PVA gel modification should also be studied. In addition to that, the presence/absence of predators

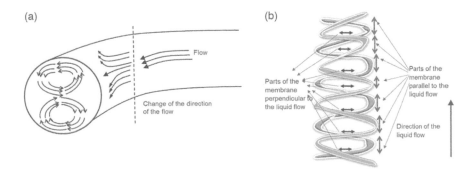

FIGURE 17.6 Advantages of spiral membrane module for MABR systems: (a) formation of secondary flow; (b) minimization of the thickness of liquid boundary layer.

such as protozoa, amoeba that feed on biofilms and create large voids at the base of the biofilm, could be verified using these advanced microbial methods. Also, measuring the oxygen gradient along the thickness of the biofilm using an oxygen microelectrode would provide an indication of the activity of nitrifying bacteria and aerobic heterotrophs.

17.2.6.6 Sludge Production in MABR Systems

According to OXYMEM (2019), MABR generates 50% less sludge content in comparison to conventional ASP. Due to biomass attachment on the membrane surface, the concentration of the suspended solids in the bulk liquid is low, so less sludge is produced. Unlike in the ASP, which is a suspended growth system, in the MABR systems, due to the attachment of biomass on the membrane surface, SRT of the sludge is detached from HRT. Therefore, biomass is retained within the system without the risk of washout causing less sludge yield values. The yield of the sludge in the system is less than 0.2 g TSS/g of COD removed. However, the sludge production in MABR has not been much investigated, which is a crucial factor in scaling up this technology. In addition to measuring the sludge content, sludge volume index (SVI) and capillary suction time (CST) experiments could be conducted for the determination of settleability and dewaterability of the produced sludgey.

17.3 ANAEROBIC MEMBRANE BIOREACTOR

Anaerobic digestion is the ideal process for the treatment of industrial wastewater with high concentrations of biodegradable organic matter. Anaerobic treatment technologies such as up-flow anaerobic sludge blanket (UASB), an expanded granular sludge bed reactor, is capable of achieving high COD removal efficiencies (Dvořák et al., 2016). However, the application of conventional anaerobic biological systems has faced challenges due to the issues of low biomass retention. Biomass retention is one of the most important factors in the anaerobic treatment process since sufficient SRT is required for the methanogenesis process (Lin et al., 2013). Although biofilm processes and granule formation could provide high retention of the biomass in high-rate anaerobic reactors, they require a long startup period and involve complex physicochemical and biological interactions (Lin et al., 2013).

Since the use of membranes in aerobic biological wastewater treatment has been already established, which can provide high microbial retention, the focus was shifted towards extending this technology for the anaerobic processes. This has resulted in the introduction of anaerobic membrane bioreactor (AnMBR), which combines the advantages of the anaerobic treatment process with membrane separation in the wastewater treatment.

17.3.1 Advantages of AnMBR Process

AnMBR has the advantages of anaerobic digestion such as high organic matter removal efficiency, low excess sludge production, stable operation, and production of

TABLE 17.2

Comparison between conventional anaerobic digestion and AnMBR process

Description	Conventional Anaerobic Digestion Process	AnMBR Process
Removal efficiency of organics	High	High
Quality of the effluent	Poor/moderate	High
Sludge production	Low	Low
Organic loading rate	High	High
Biomass retention	Low	High
Required startup time	2–4 months	< 2 weeks
Energy recovery	Yes	Yes
Footprint	High to moderate	Low

biogas as well as the advantages of membrane treatment such as high effluent quality with less total suspended solids (TSS) and bacteria contents (Dvořák et al., 2016). Produced biogas could be combusted for the production of heat and this heat could be used for maintaining the required temperatures for the anaerobic digestion process.

Around 2.02 kWh/kg COD removed, produced from AnMBR, has been approximately 7 times more than required to operate the system (Lin et al., 2013). AnMBR systems have shorter startup periods compared to the UASB reactors and according to the studies conducted by Hu & Stuckey (2006) and Lin et al. (2011), 6-day and 12-day startup periods have been reported in AnMBR studies. A comparison between anaerobic digestion and AnMBR is shown in Table 17.2.

In addition to that, since there is no requirement of oxygen for the biodegradation of organic matter, energy consumption is reduced. AnMBR systems are capable of operating under extreme conditions such as high salinity and TSS contents, as well as poor biomass granulation. Application of AnMBR for high strength wastewater with high salinity and toxicity is also highly favorable (Jeison & van Lier, 2008). While extreme conditions usually cause system failure due to lack of biomass retention, where they are washed out even in high rate anaerobic reactors, AnMBR has the ability to facilitate this due to its ability to retain biomass while adapting to the extreme conditions.

Since nitrogen and phosphorous could remain in the effluent from AnMBR, the treated water can be used for non-potable uses such as irrigation in agricultural work (Martinez-Sosa et al., 2011). Therefore, AnMBR systems could be used for achieving the goal of resource recovery.

17.3.2 Application of Thermophilic Conditions in Anaerobic MBR

Thermophilic anaerobic treatment is one of the areas with a high potential of AnMBR, and it requires more research to optimize the performances. Jeison & van Lier (2006) have provided a comparison of AnMBR operations in thermophilic and mesophilic conditions, indicating the ability to achieve high OLRs with smaller HRT in thermophilic AnMBR.

17.3.3 Types of Wastewater That Have Been Treated by AnMBR

There are four types of wastewater based on two aspects of the concentration of the pollutants and particulate nature:

a. High strength, low particulate
b. High strength, high particulate
c. Low strength, high particulate
d. Low strength, low particulate

High-strength soluble wastewater is currently being treated in high rate anaerobic sludge retaining reactors such as UASB. Therefore, the application of AnMBR in this type of wastewater is attractive only if higher suspended solids removal is required for applications such as wastewater reuse purposes. Wastewaters with high particulate concentration require higher SRT as well as a compact system with higher biomass concentration for the complete hydrolysis of slowly degrading particulates.

However, AnMBR provides an excellent environment for the degradation of wastewater with high particulate matter content. Therefore, AnMBR has an extensive opportunity to be used in the treatment of effluents from distillery, brewery, potato starch, slaughterhouses, pulp and paper, palm oil mill, tanneries, and gelatin manufacturing industries, as well as for the treatment of sludge from wastewater treatment plants.

17.3.4 Configurations of AnMBR

Either crossflow pressurized membrane modules or submerged membranes could be used in AnMBR systems depending on the pressure or vacuum applied (Chang, 2013). In the external crossflow membrane filtration, conventional plate and frame membranes or cylindrical hollow fiber membranes can be used. The cross-flow velocity of the liquid across the surface of the membrane serves as the principle mechanism to disrupt the cake formation on the membrane.

The first AnMBR study was done by Grethlein (1978), who used an external crossflow membrane for the treatment of septic tank effluent and achieved 85–95% of biochemical oxygen demand (BOD) removal efficiency and 72% of nitrate removal efficiency. The external configuration is mostly used in AnMBR because it facilitates easier cleaning of the fouled membrane modules by isolating the chamber, without having to physically remove the membrane modules. As an advancement, the external configuration of membrane modules has been studied under gravity flow instead of having a pressure pump, and it has been possible to achieve 88% of COD removal efficiency for domestic wastewater (Lew et al., 2009).

When the membrane is immersed into the bioreactor and operated under the vacuum, instead of under direct pressure, the configuration is called the submerged membrane bioreactor due to the location of the membrane. In the internal MBR configuration, a pump or gravity flow due to the elevation difference is used to withdraw permeate through the membrane. The immersed membranes include submerged flat sheet membranes and the hollow fiber membrane modules. This mode has benefits

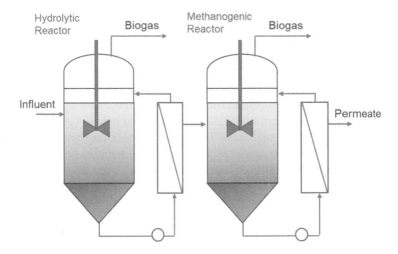

FIGURE 17.7 Two-stage anaerobic MBR systems.

including much lower energy consumption and less cleaning required (Lin et al., 2013). Kubota Corporation has developed a submerged AnMBR named KSAMBR, which has been applied in several full-scale food and beverage industries (Kanai et al., 2010). For the anaerobic MBR systems, biogas could be used for air scouring purposes. Some of the submerged AnMBR application studies have been reported by Lin et al. (2010), who did the recirculation of produced biogas for the purpose of air scouring.

In the vacuum-driven configuration, the membrane could be either directly placed in the bioreactor or immersed in a separate chamber. The latter configuration now looks like an external membrane and it will require a pump to return the retentate to the bioreactor. However, unlike the external crossflow membrane, the membrane here is operated under vacuum, instead of under the pressure.

17.3.4.1 Two-Stage Reactor Configuration for AnMBR

Another AnMBR reactor configuration is the two-stage reactor configuration. In two-stage AnMBR, the reactions of hydrolysis, acetogenesis, and acidogenesis have occurred within the first reactor, which is called the hydrolytic (acidogenic) reactor, followed by the methanogenic reactor where the methanogenesis takes place (see Figure 17.7).

In a single-stage reactor, with both of these processes taking place inside, it is difficult to provide the optimum conditions required for acid formation and methane formation such as optimum pH requirements for methane production. In a two-stage reactor, both reactors will be operating with the optimized operating conditions for the respective types of bacteria for bringing the maximum control of bacterial communities inside the reactor (Visvanathan & Abeynayaka, 2012). In a two-staged AnMBR study, for the degradation of UV-absorbing compounds in landfill leachate, it has been possible to achieve the removals of >55% of the TOC and >60% of the total COD present in the raw leachate.

17.3.5 Performance of AnMBR Systems

The instability of the attached growth and granule process at thermophilic anaerobic conditions insist on the application of the membrane process irrespective of the wastewater conditions. Due to the excellent biomass retention properties, the ability to apply in thermophilic conditions, and particulate wastewater conditions, the performance of AnMBR is quite successful. AnMBR systems have operated in a wide range of different feed concentrations, loading rates, and reactor types in mesophilic as well as in thermophilic conditions. The enhanced methane yield and biodegradability of the thermophilic anaerobic process favors the operation of anaerobic wastewater treatment in reactors at high temperatures. Apart from that, the treatment of high-temperature effluent without cooling is possible with thermophilic AnMBRs.

Most of the AnMBR studies have been conducted with synthetic wastewater at the initial stage due to the ease of process control. Several feed solutions including domestic wastewater as well as simulated high salinity wastewater have been used as the feed. It has been possible to achieve good removal efficiencies of more than 90% in almost all the studies. Better performance observed with membrane coupling in this study is attributed to higher biomass retention capacity of the AnMBR since the issue of biomass washout under extreme conditions, which cause the failure of UASB systems alone, cannot be observed in this system. The recirculation velocity in crossflow should not exceed 5 m/s to minimize biomass activity reduction. Most of the studies using biogas bubbling as a fouling reduction strategy have achieved successful performances. Pollutant removal efficiencies achieved by AnMBR have been presented in Table 17.3.

17.3.6 Challenges and Future Research Requirements in AnMBR

Methane yield from AnMBR ranges from 0.23 to 0.33 LCH_4/g COD removal (Martinez-Sosa et al., 2011; Lin et al., 2013), which is lower compared to the theoretically calculated methane yield around 0.382 LCH_4/g COD removals at 25°C. This observation could be due to the high solubility of CH_4 in liquid and a significant fraction of the generated CH_4 is wasted with the effluent in that manner. Several methods have been employed to reduce this effect such as stripping off the AnMBR effluent through post-treatment aeration, recovery of CH_4 using a degassing membrane, etc. (Lin et al., 2013). Since it is possible to recover the operational costs through the production of biogas, optimization of the recovery of methane is highly important in the AnMBR processes. In addition to that, it is required to remove the micropollutants from AnMBR effluent before introducing it for wastewater reuse applications.

Anaerobic MBR systems have not been able to achieve the higher rates that would have been possible by aerobic processes (Visvanathan & Abeynayaka, 2012). Under similar types of wastewater, AnMBR requires several times larger retention times and accommodates lower loading rates compared to aerobic MBR systems. Under aerobic conditions, significantly higher organic loading rates can be achieved. Also, the ammonia removal rates at elevated temperatures in aerobic MBR are higher due to stripping from the open reactor with high aeration rates and that type of ammonia removal rate has not been observed in AnMBR due to closed reactor conditions.

TABLE 17.3

COD removal efficiencies in AnMBR process

Scale	Type of wastewater	Type of membrane	Influent COD con. (g/L)	Effluent COD con. (g/L)	HRT (d)	SRT (d)	COD removal efficiency (%)	Reference
P	Municipal	UF	0.4	0.08	-	-	90	Martinez-Sosa et al. (2011)
L	Landfill leachate	Dynamic	13	4.91	2.5	125	62	Xie et al. (2014)
L	Landfill leachate	UF	5–19	2–9.5	45	60	50–60	Pathak et al. (2018)
L	Food	UF	0.65	0.03–0.08	1–5	50	88–95	Galib et al. (2016)
P	Food industry (salad dressings)	MF	39	0.21	-	-	99	Christian et al. (2011)
P	Food industry (snacks)	UF	11	1.7	-	-	75	Ramos et al. (2014)
P	Food industry (with high oil content)	MF	7.9	0.18	-	-	97	Diez et al. (2012)
L	Food industry (sugarcane vinasse)	MF	17.7	0.49	-	-	96	Mota et al. (2013)
P	Slaughter-house	UF	5.92	0.07	2–7	50–1000	95	Jensen et al. (2015)

P = Pilot Scale; L – Laboratory Scale.

Anaerobic MBR processes are operated at high biomass conditions due to reduced specific biodegradation rates compared to aerobic processes. However, operations at higher biomass concentrations could create issues in pumping that increases the membrane fouling, which is one of the major drawbacks in AnMBR applications. When the membrane is fouled, permeate flux is reduced and frequent cleaning is required for avoiding further fouling that can deteriorate the membrane. Therefore, membrane fouling could cause high operational costs as well as increasing membrane replacement costs. Membrane fouling in AnMBR takes place due to two major reasons: including cake layer formation by biocells and the inorganic precipitation on the membrane surface (Visvanathan & Abeynayaka, 2012).

The high mixed liquor suspended solids (MLSS) concentration in the reactor and the smaller size of biosolid particles have been identified as potential causes for increased membrane fouling. The mechanical stress applied to the biomass during the pumping has been identified as one of the reasons causing size reduction of the bioflocs, which increases the fouling in AnMBR systems. In addition to that, cross-flow velocity applied to biomass also affects the size reduction of biomass.

17.4   SUSTAINABILITY OF MBR, MABR AND AɴMBR SYSTEMS

When considering the sustainability of the wastewater treatment sector, reduction of energy consumption could be identified as a priority. Aeration is considered the largest energy consumer in a wastewater treatment plant. In a typical wastewater treatment plant, around 40–60% of energy is consumed for aeration by mechanical means or diffused air devices. However, from the energy supplied for the aeration, only 5–25% is passed to the liquid phase while the rest is wasted as air bubbles (Aybar et al., 2014).

Despite the high effluent quality provided by MBR systems, they have disadvantages such as high energy consumption for the aeration required for the treatment of pollutants as well as for the reduction of membrane fouling. In addition to that, the MBR process requires a separate anoxic tank for the denitrification process, which adds up the energy requirement for internal recirculation of produced nitrate ions in the pre-anoxic processes. In the external configuration of the MBR process, recirculation pumps are required for providing the required transmembrane pressure, which results in high energy consumption.

However, MABR could be introduced as a sustainable technology for wastewater treatment, mainly due to the reduction of the energy footprint in a typical wastewater treatment plant.

17.4.1   ENERGY NEUTRALITY AND FEWER REQUIREMENT OF THE CAPACITY IN MABR

In MABR, since oxygen can be delivered directly to bacteria using diffused aeration without the use of bubble aeration, four to five times higher the oxygen transfer efficiency could be achieved compared to a fine bubble aeration system (WaterWorld, 2017). Since MABR employs self-respiring membranes to achieve aeration, it leads to much lower energy consumption in wastewater treatment processes. MABR modules can operate at atmospheric pressure in contrast to the activated sludge process, where the air has to be forced at high pressures through fine bubble diffusers.

The airflow rate through the membrane lumen can be reduced, allowing more time for the transfer of oxygen, which results in very high oxygen transfer efficiencies around 50–90% (WaterWorld 2017). The low-pressure air combined with high oxygen transfer efficiencies leads to significant energy savings of over 75% compared to the traditional aeration technologies. In addition to the aeration energy savings, there are additional energy savings from the elimination of nitrate recirculation pumping since the process can achieve simultaneous nitrification and denitrification in the single MABR unit.

According to the observations, the energy requirement for aeration and mixing of a full-scale MABR is approximately 0.25 kWh/kg COD removed which is three to four times less intensive than the conventional activated sludge process, which consumes 1.05 kWh/kg COD removed (Syron and Casey, 2008). According to modeling conducted by Aybar et al. (2014), the reduction of power requirements by MABR is as high as 86% resulting in high oxygen transfer efficiencies. Energy consumption for a pilot-scale MABR has been reported as less than 0.1 kWh/m^3 (Syron et al., 2015), which indicates the energy efficiency of MABR technology.

Although typical bubble diffusion can transfer 1 to 2 kg O_2/kWh, in an MABR study conducted by Cote et al. (2016), aeration efficiencies of 3, 4, and 5 kgO_2/kWh have been reported and aeration efficiency of 4 kgO_2/kWh corresponds to an oxygen flux of 15 g/d/m^2. It is reported that it is possible to optimize the process to maintain an aeration efficiency greater than 6 kg O_2/kWh, which is four times greater than conventional bubble aeration.

Oxygen transfer rate is an important factor that affects the power consumption in MABR. In conventional bubble aeration, oxygen transfer efficiencies are around 12–37%, while MABR could achieve 100% oxygen transfer efficiency when operated in dead-end mode.

The ability of MABR to incrementally increase the capacity of an operating wastewater treatment plant without the need for additional tanks opens up the opportunity to increase the capacity as required and avoid long planning and construction periods. Conventional wastewater treatment technologies require large footprints. Many wastewater treatment plants that need to expand plant capacity or improve the effluent quality do not have the required footprint or capital cost requirements. MABR allows the plants to achieve more capacity from the existing space by increasing the biomass concentration on the membranes.

17.4.2 BIOGAS GENERATION IN AnMBR SYSTEMS

AnMBR systems also have sustainability aspects since they produces biogas that can be converted to heat needed to keep the required temperatures in the anaerobic digestion system. As mentioned in the earlier sections of this chapter, according to the study conducted by Lin et al. (2013), the energy content of the biogas produced from AnMBR is approximately seven times more than required to operate the system. Although membrane costs and gas scouring energy accounted for the largest percentage of total life cycle capital costs and operational costs, the cost of the operation could be completely recovered by recovering the biogas.

In addition to that, since there is no oxygen requirement for the biodegradation of organic matter, energy consumption for the system is reduced. This technology can be used for the treatment of high-strength wastewaters with minimum energy requirement, which would otherwise require a very high oxygen requirement for the operation. Also, high concentrations of organic matter contain high energy content that could be recovered in the form of biogas. To recover the maximum energy content, anaerobic MBR could be directly applied to high strength wastewater, which would increase the fraction of energy recovered as biogas.

In addition to that, to reduce the cake formation on the membranes in submerged AnMBR systems, the produced biogas could be recirculated instead of applying bubble aeration. Several studies have used biogas recirculation as a successful membrane-fouling reduction strategy (Sittisom et al., 2020).

17.5 CONCLUSION

An MBR system replaces the activated sludge process by using a membrane instead of the secondary clarifier for the filtration. High organic matter, ammonium nitrogen, and suspended solids removal could be achieved with MBR, and for the total nitrogen removal, it is required to be modified with a separate anoxic compartment. Membrane fouling is one of the challenges of the MBR process and further research should be conducted for the mitigation of this effect. MABR eliminates the disadvantages associated with the MBR process such as high energy consumption for the aeration and the requirement of the addition of an anoxic compartment for total nitrogen removal due to the ability to perform simultaneous nitrification and denitrification in a single unit. Biofilm thickness controlling in MABR is an issue that requires further consideration. AnMBR combines the advantages of both anaerobic digestion and membrane filtration. Due to the biogas generation, it has the ability to recover the operational cost of the AnMBR process, and wastewater with high organic strength could be treated by this technology. However, it is required to explore the economical ways of recovery of methane for achieving the real benefits from this treatment process. Due to the high energy efficiency of the MABR system aeration and the ability of a generation of energy in terms of biogas in the AnMBR process, they could be considered as more sustainable compared to MBR systems.

REFERENCES

Aybar, M., Pizarro, G., Boltz, J. P., Downing, L., & Nerenberg, R. (2014). Energy-efficient wastewater treatment via the air-based, hybrid membrane biofilm reactor (hybrid MfBR). *Water Science and Technology*, 69(8), 1735–1741.

Brindle, K., Stephenson, T., & Semmens, M. J. (1998). Nitrification and oxygen utilization in a membrane aeration bioreactor. *Journal of Membrane Science*, 144(1–2), 197–209.

Brindle, K., Stephenson, T., & Semmens,M. J.(1999). Pilot-plant treatment of a high strength brewery wastewater using a membrane-aeration bioreactor. *Water Environmental Research*, 71 (6), 1197–1204.

Castello, M., Díez-Montero, R., Esteban-García, A. L., & Tejero, I. (2019). Mass transfer enhancement and improved nitrification in MABR through specific membrane configuration. *Water Research*, 152, 1–11.

Chae, S., R., & Shin, H. S. (2007). Characteristics of simultaneous organic and nutrient removal in a pilot-scale vertical submerged membrane bioreactor (VSMBR) treating municipal wastewater at various temperatures. *Process Biochemistry*, 42, 193.

Chaikasem, S., Jacob, P., & Visvanathan, C. (2015). Performance improvement in a two-stage thermophilic anaerobic membrane bioreactor using PVA-gel as biocarrier. *Desalination and Water Treatment*, 53(10), 2839–2849.

Chang, S. (2013). Anaerobic membrane bioreactors (AnMBR) for wastewater treatment. *Advances in Chemical Engineering and Science*, 04(01), 56–61.

Christian, S., Grant, S., McCarthy, P., Wilson, D., & Mills, D. (2011). The first two years of full-scale anaerobic membrane bioreactor (AnMBR) operation treating high-strength industrial wastewater. *Water Practice and Technology*, 6(2).

Cote, P., Peeters, J., Adams, N., Hong, Y., Long, Z., & Ireland, J. (2016). A new membrane-aerated biofilm reactor for low energy wastewater treatment: Pilot results. *Proceedings of the Water Environment Federation*, 4226–4239.

Diez, V., Ramos, C., & Cabezas, J. L. (2012). Treating wastewater with high oil and grease content using an Anaerobic Membrane Bioreactor (AnMBR). Filtration and cleaning assays. *Water Science and Technology*, *65*(10), 1847–1853.

Downing, L., & Nerenberg, R. (2007). Performance and microbial ecology of the hybrid membrane biofilm process for concurrent nitrification and de-nitrification of wastewater. *Water Science and Technology*, *55* (8–9), 355–362.

Duvall, C. J. D. (2017). *Low-Energy Nitrification of Wastewaters using Membrane Aerated Biofilm Reactors*. Master's thesis, The University of Guelph, Ontario, Canada.

Dvořák, L., Gómez, M., Dolina, J., & Černín, A. (2016). Anaerobic membrane bioreactors—a mini review with emphasis on industrial wastewater treatment: applications, limitations and perspectives. *Desalination and Water Treatment*, *57*(41), 19062–19076.

Fudala-Ksiazek, S., Pierpaoli, M., & Luczkiewicz, A. (2018). Efficiency of landfill leachate treatment in a MBR/UF system combined with NF, with a special focus on phthalates and bisphenol A removal. *Waste Management*, *78*, 94–103.

Galib, M., Elbeshbishy, E., Reid, R., Hussain, A., & Lee, H. S. (2016). Energy-positive food wastewater treatment using an anaerobic membrane bioreactor (AnMBR). *Journal of Environmental Management*, *182*, 477–485.

Gilmore, K. R., Terada, A., Smets, B. F., Love, N. G., and Garland, J. L. (2013). Autotrophic nitrogen removal in a membrane-aerated biofilm reactor under continuous aeration: a demonstration. *Environmental Engineering Science*, *30*(1), 38–45.

Grethlein, H. E. (1978). Anaerobic digestion and membrane separation of domestic wastewater. *Journal (Water Pollution Control Federation)*, 754–763.

Grunditz, C., & Dalhammar, G. (2001). Development of nitrification inhibition assays using pure cultures of Nitrosomonas and Nitrobacter. *Water Resources*, *35*, 433–440.

Hardt, F. W., Clesceri, L. S., Nemerow, N. L., & Washington, D. R. (1970). Solids separation by ultrafiltration for concentrated activated sludge. *Journal (Water Pollution Control Federation)*, 2135–2148.

Hu, A. Y., & Stuckey, D. C. (2006). Treatment of dilute wastewaters using a novel submerged anaerobic membrane bioreactor. *Journal of Environmental Engineering*, *132*(2), 190–198.

Jeison, D. and van Lier, J.B. (2008). Anaerobic wastewater treatment and membrane filtration: a one-night Strand or a sustainable development. *Water Science and Technology*, *57*(2), 527–532.

Jeison, D., & Van Lier, J. B. (2006). Cake layer formation in anaerobic submerged membrane bioreactors (AnSMBR) for wastewater treatment. *Journal of Membrane Science*, *284*(1–2), 227–236.

Jensen, P. D., Yap, S. D., Boyle-Gotla, A., Janoschka, J., Carney, C., Pidou, M., & Batstone, D. J. (2015). Anaerobic membrane bioreactors enable high rate treatment of slaughterhouse wastewater. *Biochemical Engineering Journal*, *97*, 132–141.

Kanai, M., Ferre, V., Wakahara, S., Yamamoto, T., & Moro, M. (2010). A novel combination of methane fermentation and MBR—Kubota submerged anaerobic membrane bioreactor process. *Desalination*, *250*(3), 964–967.

Khan, S., J., Ilyas, S., Javid, S., Visvanathan, C., & Jegatheesan, V. (2011). Performance of suspended and attached growth MBR systems in treating high strength synthetic wastewater. *Bioresource Technology*, *102*(9), 5331–5336.

Kinh, C., T., Suenaga, T., Hori, T., Riya, S., Hosomi, M., Smets, B. F., et al. (2017). Counter-diffusion biofilms have lower N_2O emissions than co-diffusion biofilms during simultaneous nitrification and denitrification: Insights from depth-profile analysis. *Water Research*, 363–371.

La Para, T. M., Cole, A. C., Shanahan, J. W., & Semmens, M. J. (2006). The effects of organic carbon, ammoniacal-nitrogen, and oxygen partial pressure on the stratification of membrane-aerated biofilms. *Journal of Industrial Microbiology and Biotechnology*, *33*(4), 315–323.

Lew, B., Tarre, S., Beliavski, M., Dosoretz, C. & Green, M. (2009), "Anaerobic membrane bio-reactor (AnMBR) for domestic wastewater treatment", *Desalination*, **243**(1–3), 251–257.

Li, Q. (2018). *Pilot-Scale Plant Application of Membrane Aerated Biofilm Reactor (MABR) Technology in Wastewater Treatment*. Degree Project in Environmental Engineering, Stockholm, Sweden.

Li, M., Li, P., Du, C., Sun, L., & Li, B. (2016). Pilot-scale study of an integrated membrane-aerated biofilm reactor system on urban river remediation. *Industrial and Engineering Chemistry Research*, *55*(30), 8373–8382.

Li, P., Li, M., Zhang, Y., Zhang, H., Sun, L., & Li, B. (2016). The treatment of surface water with enhanced membrane-aerated biofilm reactor (MABR). *Chemical Engineering Science*, *144*, 267–274.

Li, P., Zhao, D., Zhang, Y., Sun, L., Zhang, H., Lian, M., & Li, B. (2015). Oil-field wastewater treatment by hybrid membrane-aerated biofilm reactor (MABR) system. *Chemical Engineering Journal*, *264*, 595–602.

Li, Y., & Zhang, K. (2017). Pilot scale treatment of polluted surface waters using membrane-aerated biofilm reactor (MABR). *Agriculture and Environmental Biotechnology*, 376–386.

Lin, H. J., Xie, K., Mahendran, B., Bagley, D. M., Leung, K. T., Liss, S. N., et al. (2010). Factors affecting sludge cake formation in a submerged anaerobic membrane bioreactor. *Journal of Membrane Science*, *361*(1–2), 126–134.

Lin, H., Liao, B. Q., Chen, J., Gao, W., Wang, L., Wang, F., & Lu, X. (2011). New insights into membrane fouling in a submerged anaerobic membrane bioreactor based on characterization of cake sludge and bulk sludge. *Bioresource Technology*, *102*(3), 2373–2379.

Lin, H., Peng, W., Zhang, M., Chen, J., Hong, H., & Zhang, Y. (2013). A review on anaerobic membrane bioreactors: applications, membrane fouling and future perspectives. *Desalination*, *314*, 169–188.

Long, Z. (2013). *Tertiary nitrification using membrane aerated biofilm reactors: Process optimization, characterization and model development*. PhD Dissertation, University of Guelph, Ontario, Canada.

Martin, K. J., & Nerenberg, R. (2012). The membrane biofilm reactor (MBfR) for water and wastewater treatment: Principles, applications, and recent developments. *Bioresource Technology*, *122*, 83–94.

Martinez-Sosa, D., Helmreich, B., Netter, T., Paris, S., Bischof, F., & Horn, H. (2011). Anaerobic submerged membrane bioreactor (AnSMBR) for municipal wastewater treatment under mesophilic and psychrophilic temperature conditions. *Bioresource Technology*, *102*(22), 10377–10385.

Metcalf & Eddy (2003). *Wastewater Engineering: Treatment and Reuse* (4th ed.). New York: McGraw-Hills.

Mota, V. T., Santos, F. S., & Amaral, M. C. (2013). Two-stage anaerobic membrane bioreactor for the treatment of sugarcane vinasse: assessment on biological activity and filtration performance. *Bioresource Technology*, *146*, 494–503.

Moulin, P., Rouch, J. C., Serra, C., Clifton, M. J., & Aptel, P. (1996). Mass transfer improvement by secondary flows: Dean vortices in coiled tubular membranes. *Journal of Membrane Science*, *114*(2), 235–244.

Onishi, H., Numazawa, R., & Takeda, H. (1980). Process and apparatus for wastewater treatment. *U.S. Patent No. 4,181,604*. Washington, DC: U.S. Patent and Trademark Office. Retrieved on November 25, 2019, from https://patents.google.com/patent/US4181604A/en

OXYMEM. (2019). *MABR and sludge reduction*. Retrieved on August 5, 2019, from https://www.oxymem.com/mabr-and-sludge-reduction

Pankhania, M., Stephenson, T., & Semmens, M. J. (1994). Hollow fibre bioreactor for wastewater treatment using bubble-less membrane aeration. *Water Research*, *28*(10), 2233–2236.

Pathak, A., Pruden, A., & Novak, J. T. (2018). Two-stage anaerobic membrane bioreactor (AnMBR) system to reduce UV absorbance in landfill leachates. *Bioresource Technology, 251*, 135–142.

Peeters, J., Long, Z., Houweling, D., Cote, P., Daigger, G., T., & Snowling, S. (2016). Nutrient removal intensification with MABR – developing a process model supported by piloting. *Resources, Conservation and Recycling, 27*, 203–215.

Pellicer-Nacher, C., Sun, S., Lackner, S., Terada, A., Schreiber, F., Zhou, Q., & Smets, B. F. (2010). Sequential aeration of membrane-aerated biofilm reactors for high-rate autotrophic nitrogen removal: experimental demonstration. *Environmental Science and Technology, 44*(19), 7628–7634.

Premarathna, N. S. M., & Visvanathan, C. (2019). Enhancement of organics and total nitrogen removal in a membrane aerated biofilm reactor using PVA-Gel bio-carriers. *Bioresource Technology Reports, 8*, 100325.

Quan, X., Huang, K., Li, M., Lan, M., & Li, B. (2018). Nitrogen removal performance of municipal reverse osmosis concentrate with low C/N ratio by membrane-aerated biofilm reactor. *Environmental Science and Engineering, 12*(6), 5.

Ramos, C., García, A., & Diez, V. (2014). Performance of an AnMBR pilot plant treating high-strength lipid wastewater: biological and filtration processes. *Water Research, 67*, 203–215.

Rouse, J. D., Burica, O., Strazar, M., & Levstek, M. (2007). A pilot-plant study of a moving-bed biofilm reactor system using PVA gel as a biocarrier for removals of organic carbon and nitrogen. *Water Science and Technology, 55*(8–9), 135–141.

Satoh, H., Ono, H., Rulin, B., Kamo, J., Okabe, S., & Fukushi, K. (2004). Macroscale and microscale analyses of nitrification and denitrification in biofilms attached on membrane aerated biofilm reactors. *Water Research, 38*(6), 1633–1641.

Semmens, M. J., Dahm, K., Shanahan, J., & Christianson, A. (2003). COD and nitrogen removal by biofilms growing on gas permeable membranes. *Water Research, 37*(18), 4343–4350.

Sittisom, P., Gotore, O., Ramaraj, R. Van Tran, G., Yuwalee, U., & Itayama, T. (2020). Energy and environmental communication: Membrane fouling issues in anaerobic membrane bioreactors (AnMBRs) for biogas production. *Energy & Environment, 1*, 15–19.

Smith, C., V., Gregorio, D., O., & Talcott, R. M. (1969). The use of ultrafiltration membranes for activated sludge separation. *24th Industrial Waste Conference*, Purdue University, Ann Arbor, MI, 1300–1310.

Sun, L., Wang, Z., Wei, X., Li, P., Zhang, H., Li, M., Li, B., & Wang, S. (2015). Enhanced biological nitrogen and phosphorus removal using sequencing batch membrane-aerated biofilm reactor. *Chemical Engineering Science, 135*, 559–565.

Sun, S. P., Nacher, C. P. I., Merkey, B., Zhou, Q., Xia, S. Q., Yang, D. H., Sun, J.H. & Smets, B. F. (2010). Effective biological nitrogen removal treatment processes for domestic wastewaters with low C/N ratios: A review. *Environmental Engineering Science, 27*(2), 111–126.

Sutton, P. M. (2003). Membrane bioreactors for industrial wastewater treatment: The state-of-the-art based on full scale commercial applications. *Proceedings of the Water Environment Federation, 6*, 23–32.

Syron, E., & Casey, E. (2013). Membrane aerated biofilm reactors. *Encyclopedia of Membrane Science and Technology*, 1–19.

Syron, E., & Casey, E. (2008). Membrane-aerated biofilms for high rate bio-treatment: Performance appraisal, engineering principles, scale-up, and development requirements. *Environmental Science and Technology, 42*(6), 1833–1844.

Syron, E., Semmens, M. J., & Casey, E. (2015). Performance analysis of a pilot-scale membrane aerated biofilm reactor for the treatment of landfill leachate. *Chemical Engineering Journal, 273*, 120–129.

Terada, A., & Tsuneda, S. (2003). Nitrogen removal characteristics and biofilm analysis of MABR applicable to high-strength nitrogenous wastewater treatment. *Journal of Bioscience and Bioengineering, 95*(2), 170–178.

Terada, A., Yamamoto, T., Igarashi, R., Tsuneda, S., & Hirata, A. (2006). Feasibility of a membrane-aerated biofilm reactor to achieve controllable nitrification. *Biochemical Engineering Journal, 28*(2), 123–130.

Visvanathan, C., & Abeynayaka, A. (2012). Developments and future potentials of anaerobic membrane bioreactors (AnMBRs). *Membrane Water Treatment, 3*(1), 1–23.

Wang, R., Zeng, X., Wang, Y., Yu, T., & Lewandowski, Z. (2019). Two-step startup improves pollutant removal in membrane-aerated biofilm reactors treating high-strength nitrogenous wastewater. *Environmental Science: Water Research and Technology, 5*(1), 39–50.

WaterWorld. (2017). *The Rise of MABR Technology: Is the Future Bubbleless?* Retrieved on July 23, 2020, from https://www.waterworld.com/international/wastewater/article/16201158/the-rise-of-mabr-technology-is-the-future-bubbleless

Wei, X., Li, B., Zhao, S., Wang, L., Zhang, H., Li, C., & Wang, S. (2012). Mixed pharmaceutical wastewater treatment by integrated membrane-aerated biofilm reactor (MABR) system – A pilot-scale study. *Bioresource Technology*, 189–195.

Xie, Z., Wang, Z., Wang, Q., Zhu, C., & Wu, Z. (2014). An anaerobic dynamic membrane bioreactor (AnDMBR) for landfill leachate treatment: performance and microbial community identification. *Bioresource Technology, 161*, 29–39.

Zhu, I. X. (2008). *Effect of oxygen partial pressure and COD loading on biofilm performance in a membrane aerated bioreactor.* University of Toronto, Toronto, Canada.

18 Membrane Hybrid System in Water and Wastewater Treatment

M. A. H. Johir
University of Technology Sydney (UTS), Broadway, Ultimo,
NSW, Australia

*Saravanamuthu Vigneswaran**
Faculty of Engineering and IT, University of Technology
Sydney (UTS), Broadway, Ultimo, NSW, Australia

CONTENTS

18.1 INTRODUCTION

The application of membranes in water and wastewater treatment has increased significantly over the last two decades due to the small footprint and production of high-quality water for different usage. Microfiltration (MF), ultrafiltration (UF), nanofiltration (NF), reverse osmosis (RO), and their hybrid systems are used in the water industry. Among these processes, membrane bioreactor (MBR) is widely used in wastewater treatment. MBR can handle high organic loading and can also be operated at a higher mixed liquor suspended solids (MLSS) concentration of around 12–18 g/L (Holler and Trösch, 2001). Although the MBR process can remove most of the organic matter and nitrogen compounds from wastewater, it fails when exposed to high and variable loads depending on the operating conditions. Other disadvantages are frequent membrane cleaning and waste sludge production when phosphorus removal is achieved by chemical precipitation. Furthermore, the research on membrane adsorption hybrid system (MAHS) is gaining attention due to its superior ability to remove organic matter. In this regard, two membrane hybrid systems—namely (a) membrane bioreactor and (b) membrane adsorption hybrid system—are discussed in this chapter.

18.2 MEMBRANE BIOREACTOR

MBR is a cost-effective and efficient process in wastewater treatment due to its ability to handle high organic loadings, small footprint, and near-zero sludge production. MBR process removes almost all suspended solids and leads to superior removal of organic matter and nutrients. The MBR system was first developed in the 1970s for the treatment of domestic wastewater. It incorporated a suspended growth biological tank together with a membrane unit, which is used in place of a secondary sedimentation tank. The growth rate of MBR is rapid (over 10% per annum) as compared to other advanced wastewater treatment systems and other membrane systems (Drews, 2010). In 1995, the annualized operation cost of MBR was $0.90/m³, which was reduced to $0.08/m³ in 2005 due to lower membrane costs and improved energy efficiency (below 0.4 kWh/m³) (Hermanowicz, 2011).

18.2.1 Different Configuration of MBR and Membranes Used in MBR

The membrane bioreactor can be configured both in side-stream or submerged systems (Figure 18.1). Each setting has relative advantages and disadvantages. For example, sludge bulking is a problem with submerged MBR, but it is not a problem with side-stream MBR (Sombatsompop, 2007). Standard membrane configurations currently used in practice are hollow fiber (HF), spiral-wound, plate-and-frame/flat sheet (FS), pleated filter cartridge, and tubular (Radjenović et al., 2008).

18.2.2 Removal of Organics and Nutrients in MBR

Biodegradation of organics and nutrients in an aerobic MBR occurs by heterotrophic and autotrophic microorganisms or bacteria. Organic matter is the primary source of

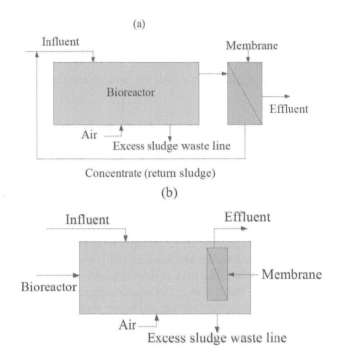

FIGURE 18.1 MBR configurations: (a) side-stream and (b) submerged.

energy. Based on the medium oxygenation, heterotrophs use either oxygen (in case of oxic conditions) or nitrate (in case of anoxic conditions) as an electron acceptor. On the other hand, oxidization of inorganic compounds provides energy to autotrophic bacteria and only utilizes oxygen as an electron acceptor (Mara and Horan, 2003).

The growth rate of autotrophs is slower than heterotrophs because there is less energy produced from the oxidation of inorganic material. In studies, a variety of toxic matter showed an inhibitory effect on autotrophic (Liu, 2007) growth rates. Autotrophic bacteria provide a favorable environment in bioreactor for heterotrophs by reducing the ammonia level.

In an MBR, nitrification takes place when ammonia is converted to nitrite and then nitrate. The conversion of ammonia to nitrite is done by ammonia-oxidizing bacteria (AOB) such as *Nitrosococcus mobilis, Nitrosomonas, Nitrosospira, Nitrosovibrio,* and *Nitrosolobus* whereas nitrate is converted to nitrate by nitrite-oxidizing bacteria (NOB) such as *Nitrobacter* and *Nitrospira* (Halling-Sørensen and Jørgensen, 1993; Schramm et al., 1998). In an aerobic MBR, the removal of phosphorus (P) occurs through the attachment of phosphorus onto bio floc and by phosphate accumulating bacteria. But, the efficient removal of P using MBR depends on various operating conditions of MBR and influent water quality parameters (Zheng et al., 2014). The removal of P using the biological process is environmentally friendly but the removal of P using the biological process is very complicated. Furthermore, the biological process also may not be able to remove P below a certain level.

18.2.3 Operating Parameters of MBR

Solid retention time (SRT), hydraulic retention time (HRT), biomass concentration, organic matter, temperature, mode of operation, and membrane morphology are the dominant operating parameters that affect the performance of an MBR (Le-Clech et al., 2006).

18.2.3.1 Solid Retention Time and Hydraulic Retention Time

SRT is defined as the total volume of the reactor divided by the volume of sludge withdrawn every day. Several researchers have suggested longer SRT for the effective operation of an MBR. For example, SRT of higher than 30 days was recommended by Adham and Gagliardo (1998). On the other hand, some other researchers used SRT of less than 10 days (Cicek et al., 2001). Furthermore, from the literature, it is found that an increase of SRT helped to reduce membrane fouling. For example, an increase of SRT by three to five times (from 2 to 10 days or 20 to 60 days) resulted in lower fouling (Ahmed et al., 2007, Trussell et al., 2006). Ke and Junxin (2009) observed the maximum fouling of MBR with an SRT of 10 days than SRT of invite days. Moreover, the effluent quality of MBR also depends on SRT. Better removal of organic materials and nutrients with higher SRT resulting from superior biodegradation and higher sludge concentration is reported by various researchers (Grelier et al., 2006; Ke and Junxin, 2009).

The operation and performance of MBR also depend on HRT, which is defined as the volume of the reactor divided by the filtration velocity. In MBR, the increase of operation flux decreases the HRT as the volume of the reactor is fixed. Research showed that the development of transmembrane pressure (TMP) is higher with shorter HRT than with longer HRTs. For example, MBR operated with lower HRT increased mixed liquor particle size, which had an adverse impact on membrane permeability and formation of fouling cake layer on the membrane surface (Meng et al., 2007). Furthermore, a study by Johir et al. (2012) showed that the higher flux (i.e. lower HRT) of 40 and 30 L/m^{2}h resulted in a higher membrane fouling than the lower flux of 20 and 25 L/m^{2}h (higher HRTs). At higher flux the suction pressure is higher than that at lower flux; hence, the depositing of floc particles on membrane surfaces in much compact resulted in the reduction of membrane permeability. Moreover, the biological activity and removal of organic matter at lower HRTs are also reduced (Meng et al., 2007; Ren et al., 2005). However, shorter HRT has some advantages and a smaller footprint, but its merit is hindered by higher membrane fouling, which reduces membrane permeability and increases both operational and maintenance costs.

18.2.3.2 Organic Loading Rate and F/M Ratio

Like SRT and HRT, organic loading rate (OLR) also plays an essential part in the sustainable operation of MBR. With higher OLR, membrane flux decreases due to an increase in membrane fouling or TMP (Trussell et al., 2006). Furthermore, they also observed 20 times higher membrane fouling development with a four-times increase to the food-to-mass ratio (F/M). Furthermore, Khoshfetrat et al. (2011) found a higher reduction of COD from 74% to 90% when OLR increased from 1 to 2.5 kg COD/m^{3}d. The degradation of organic matter was high (98%) with a lower OLR of ≤13 g CODL^{-1} d^{-1} than an OLR of 30 g CODL^{-1} d^{-1} (70%) (Shen et al., 2010).

18.2.3.3   Effect of Extracellular Polymeric Substances on Membrane Fouling

Among the different organic foulants, an extracellular polymeric substance (EPS) has the significant ability to block membrane pores and adhere to the membrane surface. EPS is a colloidal material that contains a wide range of organics, such as carbohydrates, proteins, lipids, nucleic and humic acids, etc. (Barker and Stuckey, 1999). High initial carbohydrate concentration usually results in fouling. High EPS concentration increases the viscosity (Meng et al., 2007), and the highly hydrated gel matrix of these biopolymers are responsible for the reduction of permeate flux in MBR (Cho et al., 2005). Loosely bound EPS was found to cause fouling more severely than tightly bound EPS (Ramesh et al., 2006). Dissolved organic matter rejection in sludge supernatant showed much higher EPS concentration than the effluent (Drews et al., 2006). Furthermore, the concentration of bound EPS, especially carbohydrates, influences the foaming in MBR in the sludge (Lesjean et al., 2005). This problem is further compounded when bound EPS interacts with filamentous bacteria. This result is a significant occurrence of fouling and scum growth increase. Both bound EPS and sludge characteristics should be considered together to mitigate membrane fouling.

18.2.4   Membrane Fouling

Deposition of organic and inorganic matters during operation results in the fouling of the membrane surface. Membrane fouling is due to mechanisms such as adsorption of colloids and solutes on membranes and inside their pores; deposition of sludge and formation of a cake layer on the membrane surface; the spatial and temporal changes of the foulant composition during the long-term operation; and subsequent detachment of foulants due to shear forces (Meng et al., 2009).

Fouling on membranes and the remedial measures are given below (Aryal et al., 2009):

- Reversible fouling caused by the deposition of mixed liquor solids on the membrane surface and build-up of sludge in between the membrane fibers. This can be controlled by backflushing or air scouring;
- Irreversible fouling cannot be removed by backflushing or air scouring. This can be minimized by chemical cleaning; and
- Irrecoverable fouling cannot be recovered by any cleaning measures. It takes place over a long time (more than several years) (Drews, 2010).

18.2.5   Fouling Control Strategies

A standard method to reduce sludge accumulation and fouling on the surface of the membrane is to provide aeration (air sparging) close to the surface of the membrane. This induces local shear stress, which creates favorable conditions for hydraulic distribution for sludge mixing and membrane surface scouring (Bouhabila et al., 2001). However, aeration is a major component of the operating cost of MBR (Cui et al., 2003; Judd, 2007). Optimization of the aeration is important in MBR operation, as beyond a particular aeration rate the reduction of fouling is negligible. Membrane fouling can also be minimized by introducing a medium in suspension in the reactor. Thus, it may be possible to prevent some fouling from occurring by 1) changing the operating parameters of the MBR such as aeration, 2) using an adaptive membrane cleaning method, and 3) using mechanical scouring by introducing medium in suspension in the reactor.

18.2.5.1 Aeration

Aeration is vital in MBR both for biological oxidation of organic matter and membrane defouling. Usually, air bubbling close to the membrane is one of the most efficient ways of reducing membrane fouling (Meng et al., 2008, Wicaksana et al., 2006). Governing parameters of aeration are the size and shape of bubbles, density, and viscosity of wastewater, internal circulation, temperature, and presence of surface-active compounds (Malysa et al., 2005). Bubble dissolution in a reactor is a function of the bubble size, liquid viscosity, and its aeration rate (Baral, 2003). The presence of contaminants such as surfactants can significantly change the bubble's dissolution rate, shape, and effective velocity (Lio and McLaughlin, 2000; Takemura, 2005). Recent studies separated the submerged membrane reactor from the bioreactor tanks to achieve maximum efficiency of aeration to remove foulants and to minimize the cost of aeration for biological oxidation (Lebegue et al., 2007). In theory, small bubbles are suitable for biological degradation while large bubbles assist in the reduction of membrane fouling. Sofia et al. (2004) reported that the smallest bubble could produce the best performance, whereas Madec (2000) concluded that the size of the bubble did not affect membrane performance. Further, many researchers studied the effect of rate and amount of aeration required to reduce membrane fouling (Ueda et al., 1997). They found out that after a particular aeration rate, there was no effect of aeration on membrane filtration rate. The suitable range of aeration was found to be 0.0048–0.010 $m^3 m^{-2} s^{-1}$ (for MLSS ranging from 2 to 10 g L^{-1} and filtration rate varying from 10 to 20 L $m^{-2} h^{-1}$). In addition, the suitable airflow rate per membrane surface area was reported to be 0.18 to 1.28 N m^3/m^2 h whereas the airflow rate per permeate flow produced 10 to 65 m^3/m^3 (Drews, 2010).

18.2.5.2 Mechanical Scouring

Because of the irreversible interactions between soluble compounds or bacteria and membranes, membrane fouling cannot be controlled by aeration alone. It is thus vital to find an alternative method to reduce these compounds on the membrane surface and in solution. Incorporation of supporting media/adsorbents will scour some of the foulant on the membrane surface and adsorb the organic substances that cause fouling before their contact with the membrane.

Thus, as an alternative to the application of a higher aeration rate, membrane fouling could be reduced by the incorporation of a granular medium/adsorbent in suspension in the MBR. The purpose of the suspended medium (such as activated carbon) is to adsorb the organic matter and also to provide higher shear on the membrane surface. Several studies have been conducted with adsorbents in suspension in MBR to minimize the membrane fouling (Guo et al., 2005; Jin et al., 2013, Xing et al., 2012). A pilot study by Siembida et al. (2010) with the addition of granular material (polypropylene) showed that the abrasion of granular material minimized the fouling on the membrane. When the above study was extended to more than 600 days, the abrasion did not damage the membranes. It also showed that the MBR process with the addition of granular medium could be operated at a 20% higher flux. Akram and Stuckey (2008) found that adding powdered activated carbon (PAC) in submerged anaerobic MBR helped to remove biodegradable organic compounds (both low and high molecular weight (MW) COD. They also observed a 4.5-times-higher filtration rate with the addition of PAC of 1.6 g/L than without any addition. The filtration rate increased from 2 to 9 L/m^2 h. Fang et al. (2006) investigated the effect of the activated carbon addition on the fouling reduction of an activated sludge filtration system. They observed a reduction of filtration resistance by 22% (from $6.4 \pm 0.5 \times 10^{12}$ m^{-1} to $5.0 \pm 0.1 \times 10^{12}$ m^{-1}) with the activated carbon addition. The addition of adsorbent also helps in the removal of soluble organic compounds that cause irreversible membrane fouling (Guo et al., 2005; Shanmuganathan et al., 2015a, b). Investigations by Sombatsompop et al. (2006) and Guo et al. (2008) showed that a membrane-coupled moving bed biofilm reactor (M-CMBBR) resulted in lower bio-fouling than a conventional MBR. The above studies demonstrated that the use of a suspended medium in MBR reduced membrane fouling by adsorbing organic matters and also by providing mechanical scour on the membrane surface.

18.2.5.3 Periodic Backwashing or Cleaning

A periodic backwash during a membrane filtration process is a useful tool to control reversible membrane fouling. Periodic backwashing helps to reduce pressure buildup and thus helps to prevent flux decline during membrane operation. However, it is crucial to optimize the backwash for the successful long-term maintenance of a membrane system (Smith et al., 2006). The optimum backwash can be expressed as a function of the concentration of the foulant, permeate flux, and the operational temperature (Smith et al., 2006). Optimization of backwash is required to reduce energy requirement and to increase membrane efficiency. Smith et al. (2006) reported that through an automated backwashing system, it was possible to control backwashing frequency, which reduced energy requirement and increasd productivity. Usually backflushing is done from the permeate side in the case of hollow fiber modules and relaxation is applied for flat sheet modules for approximately 15–60 seconds every 3–12 minutes of filtration. In contrast, frequent cleanings or maintenance cleanings are conducted once or twice a week and significant cleanups once or twice a year (Drews, 2010). On the other hand, Le-Clech et al. (2006) recommended less frequent, but longer, backwashing (for every 10 minutes filtration/45 seconds backwashing), which was more efficient than more frequent backwashing (for every 3 minutes filtration/15 seconds backwashing). In addition to periodic backwashing and relaxation, chemical cleaning is also needed. Chemical cleaning may include daily chemically enhanced backwash, weekly maintenance cleaning with higher chemical concentration, and half-yearly intensive chemical cleaning (Le-Clech et al., 2006).

18.2.5.4 Chemical Cleaning

If the fouling is irreversible or too high to be removed by physical means, then chemical cleaning is performed. Among the fouling control strategies used, chemical cleaning is the most effective method for the permeability recovery of highly fouled membranes. This method utilizes aggressive chemicals, such as acids (e.g., citric or volatile fatty acids), bases (e.g., caustic soda), or oxidants (mostly hypochlorite) to dissolve most irreversible inorganic foulants, as well as SMPs and EPS. However, this method weakens the integrity of the membrane, such that excessive cleaning using chemicals makes the membrane more susceptible to deterioration (Le-Clech et al., 2005). The choice of a suitable cleaning chemical is significant to clean the membrane efficiently, and at the same time, maintain the integrity of the membrane.

Acids aim to reduce inorganic fouling, which is typically caused by the chemical precipitation of inorganics and biologically induced mineralization between biopolymer and inorganic salts (Malaeb et al., 2013, Meng et al., 2009). The primary mechanisms by which acids remove the chemically and biologically induced precipitates on the membrane surface are neutralization and double-displacement. Typical acids used are hydrochloric, nitric, oxalic, citric, phosphoric, and sulfuric.

Alkaline cleaning reduces the organic fouling deposited on membrane surfaces, with the same mechanism as those of acids. Sodium hydroxide (NaOH) is the most commonly used primary cleaning agent and is more useful in removing protein-related fouling than carbohydrates because NaOH readily desorbs proteins (Kimura et al., 2013). NaOH is also used to clean membranes to treat protein-rich industrial wastewater. In highly alkaline environments, ample organics, such as microbes and colloids, can be broken into soluble organics and small particles (Yu et al., 2013).

Oxidants are also used as membrane cleaning agents due to their capability to remove organic and biological foulants through oxidation and disinfection. The common oxidants are sodium hypochlorite (NaOCl) and hydrogen peroxide (H_2O_2). H_2O_2 does not produce toxic by-products as compared with NaOCl. Ozone can be employed as well to clean submerged membranes. It is found that the intermittent ozone backwashing method is more highly efficient than the air for flux recovery (Kim et al., 2007). The primary mechanism of oxidants is to disinfect the membranes. They also oxidize the organic foulants into ketone, aldehyde, or acetyl groups. These functional groups of organic foulants have higher hydrophilicity. Thus adhesion of the foulants onto the membranes is reduced. Oxidants can also break the microbe flocs and the colloids into fine particles and soluble organics, leading to further degradation. NaOCl is a highly stable and active chemical and is easy to use together with other chemicals. Although it is beneficial in the removal of foulants, NaOCl also negatively affects membrane integrity (Arkhangelsky et al., 2007). Aside from the detrimental effects of oxidants on membrane integrity, the use of oxidants in membrane cleaning may lead to the formation of organic halogens and trihalomethanes, which are not environmentally-friendly substances (Han et al., 2010, Krause et al., 2010).

18.2.5.5 Microbiological Fouling Mitigation

Biological cleaning is widely used for cleaning MBR systems. The bioactive agents include enzymes to improve the removal of membrane fouling. Enzymatic cleaning, quorum quenching, cell wall hydrolases addition, nitric oxide-induced biofilm detachment, and bacteriophage addition are some of the biological cleaning methods reported (Xiong and Liu, 2010). Among the ways to mitigate biofouling by microbiological means is a determination of the following: cell physiology, microorganism population dynamic, quorum sensing mechanism, and biofouling mechanism. Various biological strategies have been considered as promising methods to control fouling in general and biofouling in particular. Biological approaches showed higher efficiency, lower toxicity, more sustainability, and less bacterial resistance, compared to other mitigation methods (Xiong and Liu, 2010; Yeon et al., 2009).

The use of an enzymatic agent is a standard biological cleaning method; however, enzymatic agents are specific to biopolymers. The interaction between the enzyme and the biopolymer induces the rupture of the fouling layer on the membrane, thereby inhibiting the damage by chemical and physical destruction of the membranes. Enzymes have a spectrum of activities for fouling degradation based on the selection of cleaning agents. Recent research reveals that the enzymatic cleaning method may be an alternative to remove the irreversible fouling during membrane filtration of wastewater effluent (Grélot et al., 2008). Also, enzymes do not induce membrane aging or result in the formation of the by-product. However, at present, the enzymatic agents are costlier than NaOCl. Thus, optimization of operating conditions is essential for enzymatic cleaning.

Energy uncoupling uses predatory microorganisms. Uncoupling proteins can be capable of dissipating the proton gradient and hinder the synthesis of adenosine triphosphate (ATP). Studies reported that the uncoupling of energy metabolism affects biofilm stability, which results in the dispersed growth of microorganisms (Xu and Liu, 2011).

The quorum quenching method, which focuses on the inhibition of the development of bacterial biofilm, is another biological strategy for fouling control in membranes (Lade et al., 2014). Bacteria used in quorum sensing produce and secrete specific signaling molecules, which might determine the behavior of biofilm communication. This method can be accomplished in three ways (Grandclément et al., 2016): (i) inhibition of N-acyl homoserine lactones synthesis, which commonly exists in MBR, (ii) enzymatic degradation of N-acyl homoserine lactones signal molecules, and (iii) blocking of transport or acceptance of N-acyl homoserine lactones signals. Quorum quenching is an efficient strategy to remove biofilm fouling. On the other hand, it is more likely to get clogged by inorganics, organic-inorganic complexes, and dead cells.

18.3 MEMBRANE ADSORPTION HYBRID SYSTEM

In a membrane adsorption hybrid system, the adsorbents are added to the membrane tank. The aeration used in the reactor system keeps the adsorbents in suspension. Several researchers studied the application of activated carbon in membrane-hybrid systems in reducing transmembrane pressure (TMP) development and membrane fouling (Guo et al. 2005, Johir et al. 2011, Li 2014, Stoquart et al. 2012; Vigneswaran et al. 2007). The function of activated carbon on the membrane hybrid system is twofold: firs, a majority of organic foulants can be removed by adsorption onto activated carbon; and second, activated carbon provides abrasion on the membrane surface and modifies the fouling cake layer on the membrane surface. This probably becomes more porous and helps in keeping the filtration flux stable (Pianta et al. 1998).

Many investigators also focused on the coupling of powder activated carbon (PAC) with MF/UF to improve the removal efficiency of organics and to maintain the sustainable operation of the membrane process. Vigneswaran et al. (2007) reported an 84% removal of total organic carbon (TOC) when a PAC dose of 5 g/L was added into a membrane-PAC hybrid system. PAC addition was made only at the start of the experiment, and no PAC was added then after. Guo et al. (2004) observed that a PAC addition of 5 g/L in the membrane hybrid system removed 90% of TOC from the biologically treated sewage effluent (BTSE).

In addition to the removal of organic matter, the granular activated carbon (GAC) was more effective in removing organic micropollutants compared to the flocculation process (Shanmuganathan et al. 2015a, b, 2017). The GAC removed a majority of pharmaceuticals and personal care products (PPCPs) (70% compared to more than 90%). (Shanmuganathan et al. 2015a; Shanmuganathan et al., 2017b).

18.3.1 SUBMERGED MEMBRANE FILTRATION ADSORPTION HYBRID SYSTEM FOR THE REMOVAL OF ORGANIC MICROPOLLUTANTS

In addition to removing dissolved organics, the requirement to remove micropollutants, including PPCPs, is growing due to their potential health risk to people and aquatic organisms when they are released into wastewater. Advanced oxidation processes such as ozone treatment and ultraviolet radiation are capable of removing micropollutants but may produce toxic by-products (Rossner et al. 2009). On the other hand, the adsorption process does not contribute to any undesirable by-products to water (Bonné et al. 2002; Quinlivan et al. 2005). Activated carbon is highly efficient in removing organic micropollutants (Snyder et al. 2007; Verliefde et al. 2007). The common micropollutants present in waters in Australia are presented in Table 18.1.

TABLE 18.1
Concentrations of organic micropollutants in Asia and Australian waters (data from Pal et al., 2010)

Compounds	Effluent from Wastewater Treatment Plants/Sewage Treatment Plants (ng/L)	Freshwater, rivers, canals (ng/L)
Trimethoprim	58–321	4–150
Ciprofloxacin	42–720	23–1,300
Sulfamethoxazole	3.8–1,400	1.7–2,000
Naproxen	128–548	11–181
Ibuprofen	65–1,758	28–360
Ketoprofen	-	<0.4 – 79.6
Diclofenac	8.8–127	1.1–6.8
Carbamazepine	152–226	25–34.7
Propranolol	50	-
Gemfibrozil	3.9–17	1.8–9.1

Vigneswaran et al. (2003) studied the feasibility of an MF adsorption hybrid system for dissolved organics removal. In addition to organics adsorption, the MF adsorption hybrid system was found to be effective in fouling reduction due to membrane scouring by adsorbents in suspension. The GAC/PAC in suspension provides physical abrasion on the membrane surface. This reduces the foulant accumulation on the membrane surface, which minimizes TMP development. The above two factors facilitate the reduction of the membrane fouling in the membrane adsorption hybrid system. These, in turn, extend the operation time and minimize the frequency and duration of membrane cleaning.

Earlier, PAC was used in the membrane adsorption hybrid system as an adsorbent in suspension in the tank for organics removal (Jeong et al., 2012; Kim et al., 2009; Vigneswaran et al., 2007; Thiruvenkatachari et al., 2004). The organic removal increased with the increase of PAC dose. However, at high PAC concentrations, PAC cake formation was formed on the membrane surface. This led to a flux decline. To overcome this, Johir et al. (2011) used granular activated carbon (GAC) instead of PAC. The use of GAC had several advantages over PAC such as the particle size of GAC is usually larger than PAC. GAC formed a shallowercake layer on the membrane surface than that of PAC. Also, the separation of exhausted GAC from sludge was more straightforward than that of PAC. However, one should pay attention to the selection of the size of GAC as a larger size of GAC tends to settle down on the bottom of the reactor whereas GAC of smaller size (less than 300 μm) will produce a dense cake layer on the membrane surface (Johir et al. 2011).

The membrane adsorption hybrid system has been widely researched for micropollutant removal. Löwenberg et al. (2014) studied the removal of selected organic micropollutants from sewage effluent using pressurized and submerged PAC-UF systems. They observed that the submerged system removed higher amounts of organic micropollutants as compared to the pressurized system. A submerged GAC/MF system was effective in removing most of the organic micropollutants tested in a biologically treated sewage effluent (Shanmuganathan et al. 2015a, b).

18.3.1.1 Submerged MF–GAC Adsorption Hybrid System

Shanmuganathan et al. (2015a) used a submerged MF–GAC adsorption hybrid system to remove organics and micropollutants from biologically treated sewage effluent (Figure 18.2). The organic contaminants were first adsorbed onto GAC, and the contaminant-laden adsorbent was retained by microfiltration. The MF membrane itself cannot remove micropollutants or DOC since the pore size of MF is larger than the size of the contaminants. Organic removal by MF alone was less than 10%. Daily addition of GAC of 2 g/L in the MF tank increased the removal of the organics by 65–90%.

Usually, 5–10 g/L of GAC is added into the reactor at the beginning to reduce the loading of organics to the membrane. Following this, 2–3% of GAC in the tank is replaced daily with new GAC. The daily GAC replacement (0.28 g/L) used in the study of Shanmuganathan et al. (2015a) was low compared to the one used by Wang et al. (2013) for similar or smaller removal of DOC (i.e. removal of 50–70% DOC; effluent was 5–10 mg/L of DOC). The initial amount of GAC added was high to firs, improve organics and micropollutants removals, and second, reduce the development of TMP.

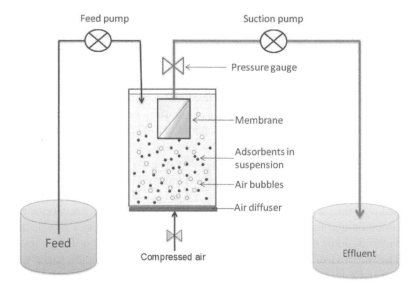

FIGURE 18.2 Schematic diagram of the submerged MF-GAC adsorption hybrid system (adapted from Shanmuganathan et al., 2015b).

In reverse osmosis concentrate (ROC), micropollutants such as carbamazepine (2240 ng/L), caffeine (1410 ng/L), atenolol (466 ng/L), trimethoprim (974 ng/L), and naproxen (443 ng/L) were detected at high concentrations. However, these concentrations are lower than their respective contaminants of emerging concern (CECs). In comparison, the concentration of a PPCP of verapamil (a cardiovascular agent) was low (83 ng/L). However, this value was higher than its CEC (24 ng/L). The amitriptyline concentration was 45 ng/L, which is near its CEC (48 ng/L). The submerged membrane adsorption hybrid system used removed 60% to more than 98.8% of micropollutants. The removal efficiency was higher on the seventh day (>81– >99%) than on the first day (60–>99%) (Table 18.2).

18.4 SUSTAINABILITY

Membrane bioreactors are widely used in wastewater treatment plants as they have a smaller footprint and produce water of reusable quality. It is a mature technology and is used in huge plants. For example, the recently built MBR plant in Stockholm, Sweden, has a design capacity of 864 million liters per day (864,000 m^3/day). The growth of MBR has been very rapid. The worldwide MBR market size rose from U.S.$0.25 to 1.17 billion from 2006 to 2010. MBR is also an excellent choice in decentralized wastewater treatment systems and provides high-quality water for reuse in toilet flushing and watering the parks and gardens. Initially, the energy requirement for air scouring of MBR has been a limiting factor. Thanks to research, air scouring for submerged hollow fiber MBR represents only 20% of the energy

TABLE 18.2

The removal of micropollutants by the MF-GAC hybrid system from ROC (Data from Shanmuganathan et al. 2017)

Micropollutants	Influent (ng/L)	Effluent (ng/L)		Removal (%)	
		Day 1	Day 7	Day 1	Day 7
Amtriptyline	45	<5	<5	>89	>89
Atenolol	466	<5	<5	>99	>99
Caffeine	1,410	31	<5	98	>99
Carbamazepine	2,240	86	<5	96	>99
Clozapine	68	<5	<5	>93	>93
DEET	68	27	<5	60	>93
Diclofenac	337	<5	<5	>99	>99
Fluoxetine	47	<5	<5	>89	>89
Gemfibrozil	344	9	<5	97	>99
Ketoprofen	377	<5	<5	>99	>99
Naproxen	443	10	<5	98	>99
Paracetamol	114	<5	<5	>96	>96
Primidone	26	<5	<5	>81	>81
Simazine	80	<5	<5	>94	>94
Sulfamethoxazole	144	35	<5	76	>97
Triclocarban	162	<10	<10	>94	>94
Triclosan	211	<5	<5	>98	>98
Trimethoprim	974	9	<5	99	>99
Verapamil	83	<5	<5	>94	>94

required for the sewage treatment plant. Energy for air scouring for submerged hollow fiber MBR has reduced 7% from 1995 to 2011. Adsorbents used in membrane adsorption hybrid systems provide mechanical scouring while providing complete micropollutant removal.

A membrane adsorption hybrid system, on the other hand, can remove organic micropollutants successfully from wastewaters. This will also increase the quantity of recycled water with useful nutrients for irrigation reuse. An economics of using the membrane adsorption hybrid system in organics and micropollutant removal was made. The GAC required for biologically treated sewage effluent is less than 50–100 g/m^3 of water produced. The significant operational cost is the periodic replacement of GAC. The cost of GAC to treat 1 m^3 of sewage effluent and ROC was US$0.05 and 0.25 per m^3, respectively. Taking into account the adverse environmental consequences of discharging the waste into waterways, the cost of US$0.25to treat a ton (or m^3) of wastewater is not high. A scheme for the beneficial use of the MF-GAC hybrid system effluent is presented in Figure 18.3.

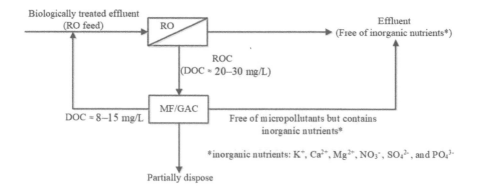

FIGURE 18.3 A scheme for the beneficial use of the MF-GAC hybrid system effluent (adapted from Shanmuganathan et al., 2017).

REFERENCES

Adham, S., & Gagliardo, P. (1998). Membrane Bioreactors for Water Repurification-Phase I. Desalination Research and Development Program Report No. 34, Project No. 1425-97-FC-81-30006J; Bureau of Reclamation: Denver, CO.

Ahmed, Z., Cho, J., Lim, B-R., Song, K-G., & Ahn, K-H. (2007). Effects of sludge retention time on membrane fouling and microbial community structure in a membrane bioreactor. Journal of Membrane Science, 287, 211–218.

Akram, A., & Stuckey, D. C. (2008). Flux and performance improvement in a submerged anaerobic membrane bioreactor (SAMBR) using powdered activated carbon (PAC). Process Biochemistry, 43(1), 93–102.

Arkhangelsky, E., Kuzmenko, D., & Gitis, V. (2007). Impact of chemical cleaning on properties and functioning of polyethersulfone membranes. Journal of Membrane Science, 305(1–2), 176–184.

Aryal, R., Lebegue, J., Vigneswaran, S., Kandasamy, J., & Grasmick, A. (2009). Identification and characterisation of biofilm formed on membrane bio-reactor. Separation and Purification Technology, 67(1), 86–94.

Baral, B. (2003), Numerical study on the bubbly plume for water purification system, Master's Thesis. University of Tokyo, Japan.

Barker, D. J., & Stuckey, D. C. (1999). A review of soluble microbial products (SMP) in wastewater treatment systems. Water Research, 33(14), 3063–3082.

Bonné, P. A. C., Hofman, J. A. M. H., & Van der Hoek, J. P. (2002). Long term capacity of biological activated carbon filtration for organics removal. Water Science and Technology: Water Supply, 2(1), 139–146.

Bouhabila, E. H., BenAïm, R., & Buisson, H. (2001). Fouling characterisation in membrane bioreactors. Separation and Purification Technology, 22–23, 123–132.

Cho, J., Song, K. G., Yun, H., Ahn, K. H., Kim, J. Y., & Chung, T. H. (2005). Quantitative analysis of biological effect on membrane fouling in submerged membrane bioreactor. Water Science and Technology, 51(6–7), 9–18.

Cicek, N., Macomber, J., Devel, J., Suidan, M. T., Audic, J., & Genestet, P. (2001). Effect of solids retention time on the performance and biological characteristics of a membrane bioreactor. Water Research, 30, 1771–1780.

Cui, Z. F., Chang, S., & Fane, A. G. (2003). The use of gas bubbling to enhance membrane processes. Journal of Membrane Science, 221, 1–35.

Drews, A., Vocks, M., Iversen, V., Lesjean, B. & Kraume, M. (2006). Influence of unsteady membrane bioreactor operation on EPS formation and filtration resistance. Desalination, 192, 1–9.

Drews, A. (2010). Membrane fouling in membrane bioreactors—Characterisation, contradictions, cause and cures. Journal of Membrane Science, 363, 1–28.

Fang, H.H.P., Shi, X., & Zhang, T. (2006). Effect of activated carbon on fouling of activated sludge filtration. Desalination, 189, 193–199.

Grandclément, C., Tannières, M., Moréra, S., Dessaux, Y., & Faure, D. (2016). Quorum quenching: role in nature and applied developments. FEMS Microbiology Reviews, 40(1), 86–116.

Grelier, P., Rosenberger, S., & Tazi-Pain, A. (2006). Influence of sludge retention time on membrane bioreactor hydraulic performance. Desalination, 192, 10–17.

Grélot, A., Machinal, C., Drouet, K., Tazi-Pain, A., Schrotter, J. C., Grasmick, A., & Grinwis, S. (2008). In the search of alternative cleaning solutions for MBR plants. Water Science and Technology, 58(10), 2041–2049.

Guo, W. S., Shim, W. G., Vigneswaran, S., & Ngo, H. H. (2005). Effect of operating parameters in a submerged membrane adsorption hybrid system: experiment and mathematical modeling. Journal of Membrane Science, 247, 65–74.

Guo, W. S., Vigneswaran, S., Ngo, H. H., & Chapman, H. (2004). Experimental investigation of adsorption–flocculation–microfiltration hybrid system in wastewater reuse. Journal of Membrane Science, 242(1–2), 27–35.

Guo, W., Vigneswaran, S., Ngo, H. H., Xing, W., & Goteti, P. (2008). Comparison of the performance of submerged membrane bioreactor (SMBR) and submerged membrane adsorption bioreactor (SMABR). Bioresource Technology, 99(5), 1012–1017.

Halling-Sørensen, B., & Jørgensen, S.E. (1993). The removal of nitrogen compounds from wastewater. Elsevier, Amsterdam, The Netherlands, 55–56.

Han, X. Y., Gu, J., Ai, B., Li, S. T., Wen, Y., & Shi, L. (2010). Analysis on MBR process operation and membrane cleaning scheme in Beixiaohe WWTP. China Water & Wastewater, 26(17), 40–43.

Hermanowicz, S.W. (2011). Membrane Bioreactors: Past, Present and Future? Water Resources Center Archives.

Holler, S., & Trösch, W. (2001). Treatment of urban wastewater in a membrane bioreactor at high organic loading rates. Journal of Biotechnology, 92, 95–101.

Jeong, S., Choi, Y. J., Nguyen, T. V., Vigneswaran, S., & Hwang, T. M. (2012). Submerged membrane hybrid systems as pretreatment in seawater reverse osmosis (SWRO): Optimisation and fouling mechanism determination. Journal of Membrane Science, 411, 173–181.

Jin, L., Ong, S.L., & Ng, H.Y. (2013). Fouling control mechanism by suspended biofilm carriers addition in submerged ceramic membrane bioreactors. Journal of Membrane Science, 427, 250–258.

Johir, M. A. H., Aryal, R., Vigneswaran, S., Kandasamy, J., & Grasmick, A. (2011). Influence of supporting media in suspension on membrane fouling reduction in submerged membrane bioreactor (SMBR). Journal of Membrane Science, 374(1–2), 121–128.

Johir, M. A., George, J., Vigneswaran, S., Kandasamy, J., Sathasivan, A., & Grasmick, A. (2012). Effect of imposed flux on fouling behavior in high rate membrane bioreactor. Bioresource Technology, 122, 42–49.

Judd, S. (2007). Membrane bioreactor technology costs. Proceedings of International Membrane Science and Technology Conference 2007, Sydney, Australia.

Ke, O., & Junxin, L. (2009). Effect of sludge retention time on sludge characteristics and membranefouling of membrane bioreactor. Journal of Environmental Science, 21, 1329–1335.

Khoshfetrat, A.B., Nikakhtari, H., Sadeghifar, M., & Khatibi, M.S. (2011). Influence of organic loading and aeration rates on performance of a lab-scale upflow aerated submerged fixed-film bioreactor. Process Safety and Environmental Protection, 89, 193–197.

Kim, J. O., Jung, J. T., Yeom, I. T., & Aoh, G. H. (2007). Effect of fouling reduction by ozone backwashing in a microfiltration system with advanced new membrane material. Desalination, 202(1–3), 361–368.

Kim, K. Y., Kim, H. S., Kim, J., Nam, J. W., Kim, J. M., & Son, S. (2009). A hybrid microfiltration–granular activated carbon system for water purification and wastewater reclamation/reuse. Desalination, 243(1–3), 132–144.

Kimura, K., Ogawa, N., & Watanabe, Y. (2013). Permeability decline in nanofiltration/reverse osmosis membranes fed with municipal wastewater treated by a membrane bioreactor. Water Science and Technology, 67(9), 1994–1999.

Krause, S., Zimmermann, B., Meyer-Blumenroth, U., Lamparter, W., Siembida, B., & Cornel, P. (2010). Enhanced membrane bioreactor process without chemical cleaning. Water Science and Technology, 61(10), 2575–2580.

Lade, H., Paul, D., & Kweon, J. H. (2014). Quorum quenching mediated approaches for control of membrane biofouling. International Journal of Biological Sciences, 10(5), 550.

Lebegue, J., Heran, M., & Grasmick, A. (2007). Proceedings of the Sixth International Membrane Science and Technology Conference, November 5–9, Sydney, Australia.

Le-Clech, P., Chen, V., & Fane, T.A.G. (2006). Fouling in membrane bioreactors used in wastewater treatment. Journal of Membrane Science, 284(1–2), 17–53.

Le-Clech, P., Fane, A., Leslie, G., & Childress, A. (2005). MBR focus: the operators' perspective. Filtration & separation, 42(5), 20–23.

Lesjean, B., Rosenberger, S., Laabs, C., Jekel, M., Gnirss, R., & Amy, G. (2005). Correlation between membrane fouling and soluble/colloidal organic substances in membrane bioreactors for municipal wastewater treatment. Water Science and Technology, 51(6–7), 1–8.

Li, W. C. (2014). Occurrence, sources, and fate of pharmaceuticals in aquatic environment and soil. Environmental Pollution, 187, 193–201.

Lio, Y., & McLaughlin, J.B. (2000). Bubble motion in aqueous surfactant solutions. Journal of Colloid and Interface Science, 224 (2), 297–310.

Liu, S. X. (2007). Food and Agricultural Wastewater Utilization and Treatment. Blackwell Publishing, West Sussex, UK, 148.

Löwenberg, J., Zenker, A., Baggenstos, M., Koch, G., Kazner, C., & Wintgens, T. (2014). Comparison of two PAC/UF processes for the removal of micropollutants from wastewater treatment plant effluent: Process performance and removal efficiency. Water Research, 56, 26–36.

Madec A. (2000). Influence of two phase flow on the filtration through submerged membrane processes. PhD thesis. INSA, Toulouse, France.

Malaeb, L., Le-Clech, P., Vrouwenvelder, J. S., Ayoub, G. M., & Saikaly, P. E. (2013). Do biological-based strategies hold promise to biofouling control in MBRs? Water Research, 47(15), 5447–5463.

Malysa, K., Krasowska, M., & Krzan, M. (2005). Influence of surface active substances on bubble motion and collision with various interfaces. Advances in Colloid and Interface Science, 114–115, 205–225.

Mara, D., & Horan, N. J. (Eds.) (2003). Handbook of Water and Wastewater Microbiology. Elsevier, London, UK, 150–151.

Meng, F., Chae, S-R., Drews, A., Kraume, M., Shin, H-S., & Yang, F. (2009). Recent advances in membrane bioreactors (MBRs): Membrane fouling and membrane material. Water Research, 43, 1489–1512.

Meng, F., Shi, B., Yang, F., & Zhang, H. (2007). Effect of hydraulic retention time on membrane fouling and biomass characteristics in submerged membrane bioreactors. Bioprocess and Biosystems Engineering, 30, 359–367.

Meng, F., Yang, F., Shi, B., & Zhang, H. (2008). A comprehensive study on membrane fouling in submerged membrane bioreactors operated under different aeration intensities. Separation and Purification Technology, 59, 91–100.

Pal, A., Gin, K. Y. H., Lin, A. Y. C., & Reinhard, M. (2010). Impacts of emerging organic contaminants on freshwater resources: review of recent occurrences, sources, fate and effects. Science of the Total Environment, 408(24), 6062–6069.

Pianta, R., Boller, M., Janex, M. L., Chappaz, A., Birou, B., Ponce, R., & Walther, J. L. (1998). Micro- and ultrafiltration of karstic spring water. Desalination, 117(1–3), 61–71.

Quinlivan, P. A., Li, L., & Knappe, D. R. (2005). Effects of activated carbon characteristics on the simultaneous adsorption of aqueous organic micropollutants and natural organic matter. Water Research, 39(8), 1663–1673.

Radjenović, J., Matošić, M., Mijatović, I., Petrović, M., & Barceló, D. (2008). Membrane bioreactor (MBR) as an advanced wastewater treatment technology. Handbook of Environmental Chemistry 5, Part S/2, 37–101.

Ramesh, A., Lee, D. J., Wang, M. L., Hsu, J. P., Juang, R. S., Hwang, K. J., Liu, J.C. & Tseng, S. J. (2006). Biofouling in membrane bioreactor. Separation Science and Technology, 41(7), 1345–1370.

Ren, N., Chen, Z., Wang, X., Hu, D., & Wang, A. (2005). Optimized operational parameters of a pilot scale membrane bioreactor for high-strength organic wastewater treatment. International Biodeterioration and Biodegradation 56, 216–223.

Rossner, A., Snyder, S. A., & Knappe, D. R. (2009). Removal of emerging contaminants of concern by alternative adsorbents. Water Research, 43(15), 3787–3796.

Schramm, A., De Beer, D., Wagner, M., & Amann. R. (1998). Identification and activities in situ of *Nitrosospira* and *Nitrospira* spp. as dominant populations in a nitrifying fluidized bed reactor. Applied and Environmental Microbiology, 64(9), 3480.

Shanmuganathan, S., Johir, M. A., Nguyen, T. V., Kandasamy, J., & Vigneswaran, S. (2015a). Experimental evaluation of microfiltration–granular activated carbon (MF–GAC)/nano filter hybrid system in high quality water reuse. Journal of Membrane Science, 476, 1–9.

Shanmuganathan, S., Loganathan, P., Kazner, C., Johir, M. A. H., & Vigneswaran, S. (2017). Submerged membrane filtration adsorption hybrid system for the removal of organic micropollutants from a water reclamation plant reverse osmosis concentrate. Desalination, 401, 134–141.

Shanmuganathan, S., Nguyen, T. V., Jeong, S., Kandasamy, J., & Vigneswaran, S. (2015b). Submerged membrane–(GAC) adsorption hybrid system in reverse osmosis concentrate treatment. Separation and Purification Technology, 146, 8–14.

Shen, L., Zhou, Y., Mahendran, B., Bagley, D. M., & Liss, S. N. (2010). Membrane fouling in a fermentative hydrogen producing membrane bioreactor at different organic loading rates. Journal of Membrane Science, 360, 226–233.

Siembida, B., Cornel, P., Krause, S., & Zimmermann, B. (2010). Effect of mechanical cleaning with granular material on the permeability of submerged membranes in the MBR process. Water Research, 44, 4037–4046.

Smith, P. J., Vigneswaran, S., Ngo, H. H., Ben-Aim, R., & Nguyen, H. (2006). A new approach to backwash initiation in membrane system. Journal of Membrane Science, 278 (1–2), 381–389.

Snyder, S. A., Adham, S., Redding, A. M., Cannon, F. S., DeCarolis, J., Oppenheimer, J., Wert, E.C., & Yoon, Y. (2007). Role of membranes and activated carbon in the removal of endocrine disruptors and pharmaceuticals. Desalination, 202(1–3), 156–181.

Sofia, A., Ng, W., & Ong, S. L. (2004). Engineering design approaches for minimum fouling in submerged MBR. Desalination, 160(1), 67–74.

Sombatsompop, K., Visvanathan, C., & BenAim, R. (2006). Evaluation of biofouling phenomenon in suspended and attached growth membrane bioreactor. Desalination, 201, 138–149.

Sombatsompop, K. (2007). Membrane fouling studies in suspended and attached growth membrane bioreactor systems. A Doctoral Dissertation Thesis. Asian Institute of Technology, School of Environment, Resources and Development, Thailand.

Stoquart, C., Servais, P., Bérubé, P. R., & Barbeau, B. (2012). Hybrid membrane processes using activated carbon treatment for drinking water: a review. Journal of Membrane Science, 411, 1–12.

Takemura, F. (2005). Adsorption of surfactants onto the surface of a spherical rising bubble and its effect on the terminal velocity of the bubble. Physics of Fluid, 17(4), 048104.

Thiruvenkatachari, R., Shim, W. G., Lee, J. W., & Moon, H. (2004). Effect of powdered activated carbon type on the performance of an adsorption-microfiltration submerged hollow fiber membrane hybrid system. Korean Journal of Chemical Engineering, 21(5), 1044–1052.

Trussell, R. S., Merlo, R. P., Hermanowicz, S. W., & Jenkins, D. (2006). The effect of organic loading on process performance and membrane fouling in a submerged membrane bioreactor treating municipal wastewater. Water Research, 40, 2675–2683.

Ueda, T., Hata, K., Kikuoka, Y., & Seino, O. (1997). Effects of aeration on suction pressure in a submerged membrane bioreactor. Water Research, 31(3), 489–494.

Verliefde, A. R., Heijman, S. G. J., Cornelissen, E. R., Amy, G., Van der Bruggen, B., & Van Dijk, J. C. (2007). Influence of electrostatic interactions on the rejection with NF and assessment of the removal efficiency during NF/GAC treatment of pharmaceutically active compounds in surface water. Water Research, 41(15), 3227–3240.

Vigneswaran, S., Chaudhary, D. S., Ngo, H. H., Shim, W. G., & Moon, H. (2003). Application of a PAC-membrane hybrid system for removal of organics from secondary sewage effluent: experiments and modelling. Separation Science and Technology, 38(10), 2183–2199.

Vigneswaran, S., Guo, W. S., Smith, P., & Ngo, H. H. (2007). Submerged membrane adsorption hybrid system (SMAHS): process control and optimization of operating parameters. Desalination, 202(1–3), 392–399.

Wang, W., Gu, P., Zhang, G., & Wang, L. (2013). Organics removal from ROC by PAC accumulative countercurrent two-stage adsorption-MF hybrid process–A laboratory-scale study. Separation and Purification Technology, 118, 342–349.

Wicaksana, F., Fane, A.G. and Chen, V. (2006). Fibre movement induced by bubbling using submerged hollow fibre membranes. Journal of Membrane Science, 271, 186–195.

Xing, W., Ngo, H. H., Guo, W. S., Listowski, A., & Cullum, P. (2012). Optimization of an integrated sponge–granular activated carbon fluidized bed bioreactor as pretreatment to microfiltration in wastewater reuse. Bioresource Technology, 113, 214–218.

Xiong, Y., & Liu, Y. (2010). Biological control of microbial attachment: a promising alternative for mitigating membrane biofouling. Applied microbiology and biotechnology, 86(3), 825–837.

Xu, H., & Liu, Y. (2011). Control and cleaning of membrane biofouling by energy uncoupling and cellular communication. Environmental Science & Technology, 45(2), 595–601.

Yeon, K. M., Cheong, W. S., Oh, H. S., Lee, W. N., Hwang, B. K., Lee, C. H., Beyenal, H., & Lewandowski, Z. (2009). Quorum sensing: a new biofouling control paradigm in a membrane bioreactor for advanced wastewater treatment. Environmental Science & Technology, 43(2), 380–385.

Yu, H., Wang, Z., Wang, Q., Wu, Z., & Ma, J. (2013). Disintegration and acidification of MBR sludge under alkaline conditions. Chemical Engineering Journal, 231, 206–213.

Zheng, X., Sun, P., Han, J., Song, Y., Hu, Z., Fan, H., & Lv, S. (2014). Inhibitory factors affecting the process of enhanced biological phosphorus removal (EBPR)–A mini-review. Process Biochemistry, 49(12), 2207–2213.

Index